T0205624

REAL AND COMPLEX CLIFFORD ANALYSIS

Advances in Complex Analysis and Its Applications

VOLUME 5

REAL AND COMPLEX CLIFFORD ANALYSIS

By

SHA HUANG
Hebei Normal University, Shijiazhuang, People's Republic of China

YU YING QIAO
Hebei Normal University, Shijiazhuang, People's Republic of China

GUO CHUN WEN
Peking University, Beijing, People's Republic of China

Springer

Library of Congress Cataloging-in-Publication Data

Huang, Sha, 1939–
Real and complex Clifford analysis / by Sha Huang, Yu Ying Qiao, Guo Chun Wen.
p. cm. — (Advances in complex analysis and its applications ; v. 5)
Includes bibliographical references and index.
ISBN-13: 978-1-4899-8655-9
ISBN-10: 1-4899-8655-3
ISBN-13: 978-0-387-24536-2 (e-book)
ISBN-10: 0-387-24536-7 (e-book)
1. Clifford algebras. 2. Functions of complex variables. 3. Differential equations, Partial. 4. Integral equations. I. Qiao, Yu Ying. II. Wen, Guo Chun. III. Title. IV. Series.

QA199.H83 2006
512´.57—dc22

2005051646

AMS Subject Classifications: 26-02, 32-02, 35-02, 45-02

Softcover re-print of the Hardcover 1st edition 2006

Printed in the United States of America.

9 8 7 6 5 4 3 2 1 SPIN 11161424

springeronline.com

Contents

Preface

Clifford analysis is a comparatively active branch of mathematics that has grown significantly over the past 30 years. It possesses both theoretical and applicable values of importance to many fields, for example in problems related to the Maxwell equation, Yang-Mills theory, quantum mechanics, and so on. Since 1965, a number of mathematicians have made great efforts in real and complex Clifford analysis, rapidly expanding our knowledge of one and multiple variable complex analysis, vector-valued analysis, generalized analytic functions, boundary value problems, singular differential and integral equations of several dimension and harmonic analysis in classical domains (see Luogeng Hua's monograph [26]1)). In recent years, more mathematicians have recognized the important role of Clifford analysis in harmonic analysis and wavelet analysis. Most of content of this book is based on the authors' research results over the past twenty years. We present the concept of quasi-permutation as a tool and establish some properties of five kinds of quasi-permutations. Moreover, we use this tool to overcome the difficulty caused by the noncommutative property of multiplication in Clifford algebra, give the sufficient and necessary condition for generalized regular functions, and discuss the solvability for some boundary value problems.

In Chapter I, we introduce the fundamentals of Clifford algebra including definitions, some properties, the Stokes theorem, Cauchy integral formulas and Pompieu formulas for generalized regular functions and harmonic functions. We also definite quasi-permutation and study the property of quasi-permutations. We state the sufficient and necessary conditions for generalized regular functions. In the last section of this chapter, we consider regular and harmonic functions in complex Clifford analysis.

In Chapters II and III, the Cauchy principle value and Plemelj formula of Cauchy type integrals are firstly studied. Next, the relation between linear and non-linear boundary value problems with Haseman shift for generalized regular and biregular functions, vector-valued functions and singular integral equations in real Clifford analysis is discussed. In addition, we discuss the existence, uniqueness and integral expressions of their solutions. By using quasi-permutation as a tool, we study the Dirichlet and mixed boundary value problems for generalized regular functions, and give the Schwarz integral formulation of hyperbolic harmonic functions in real Clifford analysis.

In Chapter IV, the theory of harmonic analysis in classic domains studied by Luogeng Hua is firstly introduced. Moreover, using quasi-permutation, we investigate two boundary value problems for four kinds of partial differential equations of second order in four kinds of classical domains of real and complex Clifford analysis, prove the existence and uniqueness of the regular solutions, and give their integral representations.

In Chapter V, we first introduce Cauchy's estimates for three kinds of integrals with parameters, and then discuss the Poincaré-Bertrand permutation formulas, inverse formulas of singular integrals with Cauchy kernel, Fredholm theory, and the regularization theorem for singular integral equations on characteristic manifolds.

In Chapter VI, we introduce the definitions of the Hadamard principle value, the Hölder continuity, recursive formulas, calculation formulas, differential formulas and Poincaré-Bertrand permutation formulas for six kinds of high order singular integrals of quasi-Bochner-Martinelli type with one and two singular points, and then prove the unique solvability of the corresponding non-linear differential integral equations in real Clifford analysis.

In Chapter VII, we use the method of Clifford analysis to solve some boundary value problems for some uniformly and degenerate elliptic systems of equations.

It is clear that when $n = 2$, the functions in real Clifford analysis are the functions in the theory of one complex variable, hence the results in this book are generalizations of the corresponding results in complex analysis of one complex variable. In this book, we introduce the history of the problems as elaborately as possible, and list many references for readers' guidance. After reading the book, it will be seen that many questions about real and complex analysis remain for further investigations. Finally the authors would like to acknowledge the support and help of NSFC, Mr. Pi-wen Yang, Lili Wang, Nanbin Cao and Yanhui Zhang.

Shijiazhuang and Beijing December, 2005

Sha Huang, Yu Ying Qiao and Guo Chun Wen
Hebei Normal University and Peking University

CHAPTER I

GENERAL REGULAR AND HARMONIC FUNCTIONS IN REAL AND COMPLEX CLIFFORD ANALYSIS

Clifford algebra is an associative and noncommutative algebraic structure, that was devised in the middle of the 1800s. Clifford analysis is an important branch of modern analysis that studies functions defined on $\mathbf{R}^n$ with values in a Clifford algebra space. In the first section of this chapter, we define a Clifford algebra. In the second section, we discuss the Cauchy type integral formula of regular functions and Plemelj formula of Cauchy type integral in real Clifford analysis. In the third section, we introduce the conception of quasi-permutation posed by Sha Huang, and from this we get an equivalent condition for regular and general regular functions. In the fourth section, we establish a new hypercomplex structure. In the last section, we discuss some properties of harmonic functions in complex Clifford analysis.

1 Real and Complex Clifford Algebra

Let $\mathcal{A}_n(R)$ (or $\mathcal{A}_n(C)$) be an real (or complex) Clifford algebra over an n-dimensional real vector space $\mathbf{R}^n$ with orthogonal basis $e := \{e_1, ..., e_n\}$, where $e_1 = 1$ is a unit element in $\mathbf{R}^n$. Then $\mathcal{A}_n(R)$ ($\mathcal{A}_n(C)$) has its basis $e_1, ..., e_n; e_2e_3, ..., e_{n-1}e_n; ...; e_2, ..., e_n$. Hence an arbitrary element of the basis may be written as $e_A = e_{\alpha_1}, ..., e_{\alpha_h}$; here $A = \{\alpha_1, ..., \alpha_h\} \subseteq \{2, ..., n\}$ and $2 \leq \alpha_1 < \alpha_2 < \cdots < \alpha_h \leq n$ and when $A = \emptyset$ (empty set) $e_A = e_1$. So real (complex) Clifford algebra is composed of elements having the type $a = \sum_A x_A e_A$, in which $x_A (\in \mathbf{R})$ are real numbers and $a = \sum_A C_A e_A$, where $C_A (\in \mathbf{C})$ are complex numbers (see [6]). In general, one has $e_i^2 = +1$, $i = 1, ..., s$, $e_i^2 = -1$, $i = s+1, ..., n$ and $e_ie_j + e_je_i = 0$, $i, j = 2, ..., n, i \neq j$. With different s we can get different partial differential equations (elliptic, hyperbolic, parabolic equations) from the regular function in Clifford analysis. In this book we let $s = 1$.

Noting that the real vector space $\mathbf{R}^n$ consists of the elements

$$z := x_1e_1 + \cdots + x_ne_n, \tag{1.1}$$

we can consider that the elements $z := x_1e_1 + \cdots + x_2e_n$ and $z = (x_1, ..., x_n)$ are identical, and denote by $\text{Re}z = x_1$ the real part x_1 of z. For $z \in \mathbf{R}^n$, we define the conjugate as

$$\overline{z} = x_1e_1 - \cdots - x_ne_n,$$

thus $z\overline{z} = e_1(x_1^2 + \cdots + x_n^2)$. The absolute value (or a norm) for an element $a = \sum_A a_Ae_A \in \mathcal{A}_n(R)$ is taken to be

$$|a| = \sqrt{|a|^2} = \sqrt{\sum_A |a_A|^2}, \tag{1.2}$$

then $|z|^2 = |\overline{z}|^2 = z\overline{z} = \overline{z}z$ for $z \in \mathbf{R}^n$. If $z \neq 0$, $z \in \mathbf{R}^n$, then we have

$$z(\frac{\overline{z}}{|z|^2}) = (\frac{\overline{z}}{|z|^2})z = 1, \tag{1.3}$$

hence, all non-zero elements of $\mathbf{R}^n$ possess the inverse operation of multiplication. However, it is not true for all Clifford elements, for instance $1 + e_{123}$.

Definition 1.1 For $a \in \mathcal{A}_n(R)$ (or $\mathcal{A}_n(C)$), we give some calculations as follows:

$$a' = \sum_A x_Ae'_A \text{ (or } a' = \sum_A c_Ae'_A),$$

where $e'_A = (-1)^{|A|}e_A$ and $|A| = n_A$(see [7]) being the cardinality of A, i.e. when $A = \emptyset$, $|A| = 0$ and when $A = \{\alpha_2, \alpha_3, ..., \alpha_h\} \neq \emptyset$, then $|A| = h$.

$$\tilde{a} = \sum_A x_A\tilde{e}_A \text{ (or } \tilde{a} = \sum_A \overline{c}_A\tilde{e}_A),$$

where $\tilde{e}_A = (-1)^{|A|(|A|-1)/2}e_A$, and

$$\overline{a} = \widetilde{a'} = \tilde{a}' = \sum_A (-1)^{\frac{|A|(|A|+1)}{2}} x_Ae_A \text{ (or } \overline{a} = \sum_A (-1)^{\frac{|A|(|A|+1)}{2}} \overline{c}_Ae_A)).$$

For any $a, b \in \mathcal{A}_n(R)$ $(\mathcal{A}_n(C))$, an inner product is defined by

$$< a, b >= [a\overline{b}]_1, \tag{1.4}$$

where $[a\bar{b}]_1$ is the coefficient of e_1 in $a\bar{b}$, and we can prove that $[a\bar{b}]_1 = [\bar{b}a]_1$, so $< a, b >= [a\bar{b}]_1 = [\bar{b}a]_1$. From (1.2), (1.4) we have

$$|a|^2 =< a, a >= \sum_A |a_A|^2. \tag{1.5}$$

Proposition 1.1 *Let $a, b \in \mathcal{A}_n(R)(\mathcal{A}_n(C))$. Then*

$$\overline{ab} = \bar{b}\bar{a},\ |a| = |a'| = |\bar{a}| = |\tilde{a}|,\ |a|^2 = [a\bar{a}]_1 = [\bar{a}a]_1.$$

Definition 1.2 A Clifford number $a \in \mathcal{A}_n(R)(\mathcal{A}_n(C))$ is said to be conjugate, if $a\bar{a} = \bar{a}a = |a|^2$, and $\bar{a}$ is called the conjugate Clifford number of a.

Denote by $\aleph\,(\aleph(C))$ the set of all conjugate Clifford numbers in $\mathcal{A}_n(R)(\mathcal{A}_n(C))$.

Proposition 1.2 *For any $a \in \mathcal{A}_n(C)$, the following conditions are equivalent:*

1) a is conjugate.

2) $|ab| = |ba| = |b||a|$ for any $b \in \mathcal{A}_n(C)$.

3) There exists $b \in \mathcal{A}_n(C)\backslash\{0\}$, such that $ab = |a||b|$ or $ba = |b||a|$.

Proof It is sufficient to show that the assertion is valid for $a \neq 0$.

We first prove 1) $\to$ 2). By Proposition 1.1, it is clear that

$$|ab|^2 = [\overline{ab}ab]_1 = [\bar{b}\bar{a}ab]_1 = |a|^2[\bar{b}b]_1 = |a|^2|b|^2.$$

Similarly we can verify $|ba| = |b||a|$.

Setting $b = \bar{a}$ and using Proposition 1.1, we can get 2) $\to$ 1), 1) $\to$ 3).

As for 3) $\to$ 1), we need to show that $ab = |a||b|$ $(b \neq 0)$, i.e. $a\bar{a} = |a|^2$. There is no harm in assuming that $|a| = |b|$, then by Proposition 1.1, we get

$$< a, \bar{b} >= [ab]_1 = ab = |a||b| = |a||\bar{b}|. \tag{1.6}$$

Hence there exists $t \in C$ such that $\bar{b} = ta$ and $|t| = 1$. From (1.6), it follows that

$$\bar{t}|a|^2 =< a, ta >= |a|^2.$$

Since $a \neq 0$, we obtain $t = 1$, that is $\bar{a} = b$, therefore

$$a\bar{a} = ab = |a||b| = |a|^2.$$

By using the above result it is easy to extend Proposition 1.1, hence we have the following corollaries.

Corollary 1.3 $a, b \in \aleph(C) \Rightarrow ab \in \aleph(C)$.

Corollary 1.4 *If* $a_1, ..., a_m \in \aleph(C)$ *and* $b \in \mathcal{A}_n(C)$, *then*

$$|a_1, ..., a_{j-1}ba_j...a_m| = |a_1, |..., |a_{j-1}||b||a_j|...|a_m|.$$

2 Cauchy Integral Formula of Regular Functions and Plemelj Formula of Cauchy Type Integrals in Real Clifford Analysis

Firstly we give the definition of $\mathcal{A}$-value function.

Definition 2.1 Let $\Omega \subset \mathbf{R}^n$ be an open connected set. The function f which is defined in Ω with values in $\mathcal{A}_n(R)$ can be expressed as

$$f(x) = \sum_A e_A f_A(x),$$

where the function f_A is a real-valued function.

The set of C^r-functions in Ω with values in $\mathcal{A}_n(R)$ is denoted by

$$F_\Omega^{(r)} = \{f | f : \Omega \to \mathcal{A}_n(R),\ f(x) = \sum_A f_A(x)e_A\}.$$

We introduce also the Dirac operator

$$\overline{\partial} = \sum_{i=1}^n e_i \frac{\partial}{\partial x_i} : F_\Omega^{(r)} \to F_\Omega^{(r-1)}, \tag{2.1}$$

that is

$$\overline{\partial} f = \sum_{i,A} e_i e_A \frac{\partial f_A}{\partial x_i}.$$

The Dirac operator ∂ is defined as

$$\partial = e_1 \frac{\partial}{\partial x_1} - e_2 \frac{\partial}{\partial x_2} - \cdots - e_n \frac{\partial}{\partial x_n}. \tag{2.2}$$

Since $\mathcal{A}_n(R)$ is not a commutative algebra, in general the expression

$$\overline{\partial} f = e_1 \frac{\partial f}{\partial x_1} + \cdots + e_n \frac{\partial f}{\partial x_n}$$

is not same as

$$f\overline{\partial} = \frac{\partial f}{\partial x_1}e_1 + \cdots + \frac{\partial f}{\partial x_n}e_n.$$

Definition 2.2 Let $\Omega \subset \mathbf{R}^n$ be a domain and $f \in F_\Omega^{(r)}$. Then f is left regular in Ω, if $\overline{\partial} f = 0$ in Ω; and f is right regular in Ω, if $f\overline{\partial} = 0$ in Ω. For $n = 2$, $\overline{\partial} f = 2\partial f/\partial \overline{z} = \partial f/\partial x_1 + e_2 \partial f/\partial x_2$, thus under this condition the Clifford algebra is just the complex space and left regular is equivalent to holomorphic. For $n = 3$, $f(x) = \sum_A e_A f_A = e_1 f_1 + e_2 f_2 + e_3 f_3 + e_{23} f_{23}$. If we transform f_1, f_2, f_3, f_{23} into Φ_1, $-\Phi_2$, $-\Phi_3$, Φ_0, then $\overline{\partial} f = 0$ is equivalent to the elliptic system in higher dimensional domains, i.e. the $M - T$ equation. Liede Huang discussed the boundary value problems of the $M - T$ equation in [28] and found a lot of applications, for example the problem of airplane wing shapes, momentum pressure, electromagnetism, laser technology and so on.

Notice that the product of Dirac operators ∂ and $\overline{\partial}$ is equivalent to the Laplace operator

$$\overline{\partial}\partial = \partial\overline{\partial} = \frac{\partial^2}{\partial x_1^2} + \cdots + \frac{\partial^2}{\partial x_n^2} = \Delta. \tag{2.3}$$

Definition 2.3 Let $\Omega \subset \mathbf{R}^n$ be a domain and $f \in F_\Omega^{(r)}$. If

$$\Delta f = 0, \tag{2.4}$$

then f is called a harmonic function.

It is clear that we have the following theorem.

Theorem 2.1 *Let $\Omega \subset \mathbf{R}^n$ be a domain and $f \in F_\Omega^{(2)}$. If f is harmonic in Ω, then ∂f is left regular in Ω, and $f\partial$ is right regular in Ω; and if f is regular in Ω, then f is harmonic in Ω.*

Let $\Omega \subset \mathbf{R}^n$ be a domain. If $U = \sum_{p=1}^n \oplus \Lambda^p\, w$ is the exterior algebra with the basis $\{dx_1, ..., dx_n\}$, we consider the pth-differential form $\psi(x) = \sum_{A,H} \psi_{A,H}(x) e_A dx^H$, where $x \in \Omega \subset \mathbf{R}^n$. Furthermore, the function $\psi_{A,H}(x)$ is assumed to be in set $C^r(\Omega)\,(r \geq 1)$. Integration of $\psi(x)$ over the p-chain $\Gamma \subseteq \Omega$ is defined as

$$\int_\Gamma \psi(x) = \sum_{A,H} \int_\Gamma \psi_{A,H}(x) e_A dx^H. \tag{2.5}$$

Now we establish the Stokes theorem for functions in $F^{(1)}$ as follows: Denote by $\Gamma \subseteq \Omega$ an n-dimensional differentiable and oriented manifold. By means of the $(n-1)$-forms

$$d\widehat{x_i} = dx_1 \wedge \cdots \wedge dx_{i-1} \wedge dx_{i+1} \wedge \cdots \wedge dx_n, \; i = 1, ..., n,$$

a $\mathcal{A}_n(R)$-valued $n-1$–form is introduced by putting

$$d\sigma = \sum_{i=1}^{n} (-1)^{i-1} e_i d\widehat{x_i}.$$

If dS stands for the "classical" surface element and

$$\overrightarrow{m} = \sum_{i=1}^{n} e_i n_i,$$

where n_i is the i-th component of the unit outward normal vector, then the $\mathcal{A}_n(R)$-valued surface element $d\sigma$ can be written as

$$d\sigma = \overrightarrow{m} dS.$$

Furthermore the volume-element $dx = dx_1 \wedge \cdots \wedge dx_n$ is used.

Now we state and prove the Stokes-Green theorem about the function in F_Ω^r.

Theorem 2.2 (see [6]) *Let Ω be as stated above, $M \subseteq \Omega$ be an n-dimensional differentiable, oriented manifold, $f, g \in F_\Omega^{(r)}$ $(r \geq 1)$, and Γ be an arbitrary n-chain on M. Then we have*

$$\int_{\partial\Gamma} f d\sigma g = \int_{\Gamma} [(f\overline{\partial})g + f(\overline{\partial} g)] dx. \tag{2.6}$$

Proof Using the Stokes theorem for real-valued functions, we successively get

$$\begin{aligned}
\int_{\partial\Gamma} f d\sigma g &= \int_{\partial\Gamma} \sum_{A,i,B} (-1)^{i-1} f_A g_B d\widehat{x_i} e_A e_i e_B \\
&= \sum_{A,i,B} (-1)^{i-1} e_A e_i e_B \int_{\Gamma} (-1)^{i-1} \partial_{x_i}(f_A g_B) dx \\
&= \int_{\Gamma} \sum_{A,i,B} [(\partial_{x_i} f_A) e_A e_i g_B e_B + f_A e_A e_i e_B (\partial_{x_i} g_B)] dx \\
&= \int_{\Gamma} [(f\overline{\partial})g + f(\overline{\partial} g)] dx.
\end{aligned}$$

Corollary 2.3 *If f is right regular in Ω and g is left regular in Ω, then for any n-chain Γ on $M \subset \Omega$, the following integral holds:*

$$\int_{\partial\Gamma} f d\sigma g = 0. \tag{2.7}$$

Corollary 2.4 *Putting $f = 1$, the formula (2.6) yields*

$$\int_{\Gamma} \bar{\partial} g dx = \int_{\partial\Gamma} d\sigma g,$$

where $g \in F_{\Omega}^{(r)}$ $(r \geq 1)$.

For any g, which is left regular in M, we have

$$\int_{\partial\Gamma} d\sigma g = 0. \tag{2.8}$$

This is called the Cauchy theorem.

Next let $\Omega \subset \mathbf{R}^n$ be a domain and the boundary $\partial\Omega$ be a differentiable, oriented, compact Liapunov surface. The so-called Liapunov surface is a kind of surface satisfying the following three conditions (see [19]).

1. Through each point in $\partial\Omega$, there is a tangent plane.

2. There exists a real constant number $d > 0$, such that for any point $N_0 \in \partial\Omega$, we can construct a sphere E with the center at N_0 and radius d, and E is divided into two parts by $\partial\Omega$, one included in the interior of E denoted by $\partial\Omega'$, the other in the exterior of E; and each straight line parallel to the normal direction of $\partial\Omega$ at N_0, with $\partial\Omega'$ intersects at most a point.

3. If the angle $\theta(N_1, N_2)$ between outward normal vectors through N_1, N_2 is an acute angle, and r_{12} is the distance between N_1 and N_2, then there exist two numbers $b, \alpha\,(0 \leq \alpha \leq 1,\ b > 0)$ independent of N_1, N_2, such that $\theta(N_1, N_2) \leq b r_{12}$ for all points $N_1, N_2 \in \partial\Omega$.

From 3 we get that θ is continuous at the point $N \in \partial\Omega$ (see [29]2)). Let $N_0 \in \partial\Omega$ be a fixed point, and establish a polar coordinate system with the origin at N_0 and the outward normal direction of $\partial\Omega$ at N_0

as the direction of the positive x_n axis. Then the surface $\partial\Omega'$ may be written in the form

$$\xi_n = \xi_n(\xi_1, ..., \xi_{n-1}), \tag{2.9}$$

and ξ_n has first order partial derivatives on ξ_i $(i = 1, ..., n-1)$.

Let $d > 0$ be small enough such that $bd^\alpha \leq 1$, and for any point $N \in \partial\Omega'$, $\theta_0 = \theta(N_0, N)$ as stated before, and $r_0 = |N_0N|$ be the distance from N_0 to N $(r_0 < d)$. We can obtain

$$\cos\theta_0 \geq 1 - \frac{1}{2}\theta_0^2 \geq 1 - \frac{1}{2}b^2 r_0^{2\alpha} \geq 0,$$

and then

$$\frac{1}{\cos\theta_0} \leq \frac{1}{1 - \frac{1}{2}b^2 r_0^{2\alpha}} \leq 1 + b^2 r_0^{2\alpha} \leq 2.$$

Thus we have the formula

$$\cos\theta_0 \geq \frac{1}{2}. \tag{2.10}$$

Moreover we introduce a local generalized spherical coordinate at N_0 as follows:

$$\xi_{n-1} = \rho_0 \cos\varphi_1 \cos\varphi_2 ... \cos\varphi_{n-3} \cos\varphi_{n-2},$$

$$\xi_{n-2} = \rho_0 \cos\varphi_1 \cos\varphi_2 ... \cos\varphi_{n-3} \sin\varphi_{n-2},$$

$$...,$$

$$\xi_2 = \rho_0 \cos\varphi_1 \sin\varphi_2,$$

$$\xi_1 = \rho_0 \sin\varphi_1,$$

where ρ_0 is the length of the projection of r_0 on the tangent plane of $\partial\Omega$ at N_0, and φ_i satisfy the conditions

$$|\varphi_j| \leq \frac{\pi}{2},\ j = 1, 2, ..., n-3,\ 0 \leq \varphi_{n-2} < 2\pi,$$

hence we get

$$\cos(\overrightarrow{m_0}, x_n) \geq \frac{1}{2},\ \left|\frac{D(\xi_1, ..., \xi_{n-1})}{D(\rho_0, \varphi_1, ..., \varphi_{n-2})}\right| \leq \rho_0^{n-2}, \tag{2.11}$$

where $\overrightarrow{m_0}$ is a normal vector through N.

A function $f(y) : \partial\Omega \to \mathcal{A}_n(R)$ is said to be Hölder continuous on $\partial\Omega$, if $f(y)$ satisfies

$$|f(y_1) - f(y_1)| \leq M_1|y_1 - y_2|^\alpha,\ y_1, y_2 \in \partial\Omega\ (0 < \alpha < 1). \tag{2.12}$$

Denote by $H^{\alpha}_{\partial\Omega}$ the set of all Hölder continuous function on $\partial\Omega$ with the index α.

Now we consider the Cauchy type integral, Cauchy singular integral and the Plemelj formula.

Definition 2.4 The integral

$$\Phi(x) = \frac{1}{\omega_n}\int_{\partial\Omega} E(x,y)n(y)f(y)ds_y, \tag{2.13}$$

or

$$\Phi(x) = \frac{1}{\omega_n}\int_{\partial\Omega} E(x,y)d\sigma_y f(y)$$

is called the Cauchy type integral, where Ω and $\partial\Omega$ are as before, $n(y)$ is the normal vector through y, dS_y is the area difference $E(x,y) = \dfrac{\overline{y}-\overline{x}}{|y-x|^n}$, and $\omega_n = \frac{2\pi^{n/2}}{\Gamma(n/2)}$ is the area of the unit sphere in $\mathbf{R}^n$. When $x \in \mathbf{R}^n\backslash\partial\Omega$, $f(x) \in H^{\alpha}_{\partial\Omega}$, it is clear that the integral is well defined.

Definition 2.5 If $x_0 \in \partial\Omega$, construct a sphere E with the center at x_0 and radius $\delta > 0$, where $\partial\Omega$ is divided into two parts by E, and the part of $\partial\Omega$ lying in the interior of E is denoted by λ_δ. If $\lim\limits_{\delta\to 0}\Phi_\delta = I$, in which

$$\Phi_\delta(x_0) = \frac{1}{\omega_n}\int_{\partial\Omega-\lambda_\delta} E(x_0,y)d\sigma_y f(y), \tag{2.14}$$

then I is called the Cauchy principal value of singular integral and denoted by $I = \Phi(x_0)$.

Now we prove the Cauchy-Pompjeu integral formula.

Theorem 2.5 *Let Ω, $\partial\Omega$ be as stated above, and $\overline{\Omega} = \Omega\cup\partial\Omega$. Then for each $x \in \Omega$, $f \in F^{r}_{\overline{\Omega}}\,(r \geq 1)$, we have*

$$f(x) = \frac{1}{\omega_n}\int_{\partial\Omega} E(x,y)d\sigma_y f(y) - \frac{1}{\omega_n}\int_{\Omega} E(x,y)(\overline{\partial}f)(y)dy, \tag{2.15}$$

and

$$f(x) = \frac{1}{\omega_n}\int_{\partial\Omega} f(y)d\sigma_y E(x,y) - \frac{1}{\omega_n}\int_{\Omega} (f\overline{\partial})(y)E(x,y)dy. \tag{2.16}$$

Proof We also use the notations as before and denote by x a point in Ω and the hypercomplex number $x = x_1e_1 + \cdots + x_ne_n$. For $n \geq 3$, set

$$E(x,y) = \frac{1}{2-n}\partial(|y-x|^{2-n}) = \frac{1}{2-n}(|y-x|^{2-n})\partial,$$

and for $n = 2$,

$$E(x, y) = \partial(ln|y - x|) = (ln|y - x|)\partial.$$

In both instances,

$$\overline{\partial}(E(x, y)) = (E(x, y))\overline{\partial} = 0,\ y \neq x.$$

So both $|y - x|^{2-n}\,(n \geq 3)$ and $\ln|y - x|\,(n = 2)$ are harmonic. If $\Omega_\varepsilon = \{y \in \Omega : |y - x| > \epsilon\}$, where $x \in \Omega$ is fixed and the ε-ball: $|y - x| < \varepsilon$ about x lies completely within Ω, by the Stokes theorem one has

$$\int_{\Omega_\varepsilon} E(x, y)(\overline{\partial} f)(y)dy = \int_{\partial\Omega} E(x, y)d\sigma_y f(y) - \int_{|y-x|=\varepsilon} E(x, y)d\sigma_y f(y).$$

About the fixed point x, $f(y)$ may be approximated as $f(y) = f(x) + O(\varepsilon)$, in which $\lim_{\varepsilon\to 0} O(\varepsilon) = 0$. Using the Stokes theorem we have

$$\int_{|y-x|=\varepsilon} E(x, y)d\sigma_y = \omega_n.$$

Hence when $\varepsilon \to 0$, we can obtain the Cauchy-Pompieu representation (2.15). A calculation analogous to the above verifies the representation (2.16).

Corollary 2.6 *Let Ω, $\partial\Omega$ be as stated above, and f be a left regular function defined on $\overline{\Omega}$. Then*

$$\Phi(x_0) = \begin{cases} f(x_0), & x_0 \in \Omega, \\ 0, & x_0 \in \mathbf{R}^n \backslash \overline{\Omega}, \end{cases} \tag{2.17}$$

here $\Phi(x_0)$ is defined by Definition 2.5.

Theorem 2.7 *Suppose that Ω, $\partial\Omega$ are as stated before, and $f \in H^\alpha_{\partial\Omega}$, $x_0 \in \partial\Omega$. Then*

$$\begin{aligned} &\frac{1}{\omega_n}\int_{\partial\Omega} E(x_0, y)d\sigma_y f(y) \\ &= \frac{1}{\omega_n}\int_{\partial\Omega} E(x_0, y)d\sigma_y[f(y) - f(x_0)] + \frac{1}{2}f(x_0), \end{aligned} \tag{2.18}$$

where the first integral is defined by Cauchy's principal value, and the second integral is defined by a generalized integral.

Proof Let λ_δ be as stated above. Then

$$\int_{\partial\Omega-\lambda_\delta} E(x_0,y)d\sigma_y f(y)$$
$$= \int_{\partial\Omega-\lambda_\delta} E(x_0,y)d\sigma_y[f(y)-f(x_0)] + \int_{\partial\Omega-\lambda_\delta} E(x_0,y)d\sigma_y f(x_0). \tag{2.19}$$

Because $f \in H^\alpha_{\partial\Omega}$ and (2.11), (2.12), we can get

$$|E(x_0,y)d\sigma_y[f(y)-f(x_0)]| \le \frac{M}{|y-x_0|^{n-1-\alpha}}|dSy|$$
$$\le \left|\frac{D(\xi_1,\dots,\xi_{n-1})}{D(\rho_0,\varphi_1,\dots,\varphi_{n-2})}\right| \frac{M}{|y-x_0|^{n-1-\alpha}}|d\rho_0 d\varphi_1 \dots d\varphi_{n-2}|$$
$$\le |\frac{M}{|y-x_0|^{n-1-\alpha}} 2\rho_0^{n-2} d\rho_0 d\varphi_1 \dots d\varphi_{n-2}| \le M' \frac{1}{\rho_0^{1-\alpha}} d\rho_0,$$

in which the integral of the last function is convergent, so

$$\lim_{\delta\to 0} \int_{\partial\Omega-\lambda_\delta} E(x_0,y)d\sigma_y[f(y)-f(x_0)] = \int_{\partial\Omega} E(x_0,y)d\sigma_y[f(y)-f(x_0)].$$

Calculating the second item, and setting $D_{out} = \{\partial[(D(x_0,\delta)) \cup \overline{\Omega}]\} \cap (R^n - \overline{\Omega})$, by the Cauchy formula (see Corollary 2.6), we have

$$\int_{\partial\Omega-\lambda_\delta+D_{out}} E(x_0,y)d\sigma_y f(x_0) = \omega_n f(x_0),$$

and

$$\int_{D_{out}} E(x_0,y)d\sigma_y f(x_0) = \int_{D_{out}} E(x_0,y)\frac{y-x_0}{|y-x_0|}dS_y f(x_0)$$
$$= \int_{D_{out}} \frac{1}{|y-x_0|^{n-1}} dS_y f(x_0),$$

and from the condition about $\partial\Omega$, we obtain

$$\lim_{\delta\to 0} \int_{D_{out}} E(x_0,y)d\sigma_y f(x_0) = \frac{1}{2}\omega_n f(x_0),$$

so

$$\lim_{\delta\to 0}\int_{\partial\Omega-\lambda_\delta} E(x_0,y)d\sigma_y f(x_0)$$
$$= \lim_{\delta\to 0}\int_{\partial\Omega-\lambda_\delta+D_{out}} E(x_0,y)d\sigma_y f(x_0)$$
$$-\lim_{\delta\to 0}\int_{D_{out}} E(x_0,y)d\sigma_y f(x_0)=\frac{1}{2}\omega_n f(x_0).$$

Summarizing the above discussion and letting $\delta \to 0$ in (2.19), we can get the required result.

Next we consider the limit of Cauchy's integral when $x \to x_0$, $x_0 \in \partial\Omega$ from $\Omega^+ = \Omega$ and $\Omega^- = R^n\backslash\Omega$; we first give a Hile's lemma.

Lemma 2.8 *Suppose that* $t, x \in \mathbf{R}^n$, $n(\geq 2)$ *and* $m(\geq 0)$ *are integers. Then*

$$\left|\frac{x}{|x|^{m+2}}-\frac{t}{|t|^{m+2}}\right| \leq \frac{P_m(x,t)}{|x|^{m+1}|t|^{m+1}}|x-t|, \tag{2.20}$$

where

$$P_m(x,t)=\begin{cases}\sum\limits_{k=0}^{m}|x|^{m-k}|t|^k, & m\neq 0,\\ 1, & m=0.\end{cases}$$

Proof The proof can be seen in [19].

We rewrite (2.13) in the form

$$\begin{aligned}\Phi(x) &= \frac{1}{\omega_n}\int_{\partial\Omega}E(x,y)d\sigma_y f(y)\\ &= \frac{1}{\omega_n}\int_{\partial\Omega}E(x,y)d\sigma_y[f(y)-f(x_0)]+\frac{1}{\omega_n}\int_{\partial\Omega}E(x,y)d\sigma_y f(x_0)\\ &= F(x)+\frac{1}{\omega_n}\int_{\partial\Omega}E(x,y)d\sigma_y f(x_0).\end{aligned} \tag{2.21}$$

From Corollary 2.6, we get

$$\frac{1}{\omega_n}\int_{\partial\Omega}E(x,y)d\sigma_y f(x_0)=\begin{cases} f(x_0), & x\in\Omega^+,\\ 0, & x\in\Omega^-.\end{cases} \tag{2.22}$$

Now we need to consider the limit of (2.14), it suffices to study the first item in (2.21).

Theorem 2.9 *Let $\Omega, \partial\Omega$ be as stated above, and $f(x) \in H^{\alpha}_{\partial\Omega}$, $0 < \alpha < 1$. Then for $x_0 \in \partial\Omega$ we have*

$$\lim_{\substack{x \to x_0 \\ x \in \mathbf{R}^n \setminus \partial\Omega}} F(x) = F(x_0). \tag{2.23}$$

Proof We first assume that $x \to x_0$ and is not along the direction of the tangent plane at $x_0 (\in \partial\Omega)$. This means that the angle between the tangent plane of $\partial\Omega$ at x_0 and line segment $\overline{xx_0}$ is greater than $2\beta_0$, then

$$\begin{aligned} & F(x) - F(x_0) \\ = & \frac{1}{\omega_n} \int_{\partial\Omega} E(x,y) d\sigma_y [f(y) - f(x_0)] - \frac{1}{\omega_n} \int_{\partial\Omega} E(x_0,y) d\sigma_y [f(y) - f(x_0)] \\ = & \frac{1}{\omega_n} \int_{\partial\Omega} [E(x,y) - E(x_0,y)] d\sigma_y [f(y) - f(x_0)] \\ = & \frac{1}{\omega_n} \int_{\lambda_\delta} [E(x,y) - E(x_0,y)] d\sigma_y [f(y) - f(x_0)] \\ & + \frac{1}{\omega_n} \int_{\partial\Omega - \lambda_\delta} [E(x,y) - E(x_0,y)] d\sigma_y [f(y) - f(x_0)] = I_1 + I_2. \end{aligned}$$

From the conditions that $\partial\Omega$ is a Liapunov surface and $x \to x_0$ is not along the direction of the tangent plane and the angle to the tangent plane of $\partial\Omega$ at x_0 is greater than $2\beta_0$, where β_0 is a constant, we can get

$$\left| \frac{y - x_0}{y - x} \right| \le M_2 < \infty, \quad \left| \frac{x - x_0}{y - x} \right| \le M_2 < \infty;$$

again by Hile's lemma, we have

$$\begin{aligned} |E(x,y) - E(x_0,y)| &= \left| \frac{\overline{y} - \overline{x}}{|y-x|^n} - \frac{\overline{y} - \overline{x_0}}{|y-x_0|^n} \right| \\ &\le \sum_{k=0}^{n-2} \frac{|x - x_0|}{|y - x|^{k+1}} \frac{1}{|y - x_0|^{n-1-k}} \\ &= \sum_{k=1}^{n-1} \frac{|x - x_0|}{|y - x|^k} \frac{1}{|y - x_0|^{n-k}} \\ &= \sum_{k=1}^{n-1} \frac{|x - x_0|}{|y - x|} \frac{|y - x_0|^{k-1}}{|y - x|^{k-1}} \frac{1}{|y - x_0|^{n-1}} \\ &\le (n-1) M_2^k \frac{1}{|y - x_0|^{n-1}} \le \frac{(n-1) M_2^k}{\rho_0^{n-1}}. \end{aligned}$$

From (2.11), it follows that

$$|dS_y| = |d\sigma_y| = \left|\frac{D(\xi_1, ..., \xi_{n-1})}{D(\rho_0, \varphi_1, ..., \varphi_{n-2})}\right| \times |d\rho_0 d\varphi_1 ... d\varphi_{n-2}| \le M_2 \rho_0^{n-2} d\rho_0,$$

and

$$|I_1| \le \frac{1}{\omega_n} \int_{\lambda_\delta} |E(x,y) - E(x_0,y)||d\sigma_y||f(y) - f(x_0)| \le M_3 \int_0^\delta \frac{1}{\rho_0^{n-1}} \rho_0^{n-2} \rho_0^\alpha d\rho_0 \le M_4 \int_0^\delta \rho_0^{\alpha-1} d\rho_0,$$

in which the constants M_3, M_4 are independent of x_0, and for arbitrary $\varepsilon > 0$, there exists a $\rho > 0$ such that when $\delta < \rho, |I_1| < \varepsilon/2$. Given a fixed δ such that $0 < \delta < \rho$, thus we can estimate I_2. By

$$|E(x,y) - E(x_0,y)| \le \left(\sum_{k=1}^{n-1} \frac{|y - x_0|^k}{|y - x|^k} \frac{1}{|y - x_0|^n} \right) |x - x_0|,$$

we get

$$|I_2| = \frac{1}{\omega_n} \int_{\partial\Omega - \lambda_\delta} |E(x,y) - E(x_0,y)||d\sigma_y||f(y) - f(x_0)| \le \frac{1}{\omega_n} \int_\delta^L \frac{(n-1)M_2^k}{\rho_0^n} 2\rho_0^{n-2} d\rho_0 M_1 |x - x_0|.$$

Here L is a positive constant because of $\partial\Omega$ being bounded. Thus when $|x - x_0|$ is small enough we can get $|I_2| < \varepsilon/2$. Hence $F(x)$ converges to $F(x_0)$ on $\partial\Omega$ as $x \to x_0$. Note that δ is independent of x_0, and $x', x_0 \in \partial\Omega$, thus it is easy to see that $F(x)$ is uniformly continuous on $\partial\Omega$.

Next consider $x \to x_0$ along the direction of the tangent plane of $\partial\Omega$ at x_0. When $|x - x_0|$ is small enough, we can choose a point $y(\in \partial\Omega)$ such that $|x - y|$, $|y - x_0|$ are also small enough, and x lies in the direction of the tangent plane of $\partial\Omega$ at y. Taking account that $F(x)$ on $\partial\Omega$ is uniformly continuous, we can derive that for any $\varepsilon > 0$, when $|x - x_0|, |y - x_0|$ are small enough, the following inequalities hold:

$$|F(y) - F(x_0)| \le \frac{\varepsilon}{2}, \ |F(x) - F(y)| \le \frac{\varepsilon}{2},$$

thus

$$|F(x) - F(x_0)| \le |F(x) - F(y)| + |F(y) - F(x_0)| \le \varepsilon.$$

This completes the proof.

Theorem 2.10 *Let* Ω, $\partial\Omega$, Ω^+, Ω^- *be as stated above,* $f(x) \in H^\alpha_{\partial\Omega}$, $0 < \alpha < 1$, $x_0 \in \partial\Omega$, *and denote by* $\Phi^+(x_0)$, $\Phi^-(x_0)$ *the limits of* $\Phi(x)$, *when* $x \to x_0$ *in* Ω^+, Ω^- *respectively. Then*

$$\begin{cases} \Phi^+(x_0) = \dfrac{1}{\omega_n}\displaystyle\int_{\partial\Omega} E(x_0, y)d\sigma_y f(y) + \dfrac{1}{2}f(x_0), \\ \Phi^-(x_0) = \dfrac{1}{\omega_n}\displaystyle\int_{\partial\Omega} E(x_0, y)d\sigma_y f(y) - \dfrac{1}{2}f(x_0), \end{cases} \tag{2.24}$$

or

$$\begin{cases} \Phi^+(x_0) - \Phi^-(x_0) = f(x_0), \\ \Phi^-(x_0) + \Phi^-(x_0) = \dfrac{2}{\omega_n}\displaystyle\int_{\partial\Omega} E(x_0, y)d\sigma_y f(y). \end{cases} \tag{2.25}$$

Proof By using Cauchy's integrals (2.13), (2.21), (2.22) and Theorem 2.5, when $x \to x_0$ in Ω^+, Ω^-, $\Phi(x)$ possesses the limits

$$\Phi^+(x_0) = \frac{1}{\omega_n}\int_{\partial\Omega} E(x_0, y)d\sigma_y[f(y) - f(x_0)] + f(x_0),$$

$$\Phi^-(x_0) = \frac{1}{\omega_n}\int_{\partial\Omega} E(x_0, y)d\sigma_y[f(y) - f(x_0)],$$

respectively. Form (2.21) it follows that (2.24) is valid. Similarly we can prove (2.25). The formula (2.25) is called the Plemelj formula for left regular functions, which is generally used to discuss boundary value problems.

3 Quasi-Permutations and Generalized Regular Functions in Real Clifford Analysis

Clifford numbers that we use in Clifford analysis do not have the commutative property, creates a lot of difficulties in handling some problems. In this section we first introduce the conception of quasi-permutation proposed by Sha Huang (see [29]4),5),6)), and then we give some equivalent conditions of regular functions and generalized regular functions by the above quasi-permutation.

3.1 The quasi-permutation and its properties

Definition 3.1 Let $A = \{h_1, h_2, ..., h_k\}$, $1 \le h_1 < h_2 < ... < h_k$, here h_i $(i = 1, ..., k)$, and m be natural numbers. For the arrangement

$$\overline{mA} = \begin{cases} A \setminus \{m\}, & m \in A, \\ \{g_1, g_2, ..., g_{k+1}\}, & m \notin A, \end{cases}$$

where $g_i \in A \cup m$, $1 \le g_1 < g_2 < ... < g_{k+1}$, we call $\overline{mA}$ the first class quasi-permutation for arrangement mA, and define $\overline{mm} = 1$, $\overline{m1} = m$. If there are p natural numbers $h_i \in [2, m]$, then the sign

$$\delta_{\overline{mA}} = (-1)^p$$

is called the sign of the first class quasi-permutation $\overline{mA}$.

Property 3.1 *1. If $\overline{mA} = B$, then $\overline{mB} = A$.*

2. $\overline{mA} = \overline{Am}$.

3. Let $m = 1$, then $\delta_{\overline{mA}} = 1$.

4. If $\overline{mA} = B$ and $m \ne 1$, then $\delta_{\overline{mA}} = -\delta_{\overline{mB}}$.

5. $e_m e_A = \delta_{\overline{mA}} e_{\overline{mA}}$.

The proof is easily obtained by Definition 3.1.

Definition 3.2 Let A be as stated in Definition 3.1 and $B = \{l_1, l_2, ..., l_m\}$, $1 \le l_1 < l_2 < ... < l_m \le n$, $l_j\,(1 \le j < m)$ be natural numbers. We call the arrangement $\overline{BA} = (B \cup A) \setminus (B \cap A)$ with the natural order as the second class quasi-permutation for arrangement BA, and

$$\delta_{\overline{BA}} = \prod_{j=1}^{m} \delta_{\overline{l_j A}}$$

as the sign of the second class quasi-permutation. And we let $\overline{BB} = 1$, $\overline{B1} = B$

Property 3.2. *1. If $\overline{BA} = C$, then $\overline{CA} = B$.*

2. $\overline{BA} = \overline{AB}$.

3. Let $A = \{h_1, h_2, ..., h_r\} \ne \{1\}$. Then $\delta_{\overline{AA}} = (-1)^{\frac{r(r+1)}{2}}$.

4. $e_B e_A = \delta_{\overline{BA}} e_{\overline{BA}}$.

5. *If* $a = \sum_C a_C e_C$, $A = \overline{CB}$, $\omega = \sum_B \omega_B e_B$, *then*

$$a\omega = \sum_A (\sum_C a_C \delta_{\overline{CB}} \omega_B) e_A.$$

Proof. Noting that

$$\begin{aligned} a\omega &= \sum_C a_C e_C \sum_B \omega_B e_B \\ &= \sum_C \sum_B a_C \omega_B e_C e_B \\ &= \sum_{B,C} a_C \omega_B \delta_{\overline{CB}} e_{\overline{CB}}, \end{aligned}$$

we can write $A = \overline{CB}$, and notice for every C, A and B monogamy, so

$$a\omega = \sum_A \sum_C a_C \delta_{\overline{CB}} \omega_B e_A.$$

Definition 3.3 Let j be a natural number and $D = \{h_1, h_2, ..., h_s\}$, $1 < h_1 < h_2 < ... < h_s \leq n$, $h_p \, (1 \leq p \leq s)$ be natural numbers. The arrangement Dj is called the third class quasi-permutation of arrangement jD, and the sign

$$\varepsilon_{jD} = \begin{cases} (-1)^s & \text{for any } p \in [1, s], \ h_p \neq j, \\ (-1)^{s-1} & \text{for some } p \in [1, s], \ h_p = j, \end{cases}$$

$$\varepsilon_{1D} = 1,$$

is called the sign of the third class quasi-permutation.

Property 3.3 *If* j, D *are as stated above, then*

$$e_j e_D = \varepsilon_{jD} e_D e_j.$$

Definition 3.4 Let the arrangement $D = \{h_1, ..., h_r\}$ and C be natural number arrangement with natural order as above. Then the arrangement CD is called the fourth class quasi-permutation of arrangement DC and the sign

$$\varepsilon_{DC} = \prod_{p=1}^{r} \varepsilon_{h_p C}$$

is called the sign of the fourth class quasi-permutation.

Property 3.4 *If C, D are as stated above, then*

$$e_D e_C = \varepsilon_{DC} e_C e_D.$$

Property 3.5 *Let $C = \{h_1, ..., h_r\}$, $D = \{k_1, ..., k_p\}$ as stated in Definition 3.1, where h_i, k_j are all not equal to 1 and q elements in C and D are the same. Then $e_D e_C = (-1)^{rp-q} e_C e_D$, $\varepsilon_{DC} = (-1)^{rp-q}$, and when r or p is an even number, then $e_D e_C = (-1)^q e_C e_D, \varepsilon_{DC} = (-1)^q$.*

Definition 3.5 Let C, D, E, M be the arrangement of natural numbers as in Definition 3.1. The arrangement $CDEM$ is called the fifth class quasi-permutation of arrangement $EDCM$, and if

$$\overline{CM} = \{h_1, ..., h_r\},\ \overline{EM} = \{b_1, ..., b_p\},\ \nu_{EDC} = \varepsilon_{ED}\varepsilon_{EC}\varepsilon_{DC},$$

then we call

$$\mu_{EDC} = (-1)^{\frac{r(r+1)+p(p+1)}{2}} \nu_{EDC}$$

the sign of the fifth class quasi-permutation.

Property 3.6 *Let E, D, M be the arrangement of natural numbers as in Definition 3.1, $\overline{ED} = \{h_1, ..., h_r\} = A\,(h_j \neq 1)$, $\overline{MA} = C$, $\overline{CD} = \{b_1, ..., b_p\} = B\,(b_j \neq 1)$. Then*

1. $\delta_{\overline{ED}}\delta_{\overline{CM}} = \mu_{EDC}\delta_{\overline{CD}}\delta_{\overline{EM}}.$
2. $\delta_{\overline{CD}}\delta_{\overline{EM}} = \mu_{EDC}\delta_{\overline{ED}}\delta_{\overline{CM}}.$
3. $e_E e_D e_C e_M = \nu_{EDC}(-1)^{\frac{p(p+1)}{2}}\delta_{\overline{CD}}\delta_{\overline{EM}}.$

Proof We calculate $e_E e_D e_C e_M$ in two ways, and obtain

$$e_E e_D e_C e_M = \delta_{\overline{ED}} e_{\overline{ED}} \delta_{\overline{CM}} e_{\overline{CM}}$$
$$= \delta_{\overline{ED}}\delta_{\overline{CM}} e_{\overline{ED}} e_{\overline{CM}} = (-1)^{\frac{r(r+1)}{2}} \delta_{\overline{ED}}\delta_{\overline{CM}},$$

where $\overline{ED} = \overline{\overline{M\overline{ED}M}} = \overline{CM}$, and then

$$e_E e_D e_C e_M = \nu_{EDC} e_C e_D e_E e_M$$
$$= \nu_{EDC}\delta_{\overline{CD}}\delta_{\overline{EM}} e_{\overline{CD}} e_{\overline{EM}}$$
$$= \nu_{EDC}(-1)^{\frac{p(p+1)}{2}} \delta_{\overline{CD}}\delta_{\overline{EM}},$$

where $\overline{CD} = \overline{\overline{\overline{ME}}D}D = \overline{EM}$. From the above equality, we can derive

$$\delta_{\overline{ED}}\delta_{\overline{CM}} = (-1)^{\frac{r(r+1)+p(p+1)}{2}}\nu_{EDC}\delta_{\overline{CD}}\delta_{\overline{EM}}$$

and the proof of 1 is finished. Similarly we can prove 2, 3.

Property 3.7 *Let D, B be the natural arrangement of natural numbers, $j(\neq m)$ be natural numbers, and*

$$\overline{jD} = A = \{h_1, ..., h_r\} = \overline{mB},\ \overline{jB} = \overline{mD} = \{b_1, ..., b_p\}.$$

Then

$$\delta_{\overline{mD}}\delta_{\overline{jB}} = -\delta_{\overline{jD}}\delta_{\overline{mB}}.$$

Proof We prove it according to the following three cases:

1. $m, j \in D$.
2. $m, j \notin D$.
3. D includes one of m and j.

If $m, j \in D$, then $m, j \notin N \setminus B$, and there exists a set E, such that $D = E \cup \{j, m\}$, $B = E \setminus \{j, m\}$, hence

$$e_m e_D e_j e_B = \delta_{\overline{mD}} e_{\overline{mD}} \delta_{\overline{jB}} e_{\overline{jB}} = \delta_{\overline{mD}}\delta_{\overline{jB}}(-1)^{\frac{r(r+1)}{2}},$$

moreover we have

$$e_m e_D e_j e_B = \nu_{mDj} e_j e_D e_m e_B = \nu_{mDj}\delta_{\overline{jD}}\delta_{\overline{mB}}$$

$$\times e_{\overline{jD}} e_{\overline{mB}} = -\delta_{\overline{jD}}\delta_{\overline{mB}}(-1)^{\frac{r(r+1)}{2}}.$$

From this, we get 1. Similarly we can prove 2, 3.

3.2 Regular functions and generalized regular functions

By the method using the sign of quasi-permutation, in the following we give equivalent conditions for regular functions and generalized regular functions.

Theorem 3.8 *If $f : \Omega \to \mathcal{A}_n(R)$, then $f(x) = \sum_A f_A(x)e_A$ is a left regular function if and only if*

$$f_{Ax_1} = \sum_{m=2}^{n} \delta_{\overline{mA}} f_{\overline{mA}x_m},$$

where $(\ )_{x_m} = \partial(\ )/\partial x_m$.

Proof By the definition of quasi-permutations, we have

$$\begin{aligned}\overline{\partial} f(x) &= \sum_{i=1}^{n} e_i \frac{\partial f}{\partial x_i} = \sum_{i=1}^{n}\sum_{A} e_i e_A \frac{\partial f_A}{\partial x_i} \\ &= \sum_{i,A} \delta_{\overline{iA}} e_{\overline{iA}} \frac{\partial f_A}{\partial x_i} = \sum_{A} \left[\delta_{\overline{1A}} e_{\overline{1A}} \frac{\partial f_A}{\partial x_1} + \sum_{i=2}^{n} \delta_{\overline{iA}} e_{\overline{iA}} \frac{\partial f_A}{\partial x_i} \right],\end{aligned}$$

when $2 \le i \le n$; if $\overline{iA} = B$, then $A = \overline{iB}$, $\delta_{\overline{iA}} = -\delta_{\overline{iB}}$. So

$$\overline{\partial} f(x) = \sum_{A} \left[\delta_{\overline{1A}} e_{\overline{1A}} \frac{\partial f_A}{\partial x_1} \right] - \sum_{B}\sum_{i=2}^{n} \left[\delta_{\overline{iB}} e_B \frac{\partial f_{\overline{iB}}}{\partial x_i} \right].$$

Again because $\overline{iA} = B$, $\overline{iB} = A\,(2 \le i \le n)$, we can verify that the elements of the sets $\{A\}$ and $\{B\}$ possess the same quantity and same form. From this formula it follows that

$$\overline{\partial} f(x) = \sum_{A} \left[e_A \frac{\partial f_A}{\partial x_1} - \sum_{i=2}^{n} \delta_{\overline{iA}} e_A \frac{\partial f_{\overline{iA}}}{\partial x_i} \right].$$

Finally by the definition of $\overline{\partial} f(x)$, the proof is completed.

Especially, when $n = 3$, the equivalent condition is the system of equations (3.0), Chapter VII.

Theorem 3.9 *Let $f : \Omega \to \mathcal{A}$. Then $f(x) = \sum_A f_A(x)e_A$ is a left regular function if and only if*

$$\sum_{m=1}^{n} \delta_{\overline{mB}} \frac{\partial f_B}{\partial x_m} = 0,$$

where $\overline{mA} = B$.

Proof By Theorem 3.8 and $\delta_{\overline{iA}} = -\delta_{\overline{iB}}\,(i \ne 1, \overline{iA} = B, \overline{iB} = A)$, the theorem is easily proved.

Let $w = \sum_A w_A e_A$. Then $\overline{w} = \sum_A w_A \overline{e}_A$, if $A = \{h_1, ..., h_r\}$, $\overline{e}_A = (-1)^{\frac{r(r+1)}{2}} e_A$ (see Definition 1.1). When $r = 0, 3 \pmod 4$, i.e. $(-1)^{\frac{r(r+1)}{2}} = 1$, we denote A by $\underline{A}$ and the A is called the $\underline{A}$-type index. When $r = 1, 2 \pmod 4$, i.e. $(-1)^{\frac{r(r+1)}{2}} = -1$, we denote A by $\underline{\underline{A}}$ and the A is called the $\underline{\underline{A}}$-type index. Hence we can write

$$\overline{\omega} = \sum_{\underline{A}} \omega_{\underline{A}} e_{\underline{A}} - \sum_{\underline{\underline{A}}} \omega_{\underline{\underline{A}}} e_{\underline{\underline{A}}}$$

Definition 3.6 Let $\Omega \subset \mathbf{R}^n$ be a bounded domain, and $w(x) = \sum_A w_A(x) e_A \in F_\Omega^{(r)}$ $(r \geq 1)$. If $w(x)$ satisfies the system of first order equations

$$\overline{\partial} w = aw + b\overline{w} + l, \; x \in \Omega, \tag{3.1}$$

then $w(x)$ is called a generalized regular function in Ω, where

$$a(x) = \sum_C a_C(x) e_C, \; b(x) = \sum_C b_C(x) e_C, \; l(x) = \sum_A l_A(x) e_A.$$

By Theorems 3.8 and 3.9, we know

$$\overline{\partial} w = \sum_{\overline{mB}=A} \sum_{m=1}^{n} \delta_{\overline{mB}} w_{Bx_m} e_A,$$

and by Property 3.2, at the same time letting $w = \sum_{\underline{M}} w_{\underline{M}} e_{\underline{M}} + \sum_{\underline{\underline{M}}} w_{\underline{\underline{M}}} e_{\underline{\underline{M}}}$, we have

$$\begin{aligned} aw + b\overline{w} + l = {} & (\sum_C a_C e_C)(\sum_{\underline{M}} w_{\underline{M}} e_{\underline{M}} + \sum_{\underline{\underline{M}}} w_{\underline{\underline{M}}} e_{\underline{\underline{M}}}) \\ & + (\sum_C b_C e_C)(\sum_{\underline{M}} w_{\underline{M}} e_{\underline{M}} - \sum_{\underline{\underline{M}}} w_{\underline{\underline{M}}} e_{\underline{\underline{M}}}) + \sum_A l_A e_A \\ = {} & \sum_{C,A} a_C w_{\underline{M}} \delta_{\overline{C\underline{M}}} e_A + \sum_{C,A} a_C w_{\underline{\underline{M}}} \delta_{\overline{C\underline{\underline{M}}}} e_A + \sum_{C,A} b_C w_{\underline{M}} \delta_{\overline{C\underline{M}}} e_A \\ & - \sum_{C,A} b_C w_{\underline{\underline{M}}} \delta_{\overline{C\underline{\underline{M}}}} e_A + \sum_A l_A e_A = \sum_{C,A} (a_C + b_C) w_{\underline{M}} \delta_{\overline{C\underline{M}}} e_A \\ & + \sum_{C,A} (a_C - b_C) w_{\underline{\underline{M}}} \delta_{\overline{C\underline{\underline{M}}}} e_A + \sum_A l_A e_A, \end{aligned}$$

where $\overline{C\underline{M}} = A$, $\overline{C\underline{\underline{M}}} = A$. Thus (3.1) becomes

$$\sum_{A}\sum_{m=1}^{n} \delta_{\overline{mB}} w_{Bx_m} e_A = \sum_{C,A} (a_C + b_C) w_{\underline{M}} \delta_{\overline{C\underline{M}}} e_A$$

$$+ \sum_{C,A} (a_C - b_C) w_{\underline{\underline{M}}} \delta_{\overline{C\underline{\underline{M}}}} e_A + \sum_{A} l_A e_A,$$

in which $\overline{mB} = A$, $\overline{C\underline{M}} = A$, $\overline{C\underline{\underline{M}}} = A$. From the above discussion, we have the following theorem. That is to say for any A, the following formula (3.2) is true.

Theorem 3.10 *Let $\Omega \subset R^n$ be a bounded domain, and $w(x) = \sum_{M} w_M(x) e_M \in F_{\Omega}^{(r)}$ $(r \geq 1)$. Then $w(x)$ is a generalized regular function in Ω if and only if $w(x)$ satisfies the elliptic system of first order equations*

$$\sum_{\substack{m=1 \\ \overline{mB}=A}}^{n} \delta_{\overline{mB}} w_{Bx_m} = \sum_{\substack{C \\ \overline{C\underline{M}}=A}} (a_C + b_C) w_{\underline{M}} \delta_{\overline{C\underline{M}}}$$

$$+ \sum_{\substack{C \\ \overline{C\underline{\underline{M}}}=A}} (a_C - b_C) w_{\underline{\underline{M}}} \delta_{\overline{C\underline{\underline{M}}}} + l_A. \tag{3.2}$$

In addition, we can also derive the relation between a generalized regular function and an elliptic system of second order equations.

Theorem 3.11 *Let $\Omega \subset \mathbf{R}^n$ be a bounded domain, and $w(x) = \sum_{M} w_M(x) e_M \in F_{\Omega}^{(r)}$ $(r \geq 2)$. Then $w(x)$ is a generalized regular function in Ω if and only if the components of $w(x)$ satisfy the elliptic system of second order equations*

$$\Delta w_D = \sum_{j,C} \delta_{\overline{jD}} \delta_{\overline{C\underline{M}}} (a_C + b_C)_{x_j} w_{\underline{M}}$$

$$+ \sum_{j,C} \delta_{\overline{jD}} \delta_{\overline{C\underline{M}}} (a_C + b_C) w_{\underline{M}x_j} + \sum_{j,C} \delta_{\overline{jD}}$$

$$\times \delta_{\overline{C\underline{\underline{M}}}} (a_C - b_C)_{x_j} w_{\underline{\underline{M}}} + \sum_{j,C} \delta_{\overline{jD}} \delta_{\overline{C\underline{\underline{M}}}} \tag{3.3}$$

$$\times (a_C - b_C) w_{\underline{\underline{M}}x_j} + \sum_{j=1}^{n} \delta_{\overline{jD}} l_{Ax_j},$$

where $j = 1, ..., n$, $\overline{jD} = A$, $\overline{C\underline{M}} = A$, $\overline{C\underline{\underline{M}}} = A$.

Proof Let $w(x) = \sum\limits_{M} w_M(x)e_M$ be a generalized regular function. From Theorem 3.10, it is clear that $w_M(x)$ satisfies (3.2), and for a fixed index D, we consider the term w_D in the left-hand side of equation (3.2), i.e. the equation corresponding to e_A satisfying that $\overline{jD} = A$. Suppose that the term corresponding to w_D is $\delta_{\overline{jD}} w_{Dk_j}$, and this equation is said to be the equation for w_D. Then we find partial derivatives with respect to x_j for the equation including W_D, and then multiply this equation by $\delta_{\overline{jD}}$. Finally sum these derivatives according to the index j, we get

$$\begin{aligned}&\sum_{j=1}^{n} \delta_{\overline{jD}} \left(\sum_{j=1}^{n} \delta_{\overline{mB}} w_{Bx_m} \right)_{x_j} \\ = &\sum_{j=1}^{n} (\delta_{\overline{jD}})^2 w_{Dx_j^2} + \sum_{j,m,j\neq m} \delta_{\overline{jD}} \delta_{\overline{mB}} w_{Bx_m x_j}.\end{aligned} \tag{3.4}$$

Noting that $\overline{mB} = A$, $A = \overline{jD}\,(j = 1, ..., n)$ in (3.4) only include a part of all indexes(real subset) of sets A, and that from Corollary 3.7 we have

$$\delta_{\overline{jD}} \delta_{\overline{mB}} w_{Bx_m x_j} = -\delta_{\overline{mD}} \delta_{\overline{jB}} w_{Bx_j x_m},$$

then

$$\sum_{j,m,j\neq m} \delta_{\overline{jD}} \delta_{\overline{mB}} w_{Bx_m x_j} = 0\,(\overline{jD} = A, \overline{mB} = A, j \neq m);$$

moreover we can derive

$$\sum_{j=1}^{n} \delta_{\overline{jD}} \left(\sum_{m=1}^{n} \delta_{\overline{mB}} w_{Bx_m} \right)_{x_j} = \Delta w_D. \tag{3.5}$$

In fact we consider the right side of equation for $w(x)$ in (3.2). By using the same method (i.e. we find the partial derivatives with respect to x_j and multiply by $\delta_{\overline{jD}}$, and then sum them by j) we can get the expression of w_D in the right side of equation (3.4):

$$\begin{aligned}&\sum_{j=1}^{n} \delta_{\overline{jD}} [\sum_{C} (a_C + b_C) w_{\underline{M}} \delta_{\overline{C\underline{M}}} \\ &+ \sum_{C} (a_C - b_C) w_{\underline{\underline{M}}} \delta_{\overline{C\underline{\underline{M}}}} + l_A]_{x_j} \\ = &\sum_{j,C} \delta_{\overline{jD}} \delta_{\overline{C\underline{M}}} [(a_C + b_C) w_{\underline{M}}]_{x_j} \\ &+ \sum_{j,C} \delta_{\overline{jD}} \delta_{\overline{C\underline{\underline{M}}}} [(a_C - b_C) w_{\underline{\underline{M}}}]_{x_j}\end{aligned}$$

$$
\begin{aligned}
&+\sum_{j=1}^{n}\delta_{\overline{jD}}l_{Ax_j}=\sum_{j,C}\delta_{\overline{jD}}\delta_{\overline{C\underline{M}}}(a_C+b_C)_{x_j}w_{\underline{M}}\\
&+\sum_{j,C}\delta_{\overline{jD}}\delta_{\overline{C\underline{M}}}(a_C+b_C)w_{\underline{M}x_j}\\
&+\sum_{j,C}\delta_{\overline{jD}}\delta_{\overline{C\underline{\underline{M}}}}(a_C-b_C)_{x_j}w_{\underline{\underline{M}}}\\
&+\sum_{j,C}\delta_{\overline{jD}}\delta_{\overline{C\underline{\underline{M}}}}(a_C-b_C)w_{\underline{\underline{M}}x_j}+\sum_{j=1}^{n}\delta_{\overline{jD}}l_{Ax_j}.
\end{aligned}
\tag{3.6}
$$

Here $\overline{C\underline{M}}=A$, $\overline{C\underline{\underline{M}}}=A$, $\overline{jD}=A$. From equations (3.4), (3.5), (3.6), it follows that the formula (3.3) is true.

4 The Chain Rule and Differentiation Rules of new Hypercomplex Differential functions in Clifford Analysis

There are three ways to discuss general holomorphic functions in complex analysis: one is Cauchy's method that is based on the differentiability; another one is the Weierstrass method that is based on power series; the third one is the Riemann method that is based on the Cauchy-Riemann equations. In the case of a quaternion, F. Sommen [72]1) uses a special differential form that avoide the differentiability in Cauchy's method. Conversely in 1990, H. Malonek [48] gave an imitation of Cauchy's classical method by using the new hypercomplex structure, first giving a relation between new hypercomplex differentiable functions and monogenic functions, a hypercomplex structure that possesses an obvious advantage. In this section, we establish a new chain rule and new differential rule in Clifford analysis from the view of the new hypercomplex structure of Malonek.

4.1. Hypercomplex Differential

Let

$$\lambda=\sum_{k=1}^{m}\lambda_k e_k,\ (m\le n,\ \lambda_k\ are\ real\ numbers)$$

be a hypercomplex number and the set of all λ make up a space $\widetilde{\mathcal{A}}_m(R)$. Then $\mathbf{R}^m \cong \widetilde{\mathcal{A}}_m(R)$, $x = (x_1, ..., x_m) \in \mathbf{R}^m$, and $z_k = x_k - x_1 e_k$, $k = 2, ..., m$, $\overrightarrow{z} = (z_2, ..., z_m)$. Denote by H^{m-1} the set of all $\overrightarrow{z}$, and denote the topological product of $m-1$ spaces $\mathcal{A}_n(R)$ by $\mathcal{A}_n^{m-1}(R)$, then $H^{m-1} \subset \mathcal{A}_n^{m-1}(R)$. For any $\overrightarrow{z} \in H^{m-1}$ define the norm of $\overrightarrow{z}$ by

$$\|\overrightarrow{z}\| = (\overrightarrow{z}, \overrightarrow{z})^{\frac{1}{2}} = (\sum_{k=2}^{m} \overline{z}_k z_k)^{\frac{1}{2}} = (m x_1^2 + \sum_{k=2}^{m} x_k^2)^{\frac{1}{2}},$$

where the inner product $(\overrightarrow{z}, \overrightarrow{\xi}) = \overline{(\overrightarrow{\xi}, \overrightarrow{z})} = \sum_{k=2}^{m} \overline{z}_k \xi_k$.

Definition 4.1 Denote by $L(\mathcal{A}_n^{m-1}(R), \mathcal{A}_n^{p-1}(R))$ the set of linear mappings from $\mathcal{A}_n^{m-1}(R)$ to $\mathcal{A}_n^{p-1}(R)$. A bounded mappings $\overrightarrow{l} \in L(\mathcal{A}_n^{m-1}(R), \mathcal{A}_n^{p-1}(R))$ is said to be $\mathcal{A}$ linear (or linear for short), if $\overrightarrow{u}$, $\overrightarrow{v} \in \mathcal{A}_n^{m-1}(R)$, λ, $\mu \in \mathcal{A}_n(R)$; then we have

$$\overrightarrow{l}(\lambda \overrightarrow{u} + \mu \overrightarrow{v}) = \lambda \overrightarrow{l}(\overrightarrow{u}) + \mu \overrightarrow{l}(\overrightarrow{v}).$$

Lemma 4.1 (see [48]) *Every $\mathcal{A}$ linear mapping $l \in L(H^{m-1}, \mathcal{A}_n(R))$ has a unique expression $l(\overrightarrow{z}) = z_2 A_2 + \cdots + z_m A_m$, where $A_k \in \mathcal{A}_n(R)$, $k = 2, ..., m$.*

Proof Let $l \in L(H^{m-1}, \mathcal{A}_n(R))$ be an $\mathcal{A}$ linear mapping. Then we have

$$l(0, 0, ..., 1, ..., 0) = A_k \in \mathcal{A}_n(R), \ k = 2, ..., m,$$

and then

$$l(\overrightarrow{z}) = z_2 A_2 + \cdots + z_m A_m.$$

If there are other A_k', $k = 2, ..., m$, such that

$$l(\overrightarrow{z}) = z_2 A_2' + \cdots + z_m A_m',$$

then

$$\begin{aligned} 0 &= l(\overrightarrow{z} - \overrightarrow{z}) = l(\overrightarrow{z}) - l(\overrightarrow{z}) \\ &= z_2(A_2 - A_2') + \cdots + z_m(A_m - A_m'), \end{aligned}$$

and we get $A_k = A_k'$, $k = 2, ..., m$.

Definition 4.2 (see [48]) Let $\overrightarrow{a} \in H^{m-1}$ and f be a continuous mapping from some domain of $\overrightarrow{a}$ to $\mathcal{A}_n(R)$. The function $f = f(\overrightarrow{z})$ is called the left hypercomplex differential, for short hypercomplex differential, if there exists an $\mathcal{A}$ linear mapping $l \in L(H^{m-1}, \mathcal{A}_n(R))$ such that

$$\lim_{\triangle \overrightarrow{z} \to 0} \frac{|f(\overrightarrow{a} + \triangle \overrightarrow{z}) - f(\overrightarrow{a}) - l(\triangle \overrightarrow{z})|}{\|\triangle \overrightarrow{z}\|} = 0,$$

and $l = l(\overrightarrow{z})$ is called the left hypercomplex differential for f (for short hypercomplex differential), written by $l = f'(\overrightarrow{a})$ or $l(\overrightarrow{z}) = f'(\overrightarrow{a})(\overrightarrow{z})$.

Lemma 4.2 (see [72]1)) *If $f(\overrightarrow{z})$ is a hypercomplex differential function from H^{m-1} to $\mathcal{A}_n(R)$, then the differential $f'(\overrightarrow{a})$ is unique.*

4.2 $\mathcal{A}$-chain rule of hypercomplex differentiation

Theorem 4.3 *If $f : H^{m-1} \to \mathcal{A}_n(R)$, $b = f(\overrightarrow{a})$, and $f'(\overrightarrow{a})$ is given, then for any $\varepsilon > 0$, there exists a number $\delta > 0$ such that when $\|\overrightarrow{z} - \overrightarrow{a}\| < \delta$, the inequality $|f(\overrightarrow{z}) - f(\overrightarrow{a})| < \varepsilon$ is valid, and we say that f is continuous at $\overrightarrow{a}$.*

Proof From the conditions, we can get

$$\lim_{\triangle \overrightarrow{z} \to 0} \frac{|f(\overrightarrow{a} + \triangle \overrightarrow{z}) - f(\overrightarrow{a}) - f'(\overrightarrow{a})(\triangle \overrightarrow{z})|}{\|\triangle \overrightarrow{z}\|} = 0,$$

and then

$$\lim_{\triangle \overrightarrow{z} \to 0} |f(\overrightarrow{a} + \triangle \overrightarrow{z}) - f(\overrightarrow{a}) - f'(\overrightarrow{a})(\triangle \overrightarrow{z})| = 0.$$

On the basis of Lemma 4.1, it is easy to see that $f'(\overrightarrow{a})(\Delta \overrightarrow{z}) = \triangle z_2 A_2 + \cdots + \triangle z_m A_m$, where A_k $(k = 2, ..., m)$ are fixed Clifford numbers that are independent with $\Delta \overrightarrow{z}$, hence $\lim\limits_{\Delta \overrightarrow{z} \to 0} f'(\overrightarrow{a})(\Delta \overrightarrow{z}) = 0$ and

$$\lim_{\triangle \overrightarrow{z} \to 0} |f(\overrightarrow{a} + \triangle \overrightarrow{z}) - f(\overrightarrow{a})|$$

$$= \lim_{\triangle \overrightarrow{z} \to 0} |f(\overrightarrow{a} + \triangle \overrightarrow{z}) - f(\overrightarrow{a}) - f'(\overrightarrow{a})(\triangle \overrightarrow{z}) + f'(\overrightarrow{a})(\triangle \overrightarrow{z})| = 0.$$

Theorem 4.4 *Let $\overrightarrow{l} \in L(\mathcal{A}_n^{m-1}(R), \mathcal{A}_n^{p-1}(R))$. Then there exists an $M > 0$ such that*

$$\|\overrightarrow{l}(\overrightarrow{u})\| \le M\|\overrightarrow{u}\|, \ \overrightarrow{u} \in \mathcal{A}_n^{m-1}(R),$$

where $\|\overrightarrow{u}\| = \sqrt{\sum\limits_{i=2}^{m} |u_i|^2}$, $\overrightarrow{u} = (u_2, ..., u_m)$, $u_i \in \mathcal{A}_n(R)$, $i = 2, ..., m$.

Definition 4.3 Let $\overrightarrow{a} \in \mathcal{A}_n^{p-1}(R)$ and $\overrightarrow{f}$ be a continuous mapping from some domain of $\overrightarrow{a}$ to $\mathcal{A}_n^{m-1}(R)$. Then $\overrightarrow{f} = \overrightarrow{f}(\overrightarrow{z})$ is said to be $\mathcal{A}$ differential at the point $\overrightarrow{a}$; if there exists an $\mathcal{A}$ linear mapping $l \in L(\mathcal{A}_n^{p-1}(R), \mathcal{A}_n^{m-1}(R))$ such that

$$\lim_{\triangle \overrightarrow{z} \to 0} \frac{\|\overrightarrow{f}(\overrightarrow{a}+\triangle\overrightarrow{z}) - \overrightarrow{f}(\overrightarrow{a}) - \overrightarrow{l}(\triangle\overrightarrow{z})\|}{\|\triangle\overrightarrow{z}\|} = 0,$$

then $\overrightarrow{l} = \overrightarrow{l}(\overrightarrow{z})$ is called the left differential, written by $\overrightarrow{l}(\overrightarrow{z}) = \overrightarrow{(f')}(\overrightarrow{a})$ or $\overrightarrow{l} = \overrightarrow{(f')}(\overrightarrow{a})$.

Theorem 4.5 *Let $\overrightarrow{f} : \mathcal{A}_n^{p-1}(R) \to \mathcal{A}_n^{m-1}(R)$ and $\overrightarrow{f'}(\overrightarrow{a})$ be given. Then $\overrightarrow{f}(\overrightarrow{z})$ is continuous at $\overrightarrow{a}$.*

Lemma 4.6 (see [72]1)) *If $\overrightarrow{f} : \mathcal{A}_n^{p-1}(R) \to \mathcal{A}_n^{m-1}(R)$ is $\mathcal{A}$ differentiable, then $\overrightarrow{f'}(\overrightarrow{a})$ is unique.*

Theorem 4.7 *Let $\overrightarrow{f} : \mathcal{A}_n^{p-1}(R) \to \mathcal{A}_n^{m-1}(R)$ be $\mathcal{A}$ differentiable at $\overrightarrow{a}$ and $g : \mathcal{A}_n^{m-1}(R) \to \mathcal{A}_n(R)$ be $\mathcal{A}_n(R)$ differentiable at $\overrightarrow{f}(\overrightarrow{a})$. Then the composite function $g \circ \overrightarrow{f} : \mathcal{A}_n^{p-1}(R) \to \mathcal{A}_n(R)$ is differentiable at $\overrightarrow{a}$, and*

$$(g \circ \overrightarrow{f})'(\overrightarrow{a}) = g'(\overrightarrow{f}(\overrightarrow{a})) \circ (\overrightarrow{(f')}(\overrightarrow{a})).$$

Proof Suppose that $\overrightarrow{b} = \overrightarrow{f}(\overrightarrow{a})$, $\overrightarrow{w} = \overrightarrow{f}(\overrightarrow{z})$, $\overrightarrow{\lambda} = \overrightarrow{f'}(\overrightarrow{a})$, $\mu = g'(\overrightarrow{f}(\overrightarrow{a}))$, and let

$$\triangle\overrightarrow{z} = \overrightarrow{z} - \overrightarrow{a}, \triangle\overrightarrow{w} = \overrightarrow{w} - \overrightarrow{b},$$

$$\overrightarrow{\varphi}(\overrightarrow{z}) = \overrightarrow{f}(\overrightarrow{z}) - \overrightarrow{f}(\overrightarrow{a}) - \overrightarrow{\lambda}(\overrightarrow{z} - \overrightarrow{a}),$$

$$\psi(\overrightarrow{w}) = g(\overrightarrow{w}) - g(\overrightarrow{b}) - \mu(\overrightarrow{w} - \overrightarrow{b}),$$

$$\rho(\overrightarrow{z}) = g \circ \overrightarrow{f}(\overrightarrow{z}) - g \circ \overrightarrow{f}(\overrightarrow{a}) - \mu \circ \overrightarrow{\lambda}(\overrightarrow{z} - \overrightarrow{a}),$$

where

$$\mu \circ \overrightarrow{\lambda}(\overrightarrow{z} - \overrightarrow{a}) = g'(\overrightarrow{f}(\overrightarrow{a})) \circ (\overrightarrow{f'}(\overrightarrow{a})).$$

Then we have

$$\lim_{(\overrightarrow{z} - \overrightarrow{a}) \to 0} \frac{\|\overrightarrow{\varphi}(\overrightarrow{z})\|}{\|\overrightarrow{z} - \overrightarrow{a}\|} = 0, \tag{4.1}$$

$$\lim_{(\overrightarrow{w} - \overrightarrow{b}) \to 0} \frac{|\psi(\overrightarrow{w})|}{\|\overrightarrow{w} - \overrightarrow{b}\|} = 0. \tag{4.2}$$

In order to prove

$$\lim_{(\overrightarrow{z} - \overrightarrow{a}) \to 0} \frac{|\rho(\overrightarrow{z})|}{\|\overrightarrow{z} - \overrightarrow{a}\|} = 0, \tag{4.3}$$

we consider

$$\begin{aligned}\rho(\vec{z}) &= g \circ \vec{f}(\vec{z}) - g \circ \vec{f}(\vec{a}) - \mu \circ [\vec{f}(\vec{z}) - \vec{f}(\vec{a}) - \vec{\varphi}(\vec{z})] \\ &= g \circ \vec{f}(\vec{z}) - g \circ \vec{f}(\vec{a}) - \mu \circ [\vec{f}(\vec{z}) - \vec{f}(\vec{a})] + \mu \circ [\vec{\varphi}(\vec{z})] \\ &= \psi \circ \vec{f}(\vec{z}) + \mu(\vec{\varphi}(\vec{z})).\end{aligned} \tag{4.4}$$

It is sufficient to prove

$$\lim_{(\vec{z}-\vec{a})\to 0} \frac{|\psi \circ \vec{f}(\vec{z})|}{\|\vec{z}-\vec{a}\|} = 0, \tag{4.5}$$

$$\lim_{(\vec{z}-\vec{a})\to 0} \frac{|\mu(\vec{\varphi}(\vec{z}))|}{\|\vec{z}-\vec{a}\|} = 0. \tag{4.6}$$

For any fixed number $\varepsilon > 0$, from (4.1) there exists a $\delta > 0$ such that when

$$\|\vec{w} - \vec{b}\| = \|\vec{f}(\vec{z}) - \vec{f}(\vec{a})\| < \delta,$$

the following inequality holds:

$$|\psi \circ \vec{f}(\vec{z})| \le \varepsilon \|\vec{f}(\vec{z}) - \vec{f}(\vec{a})\|.$$

Moreover from Theorem 4.3, we see that for the δ as stated above there exists a $\delta_1 > 0$, such that when $\|\vec{z} - \vec{a}\| < \delta_1$, we have

$$\|\vec{f}(\vec{z}) - \vec{f}(\vec{a})\| < \delta.$$

Again by Theorem 4.2, there exists $M > 0$, such that

$$\|\vec{\lambda}(\vec{z} - \vec{a})\| \le M\|\vec{z} - \vec{a}\|.$$

Hence

$$\begin{aligned}|\psi \circ \vec{f}(\vec{z})| &\le \varepsilon \|\vec{f}\vec{z}) - \vec{f}(\vec{a})\| \le \varepsilon\|\vec{\varphi}(\vec{z})\| + \varepsilon\|\vec{\lambda}(\vec{z} - \vec{a})\| \\ &\le \varepsilon\|\vec{\varphi}(\vec{z})\| + M\varepsilon\|\vec{z} - \vec{a}\|.\end{aligned} \tag{4.7}$$

In addition from (4.1), (4.7), it follows that (4.5) holds. Again from Theorem 4.2, we get

$$\frac{\|\mu(\vec{\varphi}(\vec{z}))|}{\|\vec{z}-\vec{a}\|} \le \frac{M\|\vec{\varphi}(\vec{z})\|}{\|\vec{z}-\vec{a}\|}.$$

Noting (4.1) and (4.4), we see that (4.6) and (4.3) are true, i.e.

$$(g \circ \vec{f})'(\vec{a}) = \mu \circ \vec{\lambda}(\vec{z} - \vec{a}) = g'(\vec{f}(\vec{a})) \circ (\vec{f}'(\vec{a})).$$

4.3. Differentiation rules

First of all, we give a notation for a project function. A function $\overrightarrow{f}\,(=\overrightarrow{f}(\overrightarrow{z}) = (f_2(\overrightarrow{z}), ..., f_m(\overrightarrow{z})) : H^{p-1} \to \mathcal{A}_n^{m-1}(R)$ can uniquely determine $m-1$ component functions $f_i(\overrightarrow{z}) : H^{p-1} \to \mathcal{A}_n(R)$, $i = 2, ..., m$. If $\overrightarrow{\pi} : \mathcal{A}_n^{m-1}(R) \to \mathcal{A}_n^{m-1}(R)$ is an identical mapping $\overrightarrow{\pi}(\overrightarrow{u}) = \overrightarrow{u} = (u_2, ..., u_m)$, then $\pi_i : \mathcal{A}_n^{m-1}(R) \to \mathcal{A}_n(R)$, $\pi_i(\overrightarrow{u}) = u_i$, $i = 2, ..., m$ is called the i-th project function. From this we can get $f_i = \pi_i \circ \overrightarrow{f}$, $i = 2, ..., m$.

Theorem 4.8 *If $f : \mathcal{A}_n^{m-1}(R) \to \mathcal{A}_n(R)$ is an $\mathcal{A}$ linear mapping, then $f'(\overrightarrow{a})(\overrightarrow{z}) = f(\overrightarrow{z})$, for short $f'(\overrightarrow{a}) = f(\overrightarrow{z})$.*

Theorem 4.9 *Let $\overrightarrow{f} = (f_2, ..., f_m) : H^{p-1} \to \mathcal{A}_n^{m-1}(R)$. Then $\overrightarrow{f}$ is $\mathcal{A}$ differentiable at $\overrightarrow{a} \in H^{p-1}$, if and only if every f_i $(i = 2, ..., m)$ is hypercomplex differentiable at $\overrightarrow{a}$, and*

$$\overrightarrow{f}'(\overrightarrow{a}) = (f_2'(\overrightarrow{a}), ..., f_m'(\overrightarrow{a})).$$

Proof Suppose that every f_i $(i = 2, ..., m)$ is hypercomplex differentiable at $\overrightarrow{a}$. If $\overrightarrow{\lambda}(\overrightarrow{z}) = (f_2'(\overrightarrow{a}), ..., f_m'(\overrightarrow{a}))(\overrightarrow{z})$, then $\overrightarrow{f}(\overrightarrow{a} + \triangle\overrightarrow{z}) - \overrightarrow{f}(\overrightarrow{a}) - \overrightarrow{\lambda}(\triangle\overrightarrow{z}) = (f_2(\overrightarrow{a} + \triangle\overrightarrow{z}) - f_2(\overrightarrow{a}) - f_2'(\overrightarrow{a})(\triangle\overrightarrow{z}), ..., f_m(\overrightarrow{a} + \triangle\overrightarrow{z}) - f_m(\overrightarrow{a}) - f_m'(\overrightarrow{a})(\triangle\overrightarrow{z}))$, and then

$$\lim_{\triangle\overrightarrow{z}\to 0} \frac{\|\overrightarrow{f}(\overrightarrow{a}+\triangle\overrightarrow{z}) - \overrightarrow{f}(\overrightarrow{a}) - \overrightarrow{\lambda}(\triangle\overrightarrow{z})\|}{\|\triangle\overrightarrow{z}\|}$$

$$\leq \lim_{\triangle\overrightarrow{z}\to 0} \sum_{i=2}^{m} \frac{\|f_i(\overrightarrow{a}+\triangle\overrightarrow{z}) - f_i(\overrightarrow{a}) - f_i'(\overrightarrow{a})(\triangle\overrightarrow{z})\|}{\|\triangle\overrightarrow{z}\|},$$

namely $\overrightarrow{f}$ is $\mathcal{A}$ differentiable at $\overrightarrow{a}$, and the equality in this theorem is true. Conversely if $\overrightarrow{f}$ is differentiable at $\overrightarrow{a} \in H^{m-1}$, then from Theorems 4.4 and 4.5, we see that $f_i = \pi_i \circ \overrightarrow{f}$ is hypercomplex differentiable at $\overrightarrow{a}_i$.

Theorem 4.10 *If $S : \mathcal{A}_n^2(R) \to \mathcal{A}_n(R)$, $S(u, v) = u + v$, then $S'(a, b) = S(u, v)$.*

Theorem 4.11 *Let $q : \mathcal{A}_n^2(R) \to \mathcal{A}_n(R)$, $q(u, v) = uv$. Then $q'(a, b)(u, v) = ub + av$.*

Proof Suppose $\lambda(u,v) = ub + av$. Then

$$\lim_{(\triangle u,\triangle v)\to 0} \frac{\|q(a+\triangle u, b+\triangle v) - q(a,b) - \lambda(\triangle u,\triangle v)\|}{\|(\triangle u,\triangle v)\|}$$

$$= \lim_{(\triangle u,\triangle v)\to 0} \frac{\|\triangle u \triangle v\|}{\|(\triangle u,\triangle v)\|} \leq \lim_{(\triangle u,\triangle v)\to 0} \frac{B\|\triangle u\|\|\triangle v\|}{\sqrt{(\|\triangle u\|)^2 + (\|\triangle v\|)^2}}$$

$$\leq \lim_{(\triangle u,\triangle v)\to 0} \frac{B}{2}\frac{\|\triangle u\|^2 + \|\triangle v\|^2}{\sqrt{(\|\triangle u\|)^2 + (\|\triangle v\|)^2}} = 0,$$

where $B > 0$ is a constant, hence $q'(a,b)(u,v) = ub + av$.

Theorem 4.12 *Let $f, g : H^{m-1} \to \mathcal{A}_n(R)$ be hypercomplex differentiable at $\overrightarrow{a}$. Then*

1. $(f+g)'(\overrightarrow{a}) = (f)'(\overrightarrow{a}) + (g)'(\overrightarrow{a})$.
2. $(fg)'(\overrightarrow{a}) = (f)'(\overrightarrow{a})g(\overrightarrow{a}) + f(\overrightarrow{a})g'(\overrightarrow{a})$.

Proof On the basis of Theorem 4.4, Theorem4.6 and Theorem4.7, we can get 1. Moreover from Theorems 4.4, Lemma 4.6 and Theorem 4.8, we can derive 2.

5 Regular and Harmonic Functions in Complex Clifford Analysis

Let $e := \{e_1, ..., e_n\}$ be the orthogonal basis of $\mathbf{R}^n$. According to the structure of real Clifford analysis, we use the complex numbers to construct the complex Clifford algebra $\mathcal{A}(C)$; its elements possess the form

$$Z = \sum_A c_A e_A, \tag{5.1}$$

where c_A are the complex numbers and e_A is as above. The norm of the element is defined as

$$|Z| = \sqrt{\sum_A |c_A|^2}. \tag{5.2}$$

Complex Clifford analysis studies the functions defined on $\mathbf{C}^n$ and taking value on a complex Clifford algebra. About Complex Clifford analysis, F. Sommen and J. Ryan have obtained some results. In 1982,

F. Sommen discussed the relation between functions of several variables and monogenic functions by using the Fourier-Borel transformation (see [72]2)). In 1982, J. Ryan gave a simulation of Cauchy's theorem, proved the invariant property of the complex left regular function under the action of a Lie group and structured the holomorphic function by complex harmonic functions (see [68]1)). In 1983, J. Ryan also gave a general Cauchy integral formula for a kind of special holomorphic function and used this formula to prove that all these special holomorphic functions form a Fréchet model (see [68]2).

5.1 Complex regular functions

Now we give the concept of a complex regular function by using the Dirac operator.

Definition 5.1. Let $U \subset \mathbf{C}^n$ be a domain and $f : U \to \mathcal{A}_n(C)$ be a holomorphic function. If for every point $z \in U$, we have

$$\overline{\partial} f = \sum_{j=1}^{n} e_j \frac{\partial f(z)}{\partial z_j} = 0,$$

then $f(z)$ is called a complex left regular function in U. If for every point $z \in U$, we have

$$f \overline{\partial} = \sum_{j=1}^{n} \frac{\partial f(z)}{\partial z_j} e_j = 0,$$

then $f(z)$ is called a complex right regular function in U.

In the following, we mainly discuss the complex left regular function, and the complex right regular function can be analogously discussed. Let

$$A = \{j_1, ..., j_r\}, \ \widehat{A}_k = \{j_1, ...j_{k-1}, j_{k+1}, ..., j_r\},$$

$$2 \leq j_1 < j_2 < ... < j_r \leq n,$$

$$KA = Kj_1...j_r, \ AK = j_1...j_rK,$$

and denote the sign of permutation $(l, j_1, ..., j_r)$ by $\operatorname{sgn}(l, A)$. And by a direct calculation, we get

$$\begin{aligned}
& e_{j_k}\frac{\partial f_{\widehat{A}_k}}{\partial z_{j_k}}e_{\widehat{A}_k} = \frac{\partial f_{\widehat{A}_k}}{\partial z_{j_k}}(-1)^{k-1}e_A\,(k=1,2,...,r),\\
& e_k\frac{\partial f_{kA}}{\partial z_k}e_{kA} = \frac{\partial f_{kA}}{\partial z_k}e_k^2 e_A = -\frac{\partial f_{kA}}{\partial z_k}e_A\\
= \;& -\mathrm{sgn}(k,A)\frac{\partial f_{kA}}{\partial z_k}e_A\,(k<j_1, k=1,2,...,j_1-1),\\
& e_k\frac{\partial f_{Ak}}{\partial z_k}e_{Ak} = (-1)^r\frac{\partial f_{Ak}}{\partial z_k}e_k^2 e_A = (-1)^{r+1}\frac{\partial f_{Ak}}{\partial z_k}e_A\\
= \;& -\mathrm{sgn}(k,A)\tfrac{\partial f_{Ak}}{\partial z_k}e_A\,(j_r<k, k=j_r+1, j_r+2,...,n),\\
& e_k\frac{\partial f_{j_1,...,k,...,j_r}}{\partial z_k}e_{j_1,...,k,...,j_r} = -\mathrm{sgn}(k,A)\frac{\partial f_{\{k\}\cup A}}{\partial z_k}e_A,\\
& (j_1<k<j_r,\ k\notin A),
\end{aligned}$$

and then

$$\begin{aligned}
& \sum_{j=1}^n e_j\frac{\partial f(z)}{\partial z_j} = \sum_{j=1}^n e_j\frac{\partial}{\partial z_j}\sum_A f_A e_A\\
= \;& \sum_A\left\{\sum_{k=1}^r(-1)^{k-1}\frac{\partial f_{\widehat{A}_k}}{\partial z_{j_k}} + \sum_{k=1}^{j_1-1}\frac{-\partial f_{kA}}{\partial z_k} + \sum_{k=j_r+1}^n(-1)^{r+1}\frac{\partial f_{Ak}}{\partial z_k}\right.\\
& \left. + \sum_{j_1<k<j_r, k\notin A,} -sgn(k,A)\frac{\partial f_{\{k\}\cup A}}{\partial z_k}\right\}e_A\\
= \;& \sum_A\left\{\sum_{k=1}^r(-1)^{k+1}\frac{\partial f_{\widehat{A}_k}}{\partial z_{j_k}} + \sum_{k=1}^{j_1-1}(-sgn(k,A)\frac{\partial f_{kA}}{\partial z_k})\right.\\
& \left. + \sum_{k=j_r+1}^n(-sgn(k,A))\frac{\partial f_{Ak}}{\partial z_k} + \sum_{j_1<k<j_r,k\notin A}(-sgn(k,A)\frac{\partial f_{\{k\}\cup A}}{\partial z_k})\right\}e_A\\
= \;& \sum_A\left[\sum_{k=1}^r(-1)^{k-1}\frac{\partial f_{\widehat{A}_k}}{\partial z_{j_k}} - \sum_{k\notin A} sgn(k,A)\frac{\partial f_{\{k\}\cup A}}{\partial z_k}\right]e_A,
\end{aligned}$$

thus we have the theorem

Theorem 5.1 *Let $f: U \to \mathcal{A}_n(C)$ be an analytic function. Then f is* *complex left regular if and only if*

$$\sum_{k=1}^{r}(-1)^{k-1}\frac{\partial f_{\widehat{A_k}}}{\partial z_{j_k}} - \sum_{k\notin A} sgn(k,A)\frac{\partial f_{\{k\}\cup A}}{\partial z_k} = 0.$$

Similarly to the M-T system in real Clifford analysis (see [28]), we think it is useful to study the system which is equivalent to complex regular function when $n = 3$.

Definition 5.2. If $f : U \subset \mathbf{C}^n \to \mathcal{A}_n(C)$ is a holomorphic function and satisfies the equation

$$\Box_\varepsilon f(z) \equiv \sum_{j=1}^{n} \frac{\partial^2 f}{\partial z_j^2}(z) = 0,$$

then we call $f(z)$ a complex harmonic function.

From this we can get the following result.

Theorem 5.2 *Suppose that $f(z)$ is a holomorphic function, then f is a complex harmonic function if and only if f_A satisfies*

$$\Box_\varepsilon f_A(z) = 0.$$

Theorem 5.3. *Let $f(z)$ be a complex left regular function. Then $f(z)$ is a complex harmonic function.*

The inverse of Theorem 5.3 is not true, for example, if $f : U \subseteq \mathbf{C}^3 \to \mathcal{A}_2(C)$, $f(z) = (z_2 + z_3)e_1 + z_2e_2 + z_3e_3$ is a complex harmonic function, but it is not the complex left regular function, therefore we have the following theorem.

Theorem 5.4. *If $f(z)$ is a complex harmonic function, then the function*

$$e_1\frac{\partial f(z)}{\partial z_1} - \sum_{j=2}^{n} e_j\frac{\partial f}{\partial z_j} = 2e_1\frac{\partial f(z)}{\partial z_1} - \sum_{j=1}^{n} e_j\frac{\partial f(z)}{\partial z_j}$$

is a complex left regular function.

5.2 The structure of complex left regular functions

From Theorem 5.2 and 5.3, we easily obtain the following theorem.

Theorem 5.5 *Let $f(z)$ be a complex left regular function. Then for every A, f_A is a complex harmonic function.*

The inverse of Theorem 5.5 is not true. It is interesting that when condition f_A is satisfied, the function $f(z)$ is complex left regular. For this we first construct a complex left regular function from the complex harmonic function.

Theorem 5.6. *Let $U \subseteq \mathbf{C}^n$ be a star domain, $A = \{j_1, j_2, ..., j_r\}$, and $u_A : U \to C$ be a complex harmonic function. Then there exists a complex left regular function*

$$f^A : U \to \mathcal{A}(C),$$

such that the component of $f^A(z)$ about the basic element e_A is equal to $u_A(z)$.

Proof By using the transformation of coordinates, we can consider that the domain includes the origin and is a star domain about the origin. Under the conditions, we discuss the function

$$\begin{aligned} f^A(z) \equiv\ & u_A(z)e_A \\ & + Vec^{(A)}\left[\int_0^1 \left\{ s^{n-3}\left(e_1\frac{\partial}{\partial z_1} - \sum_{j=2}^n e_j \frac{\partial}{\partial z_j}\right) u_A(sz)\right\} ze_A ds\right], \end{aligned}$$

where the sign $Vec^A[z] = z - z_A e_A$, $z \in \mathcal{A}_n(C)$. It is clear that the component of f^A about e_A is just $u_A(z)$. Now we only consider the complex left regularity.

Because $z = \sum_{k=1}^n z_k e_k$ and when $j = k$ we have $e_k^2 = -1$, so the component of the function

$$\int_0^1 \left\{ s^{n-3}\left(e_1\frac{\partial}{\partial z_1} - \sum_{j=2}^n e_j \frac{\partial}{\partial z_j}\right) u_A(sz)\right\} ze_A ds$$

about e_A is

$$\begin{aligned} & \int_0^1 s^{n-3} \sum_{k=1}^n z_k \frac{\partial}{\partial z_k} u_A(sz) ds \\ = & \int_0^1 s^{n-2} \frac{d}{ds} u_A(sz) ds = u_A - \int_0^1 (n-2)s^{n-3} u_A(sz) ds, \end{aligned}$$

and then

$$f^A(z)=\int_0^1 s^{n-3}\left[\left(e_1\frac{\partial}{\partial z_1}-\sum_{j=2}^n e_j\frac{\partial}{\partial z_j}\right)u_A(sz)\right]ze_A ds$$
$$+\int_0^1 (n-2)s^{n-3}u_A(sz)e_A ds.$$

Moreover we have

$$\sum_{k=1}^n e_A\frac{\partial f^A(z)}{\partial z_k}$$
$$= \int_0^1 s^{n-3}\left[\sum_{k=1}^n e_k\frac{\partial}{\partial z_k}\left(e_1\frac{\partial}{\partial z_1}-\sum_{j=2}^n e_j\frac{\partial}{\partial z_j}\right)u_A(sz)\right]ze_A ds$$
$$+\int_0^1 s^{n-3}\left\{\sum_{k=1}^n e_k\left[\left(e_1\frac{\partial}{\partial z_1}-\sum_{j=2}^n e_j\frac{\partial}{\partial z_j}\right)u_A(sz)\right]\frac{\partial z}{\partial z_k}\right\}e_A ds$$
$$+\int_0^1 (n-2)s^{n-3}\sum_{k=1}^n e_k\frac{\partial u_A(sz)}{\partial z_k}e_A ds.$$

From Theorem 5.4 we obtain

$$\sum_{k=1}^n e_k\frac{\partial}{\partial z_k}\left(e_1\frac{\partial}{\partial z_1}-\sum_{j=2}^n e_j\frac{\partial}{\partial z_j}\right)u_A(sz)=0,$$

hence

$$\sum_{k=1}^n e_k\frac{\partial f^A(z)}{\partial z_k}$$
$$= \int_0^1 s^{n-3}\left\{\sum_{k=1}^n e_k\left[\left(e_1\frac{\partial}{\partial z_1}-\sum_{j=2}^n e_j\frac{\partial}{\partial z_j}\right)u_A(sz)\right]e_k\right\}e_A ds$$
$$+\int_0^1 (n-2)s^{n-3}\sum_{k=1}^n e_k\frac{\partial u_A(sz)}{\partial z_k}e_A ds,$$

and by direct calculation, we get

$$\sum_{k=1}^n e_k\left[\left(e_1\frac{\partial}{\partial z_1}-\sum_{j=2}^n e_j\frac{\partial}{\partial z_j}\right)u_A(sz)\right]e_k$$
$$= \left(e_1\frac{\partial}{\partial z_1}-\sum_{j=2}^n e_j\frac{\partial}{\partial z_j}\right)u_A(sz)$$

$$
\begin{aligned}
&+\sum_{k=2}^{n} e_k \left[\left(e_1\frac{\partial}{\partial z_1}-\sum_{j=2}^{n} e_j\frac{\partial}{\partial z_j}\right) u_A(sz)\right] e_k\\
=\ & \left(e_1\frac{\partial}{\partial z_1}-\sum_{j=2}^{n} e_j\frac{\partial}{\partial z_j}\right) u_A(sz)\\
&+\sum_{k=2}^{n} e_k^2\frac{\partial u_A}{\partial z_1}-\sum_{k=2}^{n} e_k\left(\sum_{j=2}^{n} e_j\frac{\partial u_A}{\partial z_j}\right) e_k\\
=\ & \left(e_1\frac{\partial}{\partial z_1}-\sum_{j=2}^{n} e_j\frac{\partial}{\partial z_j}\right) u_A(sz)-\sum_{k=2}^{n} e_1\frac{\partial u_A}{\partial z_1}\\
&+\sum_{k=2,j\neq k}^{n} e_k^2\sum_{j=2}^{n} e_j\frac{\partial u_A}{\partial z_j}-\sum_{k=2,j=k}^{n} e_k^3\frac{\partial u_A}{\partial z_k}\\
=\ & -(n-2)e_1\frac{\partial u_A}{\partial z_1}-\sum_{j=2}^{n} e_j\frac{\partial u_A}{\partial z_j}-\sum_{k=2,j\neq k}^{n} e_1\sum_{j=2}^{n} e_j\frac{\partial u_A}{\partial z_j}+\sum_{k=2}^{n} e_k\frac{\partial u_A}{\partial z_k}\\
=\ & -(n-2)e_1\frac{\partial u_A}{\partial z_1}-(n-2)\sum_{j=2}^{n} e_j\frac{\partial u_A}{\partial z_j}=-(n-2)\sum_{j=1}^{n} e_j\frac{\partial u_A(sz)}{\partial z_j},
\end{aligned}
$$

and then

$$\sum_{j=1}^{n} e_A\frac{\partial f^A(z)}{\partial z_j}=0.$$

This shows that f^A is a complex left regular function.

Theorem 5.7 *If $u_A : U \subseteq \mathbf{C}^n \to C$ is a complex harmonic function defined in a star domain $U \in \mathbf{C}^n$, moreover, there exists the following relation:*

$$\int_0^1 s^{n-3}\left\{\sum_A\sum_{j=1}^{n}[\frac{\partial}{\partial z_j}(\sum_{l=2}^{n} z_l\frac{\partial u_A(sz)}{\partial z_l})+(n-1)\frac{\partial u_A(sz)}{\partial z_j}]e_je_A\right\} ds=0,$$

$$A=\{j_1...j_r\},$$

then the function $F(z)=\sum_A u_Ae_A$ in U is a complex left regular function.

Proof According to Theorem 5.6, we know that for every u_A, there exists a complex left regular function $f^{(A)}$ such that

$$f^{(A)}(z)=u_Ae_A+Vec^{(A)}[g^{(A)}(z)],$$

where

$$g^{(A)}(z) = \int_0^1 s^{n-3}[(e_1\frac{\partial}{\partial z_1} - \sum_{j=2}^n e_j\frac{\partial}{\partial z_j})u_A(sz)]ze_A ds,$$

and

$$F(z) = \sum_A u_A e_A = \sum_A f^{(A)}(z) - \sum_A Vec^{(A)} g^{(A)}.$$

Because $f^{(A)}$ is a complex left regular function, so

$$\sum_{l=1}^n e_l \frac{\partial \sum_A f^{(A)}}{\partial z_l} = \sum_{l=1}^n e_l \sum_A \frac{\partial f^{(A)}}{\partial z_l} = \sum_A \sum_{l=1}^n e_l \frac{\partial f^{(A)}}{\partial z_l} = 0,$$

and

$$\sum_{l=1}^n e_l \frac{\partial F}{\partial z_l} = -\sum_{l=1}^n e_l \sum_A Vec^{(A)} \frac{\partial g^{(A)}}{\partial z_l}$$

$$= -\sum_{l=1}^n e_l \sum_A Vec^{(A)} \int_0^1 s^{n-3} \frac{\partial}{\partial z_l}\left\{\left[\left(e_1\frac{\partial}{\partial z_1} - \sum_{j=2}^n e_j\frac{\partial}{\partial z_j}\right)u_A(sz)\right]z\right\}e_A ds$$

$$= -\sum_{l=1}^n e_l \sum_A Vec^{(A)} \int_0^1 s^{n-3}\left\{\left(e_1\frac{\partial}{\partial z_1} - \sum_{j=2}^n e_j\frac{\partial}{\partial z_j}\right)\frac{\partial u_A(sz)}{\partial z_l} z\right.$$

$$\left. + \left[\left(e_1\frac{\partial}{\partial z_1} - \sum_{j=2}^n e_j\frac{\partial}{\partial z_j}\right)u_A(sz)\right]\frac{\partial z}{\partial z_l}\right\} e_A ds$$

$$= -\sum_{l=1}^n e_l \sum_A Vec^{(A)}\left\{\int_0^1 s^{n-3}\left[\left(e_1\frac{\partial}{\partial z_1} - \sum_{j=2}^n e_j\frac{\partial}{\partial z_j}\right)\frac{\partial u_A(sz)}{\partial z_l}\right] ze_A ds\right\}$$

$$-\sum_{l=1}^n e_l \sum_A Vec^{(A)}\left\{\int_0^1 s^{n-3}\left[\left(e_1\frac{\partial}{\partial z_1} - \sum_{j=2}^n e_j\frac{\partial}{\partial z_j}\right)u_A(sz)\right]e_l e_A ds\right\}.$$

Noting that

$$Vec^{(A)}\left[\sum_A z_A e_A\right] = \sum_A z_A e_A - z_A e_A,$$

and the component about e_A of

$$\int_0^1 s^{n-3}\left[\left(e_1\frac{\partial}{\partial z_1} - \sum_{j=2}^n e_j\frac{\partial}{\partial z_j}\right)\frac{\partial u_A(sz)}{\partial z_l}\right] ze_A ds$$

is

$$\int_0^1 s^{n-3} \sum_{k=1}^n z_k \frac{\partial^2 u_A(sz)}{\partial z_k \partial z_l} ds,$$

we have

$$Vec^{(A)} \left\{ \int_0^1 s^{n-3} \left[\left(e_1 \frac{\partial}{\partial z_1} - \sum_{j=2}^n e_j \frac{\partial}{\partial z_j} \right) \frac{\partial u_A(sz)}{\partial z_l} \right] z e_A ds \right\}$$
$$= \int_0^1 s^{n-3} \left[\left(e_1 \frac{\partial}{\partial z_1} - \sum_{j=2}^n e_j \frac{\partial}{\partial z_j} \right) \frac{\partial u_A(sz)}{\partial z_l} \right] z e_A ds$$
$$- \int_0^1 s^{n-3} \sum_{k=1}^n z_k \frac{\partial^2 u_A(sz)}{\partial z_k z_l} e_A ds.$$

Because

$$\left(e_1 \frac{\partial}{\partial z_1} - \sum_{j=2}^n e_j \frac{\partial}{\partial z_j} \right) u_A(sz)$$

is a complex left regular function, we get

$$-\sum_{l=1}^n e_l \sum_A \int_0^1 s^{n-3} \left[\left(e_1 \frac{\partial}{\partial z_1} - \sum_{j=2}^n e_j \frac{\partial}{\partial z_j} \right) \frac{\partial u_A(sz)}{\partial z_l} \right] z e_A ds$$
$$= - \int_0^1 s^{n-3} \sum_A \left\{ \sum_{l=1}^n e_l \left[\frac{\partial}{\partial z_l} \left(e_1 \frac{\partial}{\partial z_1} - \sum_{j=2}^n e_j \frac{\partial}{\partial z_j} \right) u_A(sz) \right] \right\} z e_A ds = 0,$$

and

$$\sum_{l=1}^n e_l \frac{\partial F}{\partial z_l} = \sum_{l=1}^n e_l \sum_A \int_0^1 s^{n-3} \sum_{k=1}^n z_k \frac{\partial^2 u_A(sz)}{\partial z_k z_l} e_A ds$$
$$- \sum_{l=1}^n e_l \sum_A Vec^{(A)} \left\{ \int_0^1 s^{n-3} \left[\left(e_1 \frac{\partial}{\partial z_1} - \sum_{j=2}^n e_j \frac{\partial}{\partial z_j} \right) u_A(sz) \right] e_l e_A ds \right\}$$
$$= \int_0^1 s^{n-3} \sum_A \sum_{l,k=1}^n e_l e_k \frac{\partial^2 u_A(sz)}{\partial z_k z_l} e_A ds$$

$$-\int_0^1 s^{n-3}\sum_A\sum_{l=1}^n e_l Vec^{(A)}\left\{\left[\left(e_1\frac{\partial}{\partial z_1}-\sum_{j=2}^n e_j\frac{\partial}{\partial z_j}\right)u_A(sz)\right]e_le_A\right\}ds$$

$$=\int_0^1 s^{n-3}\sum_A\sum_{j,l=1}^n\frac{\partial}{\partial z_j}\left[z_l\frac{\partial u_A(sz)}{\partial z_l}\right]e_je_A ds$$

$$-\int_0^1 s^{n-3}\sum_A\sum_{l=1}^n e_l Vec^{(A)}\left\{\left[\left(e_1\frac{\partial}{\partial z_1}-\sum_{j=2}^n e_j\frac{\partial}{\partial z_j}\right)u_A(sz)\right]e_le_A\right\}ds.$$

Taking into account

$$\begin{aligned}
&\sum_{l=1}^n e_l Vec^{(A)}\left\{\left[\left(e_1\frac{\partial}{\partial z_1}-\sum_{j=2}^n e_j\frac{\partial}{\partial z_j}\right)u_A(sz)\right]e_le_A\right\}\\
=\;& -\sum_{\substack{j=2\\ l=1}}^n e_j\frac{\partial}{\partial z_j}u_A(sz)e_A+\sum_{l=2}^n e_l\frac{\partial}{\partial z_1}u_A(sz)e_le_A\\
&+\sum_{l=2}^n e_l\left(-\sum_{\substack{j=2\\ j\neq l}}^n e_j\frac{\partial}{\partial z_j}u_A(sz)\right)e_le_A\\
=\;& -\left(\sum_{j=2}^n e_j\frac{\partial}{\partial z_j}u_A(sz)\right)e_A-(n-1)\frac{\partial u_A(sz)}{\partial z_1}e_A\\
&+\sum_{l=2}^n e_l^2\left(\sum_{\substack{j=2\\ j\neq l}}^n e_j\frac{\partial}{\partial z_j}u_A(sz)e_A\right)\\
=\;& \left(-\sum_{j=2}^n e_j\frac{\partial}{\partial z_j}u_A(sz)\right)e_A-(n-1)\frac{\partial u_A(sz)}{\partial z_1}e_A\\
&-(n-2)\left(\sum_{j=2}^n e_j\frac{\partial}{\partial z_j}u_A(sz)\right)e_A\\
=\;& -(n-1)\left[\left(e_1\frac{\partial}{\partial z_1}+\sum_{j=2}^n e_j\frac{\partial}{\partial z_j}\right)u_A(sz)\right]e_A\\
=\;& -(n-1)\sum_{j=1}^n e_j\frac{\partial}{\partial z_j}u_A(sz)e_A=-(n-1)\sum_{j=1}^n\frac{\partial}{\partial z_j}u_A(sz)e_je_A,
\end{aligned}$$

we have

$$\sum_{l=1}^{n} e_l \frac{\partial F}{\partial z_l} = \int_0^1 s^{n-3} \sum_A \sum_{j,l=1}^{n} \frac{\partial}{\partial z_j} \left[z_l \frac{\partial u_A(sz)}{\partial z_l} \right] e_j e_A ds$$

$$- \int_0^1 s^{n-3} \sum_A \left[(1-n) \sum_{j=1}^{n} \frac{\partial}{\partial z_j} u_A(sz) e_j e_A \right] ds$$

$$= \int_0^1 s^{n-3} \left\{ \sum_A \sum_{j=1}^{n} \left[\frac{\partial}{\partial z_j} \sum_{l=1}^{n} z_l \frac{\partial u_A(sz)}{\partial z_l} + (n-1) \frac{\partial}{\partial z_j} u_A(sz) \right] e_j e_A \right\} ds = 0.$$

Consequently, $F(z)$ in U is a complex left regular function.

According to Theorem 5.6 and 5.7, we see that there exists a similar characteristic between the relation of complex left regular functions and complex harmonic functions in complex Clifford analysis and the relation of analytic functions and harmonic functions in one complex variable function theory, but they aren't the same. The results of Theorem 5.6 and 5.7 are generalizations of those in [19].

CHAPTER II

BOUNDARY VALUE PROBLEMS OF GENERALIZED REGULAR FUNCTIONS AND HYPERBOLIC HARMONIC FUNCTIONS IN REAL CLIFFORD ANALYSIS

This chapter deals with boundary value problems of some functions in real Clifford analysis. In the first three sections, the problems of regular and generalized regular functions are considered, and in the last section, the Dirichlet problem of hyperbolic harmonic functions is discussed. Most results in this chapter have been obtained by us in recent years.

1 The Dirichlet Problem of Regular Functions for a ball in Real Clifford Analysis

In this section, we discuss two boundary value problems of regular functions for a ball in real Clifford analysis, which are obtained from the papers of Luogeng Hua[26]1) and Sha Huang[29]4),5).

Firstly, we give definitions of some differential operators

$$\overline{\partial}_{i-1,i}=e_{i-1}\frac{\partial}{\partial x_{i-1}}+e_i\frac{\partial}{\partial x_i},\ i=2,3,...,n,$$
$$\partial_{i-1,i}=e_{i-1}\frac{\partial}{\partial x_{i-1}}-e_i\frac{\partial}{\partial x_i},\ i=2,3,...,n,$$

and then we have

$$\overline{\partial}=\frac{1}{2}[\overline{\partial}_{12}+\partial_{12}+\sum_{i=2}^{n}(\overline{\partial}_{i-1,i}-\partial_{i-1,i})],$$
$$\partial=\frac{1}{2}[\overline{\partial}_{12}+\partial_{12}-\sum_{i=2}^{n}(\overline{\partial}_{i-1,i}-\partial_{i-1,i})],\quad \frac{\partial}{\partial x_1}=\frac{1}{2}(\overline{\partial}_{12}+\partial_{12}),$$

$$\frac{\partial}{\partial x_i} = \frac{-e_i}{2}(\overline{\partial}_{i-1,i} - \partial_{i-1,i}),\ i = 2, 3, ..., n.$$

By using the quasi-permutation signs introduced in Chapter I, we can give some regular conditions of functions.

If we write the element $\sum_A a_A e_A$ in $\mathcal{A}$ as

$$\sum_A a_A e_A = \sum_B (a_B + a_{2B} e_2) e_B = \sum_B I_B e_B,$$

where $B = \{\alpha_1, ..., \alpha_k\} \subseteq \{3, 4, ..., n\}, 3 \le \alpha_1 < \cdots < \alpha_k \le n$, $I_B \in \mathbf{C}$ (the complex plane). It is evident that we can obtain the following theorem.

Theorem 1.1 *The sufficient and necessary condition for* $\sum_A a_A e_A = 0$ *is that for all* $B = \{\alpha_1, ..., \alpha_k\} \subseteq \{3, ..., n\}$, $3 \le \alpha_1 < \cdots < \alpha_k \le n$, *the following equality holds:*

$$I_B = a_B + a_{2B} e_2 = 0. \tag{1.1}$$

Moreover, for a function whose value is in the real Clifford algebra $\mathcal{A}_n(\mathbf{R})$*:*

$$f(x) = \sum_A f_A(x) e_A :\ \Omega \to \mathcal{A}_n(\mathbf{R}),$$

we can write it as

$$f(x) = \sum_B I_B e_B :\ \Omega \to \mathcal{A}_{n-1}(\mathbf{C}),$$

where $I_B : \Omega \to \mathbf{C}$, $\mathcal{A}_{n-1}(\mathbf{C})$ *is the complex Clifford algebra.*

Theorem 1.2 *A function whose value is in the real Clifford algebra* $\mathcal{A}$*:*

$$\begin{gathered} f(x) = \sum_A f_A(x) e_A = \sum_B I_B e_B, \\ (A = \{\beta_1, ..., \beta_k\} \subseteq \{2, 3, ..., n\},\ 2 \le \beta_1 < \cdots < \beta_k \le n, \\ B = \{\alpha_1, ..., \alpha_k\} \subseteq \{3, 4, ..., n\}, 3 \le \alpha_1 < \cdots < \alpha_k \le n) \end{gathered} \tag{1.2}$$

is regular in Ω *if and only if*

$$\overline{\partial}_{12} I_B = \sum_{m=3}^{n} \delta_{\overline{mB}} I_{\overline{mB}x_m},$$

where $I_{\overline{mB}x_m} = \partial I_{\overline{mB}}/\partial x_m$, $\overline{I}_{\overline{mB}x_m}$ *is the conjugate of* $I_{\overline{mB}x_m}$, *and* $\overline{mB}$ *is the quasi-permutation for* mB, $\delta_{\overline{mB}}$ *is the sign of quasi-permutation* $\overline{mB}$. *In addition, a function* f *is harmonic in* Ω *if and only if every* I_B *is harmonic.*

Proof It is clear that $e_m I_B = a_B e_m - a_{2B} e_2 e_m = \overline{I}_B e_m$, so

$$\overline{\partial} f(x) = \sum_B \overline{\partial}_{12} I_B e_B + \sum_B \sum_{m=3}^{n} \overline{I}_{Bx_m} e_m e_B.$$

Denote $B = \{\alpha_1, ..., \alpha_k\}$, $\hat{B}_p = \{\alpha_1, ..., \alpha_{p-1}, \alpha_{p+1}, ..., \alpha_k\}$, $mB = m\alpha_1 ...\alpha_k$, $Bm = \alpha_1 ...\alpha_k m$. When m is some α_p among B, we have

$$e_{\alpha_p} \overline{I}_{\hat{B}_p x_{\alpha_p}} e_{\hat{B}_p} = -\overline{I}_{\hat{B}_p x_{\alpha_p}} (-1)^p e_B = -\overline{I}_{\overline{\alpha_p B} x_{\alpha_p}} \delta_{\overline{\alpha_p B}} e_B.$$

When $m < \alpha_1$, we have

$$e_m \overline{I}_{mBx_m} e_{mB} = -\overline{I}_{\overline{mB}x_m} e_B = -\overline{I}_{\overline{mB}x_m} \delta_{\overline{mB}} e_B.$$

Similarly, when $m > \alpha_k$, we get

$$e_m \overline{I}_{Bmx_m} e_{Bm} = -\overline{I}_{\overline{mB}x_m} \delta_{\overline{mB}} e_B,$$

and

$$\sum_B \sum_{m=3}^{n} \overline{I}_{Bm\,x_m} e_m e_B = -\sum_B \sum_{m=3}^{n} [\overline{I}_{\overline{mB}x_m} \delta_{\overline{mB}}] e_B,$$

hence

$$\overline{\partial} f(x) = \sum_B \overline{\partial}_{12} I_B e_B - \sum_B \sum_{m=3}^{n} \overline{I}_{\overline{mB}x_m} \delta_{\overline{mB}} e_B.$$

Thus according to Theorem 1.1, the function f is regular if and only if

$$\overline{\partial}_{12} I_B = \sum_{m=3}^{n} \overline{I}_{\overline{mB}x_m} \delta_{\overline{mB}}.$$

This completes the proof.

In order to derive another sufficient and necessary requirement of the generalized Cauchy-Riemann condition, we divide the function

$$f(x) = \sum_A f_A e_A = \sum_B I_B e_B$$

into the two parts

$$f(x) = f^{(1)} + f^{(2)} = \sum_B{}' I'_B e'_B + \sum_B{}'' I''_B e''_B,$$

where B in the sum $\sum\limits_{B}'$ obtained from $(3,4,...,n)$ is a combination with odd integers; it is called the first suffix, the rest is called the second suffix. The corresponding sum is denoted by $\sum\limits_{B}''$, $I_B(x)$ whose B is got from the first suffix and is denoted by $I'_B(x)$, e_B whose B is derived from the second suffix is denoted by e'_B, and $I_B(x), e_B$ whose B is derived from the second suffix are denoted by I''_B, e''_B respectively. In addition, we call $f^{(i)}$ the ith part of f, and denote $f^{(i)} = J_i f (i = 1.2)$. According to this and the above theorem, we can get the following theorem.

Theorem 1.3 *A function f whose value is in real Clifford algebra $\mathcal{A}_n(R)$ is regular in domain Ω, if and only if*

$$\begin{cases} \overline{\partial}_{12} I''_B = \sum\limits_{m=3}^{n} \delta'_{\overline{mB}} \overline{I}'_{\overline{mB}x_m}, \\ \overline{\partial}_{12} I'_B = \sum\limits_{m=3}^{n} \delta''_{\overline{mB}} \overline{I}''_{\overline{mB}x_m}, \end{cases} \tag{1.3}$$

in which $f(x) = \sum\limits_{A} f_A e_A = \sum\limits_{B} I_B e_B = \sum\limits_{B}' I'_B e'_B + \sum\limits_{B}'' I''_B e''_B$, $B = \{\alpha_1, ..., \alpha_k\} \subset \{3, 4, ..., n\}$, *and* $\overline{\partial}_{12} I'_B$, $\overline{I}'_{\overline{mB}x_m}$, $\delta'_{\overline{mB}}$ *denote the corresponding part of* $\overline{\partial}_{12} I_B$, $\overline{I}_{\overline{mB}x_m}$, $\delta_{\overline{mB}}$ *respectively, when* B, $\overline{mB}$ *are derived from the first suffix. The rest is the corresponding part which is derived from the second suffix.*

Let $x = (x_1, x_2, ..., x_n) \subset R^n$, and x^T be the transpose of x, Ω : $xx^T = \sum\limits_{i=1}^{n} x_i^2 < 1$ represent a unit ball, and $\partial\Omega : xx^T = 1$ be a unit sphere, whose area $\omega_n = 2\pi^{n/2}/\Gamma(n/2)$.

Definition 1.1 If $\sum\limits_{B}' u'_B e'_B$ is continuous in $\partial\Omega$, we find a function $f(x)$ to be regular in Ω, and continuous in $\overline{\Omega} = \Omega \cup \partial\Omega$ with the condition

$$J_1 f(\xi) = \sum_{B}{}' u'_B(\xi) e'_B, \ \xi \in \partial\Omega. \tag{1.4}$$

The above problem is called the Dirichlet boundary value problem in the unit ball, and we denote it by Problem D.

Theorem 1.4 *Let $\sum\limits_{B}' u'_B e'_B$ be continuous on the sphere $\partial\Omega$. Then Problem D in the ball Ω is solvable, and the solution can be represented by*

$$f(x) = \sum_{B}{}' I'_B(x) e'_B + \sum_{B}{}'' I''_B(x) e''_B, \tag{1.5}$$

where

$$I'_B(x) = \frac{1}{\omega_n} \underbrace{\int \cdots \int}_{\xi\xi^T=1} P(x,\xi) u'_B(\xi)\dot{\xi}, \tag{1.6}$$

in which $P(x,\xi) = (1 - xx^T)/(1 - 2\xi x^T + xx^T)^{\frac{n}{2}}$ *is called the Poisson kernel,* $\dot{\xi}$ *is the area element of the sphere* $\xi\xi^T = 1$,

$$I''_B(x) = T_{12} R''_B(x) + Q''_B(x), \tag{1.7}$$

$$R''_B(x) = \frac{1}{2\omega_n} \underbrace{\int \cdots \int}_{\xi\xi^T=1} \sum_{m=3}^{n} P_{x_m} \delta'_{\overline{mB}} \overline{u'_{mB}(\xi)}\dot{\xi}, \tag{1.8}$$

and $Q''_B(x)$ *satisfies the following relations:*

$$\partial_{\overline{z}_{12}} Q''_B(x) = 0, \tag{1.9}$$

$$\overline{\partial}_{12} I'_B(x) = \sum_{k=3}^{n} \delta''_{\overline{kB}} [\overline{T}_{12} \overline{R''_{\overline{kB}x_k}(x)} + \overline{Q''_{\overline{kB}x_k}(x)}]. \tag{1.10}$$

The operators T_{12}, $\overline{T}_{12}$ *and* $\overline{\partial}_{\overline{z}_{12}}$ *can be seen below.*

Proof Firstly, we find an expression of the solution. Suppose that $f(x)$ is a solution of Problem D in the ball Ω. Then from Theorem 2.1, Chapter I, and Theorem 1.2, we can derive that $I'_B(x)$ is harmonic in Ω. By [26]1) we obtain

$$I'_B(x) = \frac{1}{\omega_n} \underbrace{\int \cdots \int}_{\xi\xi^T=1} \frac{1 - xx^T}{(1 - 2\xi x^T + xx^T)^{\frac{n}{2}}} u'_B(\xi)\dot{\xi},$$

which satisfies $I'_B(\xi) = u'_B(\xi)$ $(\xi \in \partial\Omega)$, where $\underbrace{\int \cdots \int}_{\xi\xi^T=1} \dot{\xi} = \omega_n$.

Denote $z_{12} = x_1 + x_2 e_2, \zeta_{12} = \xi_1 + \xi_2 e_2, \partial_{z_{12}} = \frac{1}{2}\left(\frac{\partial}{\partial x_1} - e_2 \frac{\partial}{\partial x_2}\right), \partial_{\overline{z}_{12}} = \frac{1}{2}\left(\frac{\partial}{\partial x_1} + e_2 \frac{\partial}{\partial x_2}\right)$. It follows that $\partial_{12} = 2\partial_{z_{12}}$, $\overline{\partial}_{12} = 2\partial_{\overline{z}_{12}}$. When $xx^T < 1$ we introduce two operators:

$$T_{12} f_B(x) = \frac{1}{\pi} \underbrace{\int \cdots \int}_{\xi_1^2+\xi_2^2<1-x_3^2-\cdots-x_n^2} \frac{f_B(\xi_1, \xi_2, x_3, x_4, \ldots, x_n)}{z_{12} - \zeta_{12}} d\xi_1 d\xi_2,$$

$$\overline{T}_{12} f_B(x) = \frac{1}{\pi} \underbrace{\int \cdots \int}_{\xi_1^2+\xi_2^2<1-x_3^2-\cdots-x_n^2} \frac{f_B(\xi_1, \xi_2, x_3, x_4, \ldots, x_n)}{\overline{z}_{12} - \overline{\zeta}_{12}} d\xi_1 d\xi_2.$$

By using the results in [77] and the first expression of Theorem 1.3, we can obtain

$$I''_B(x)=\frac{1}{2}T_{12}\sum_{m=3}^{n}\delta'_{\overline{mB}}\overline{I'_{\overline{mB}x_m}}+Q''_B(x),$$

where $\partial_{\overline{z}_{12}}Q''_B(x)=0\ (x\in\Omega)$. By using [26]1), we get

$$R''_B(x)=\frac{1}{2\omega_n}\underbrace{\int\cdots\int}_{\xi\xi^T=1}\sum_{m=3}^{n}\frac{-2x_m(1-2\xi x^T+xx^T)-2^{-1}n\xi_m}{(1-2\xi x^T+xx^T)^{\frac{n}{2}}}\delta_{\overline{mB}}\overline{u'_{\overline{mB}}}(\xi)\dot{\xi}.$$

Therefore,

$$I''_B(x)=T_{12}R''_B(x)+Q''_B(x).$$

Substituting (1.7) into the second expression of Theorem 1.3, we can derive

$$\overline{\partial}_{12}I'_B(x)=\sum_{k=3}^{n}\delta''_{\overline{kB}}[\overline{T}_{12}\overline{R''_{\overline{kB}x_k}(x)}+\overline{Q''_{\overline{kB}x_k}(x)}].$$

From the overdetermined system (1.10) and $\partial_{\overline{z}_{12}}Q''_B(x)=0\ (x\in\Omega)$ and by [76], we can find $Q''_B(x)$. Thus, from (1.7) again, $I''_B(x)$ can also be found. That is to say, if $f(x)$ is a solution of Problem D, then the expressions (1.5) – (1.10) hold.

Moreover, we verify that the function satisfying expressions (1.5) – (1.10) is a solution of Problem D. In fact, since $f^{(1)}|_{\partial\Omega}=\sum_B{}' I'_B e'_B$, $\xi\in\partial\Omega$, by using [27], we have $I'_B(\xi)=u'_B(\xi)$, and then

$$f^{(1)}|_{\partial\Omega}=\sum_B{}' u'_B(\xi)e'_B.$$

From (1.7), (1.9) and $\overline{\partial}_{12}=2\partial_{\overline{z}_{12}}$, we immediately derive

$$\overline{\partial}_{12}I''_B=\overline{\partial}_{12}T_{12}R''_B+2\partial\overline{z}_{12}Q''_B=\overline{\partial}_{12}T_{12}R''_B=2\partial_{\overline{z}_{12}}T_{12}R''_B.$$

In addition, from (1.6) and (1.8), we have

$$\begin{aligned}2\partial_{\overline{z}_{12}}T_{12}R''_B &= \sum_{m=3}^{n}\frac{\delta'_{\overline{mB}}}{\omega_n}\underbrace{\int\cdots\int}_{\xi\xi^T=1}P_{x_m}(x,\xi)\overline{u'_{\overline{mB}}(\xi)}\dot{\xi}\\ &= \sum_{m=3}^{n}\delta'_{\overline{mB}}\overline{I'_{\overline{mB}}}x_m.\end{aligned}$$

Thus, from (1.10) and (1.7), we get

$$\overline{\partial_{12}} I'_B = \sum_{m=3}^{n} \delta''_{\overline{mB}} \overline{I}''_{\overline{mB}x_m}.$$

From Theorem 1.3 again, it is easy to see that the function $f(x)$ satisfying (1.5) – (1.10) is regular in Ω. To sum up, $f(x)$ is a solution of Problem D. This proof is completed.

Next, we discuss the pseudo-modified Dirichlet problem. In order to discuss the uniqueness of its solution, we first consider the sectional domains of Ω. Cutting Ω by "the planes":

$$\begin{cases} x_3 = a_3, \\ x_4 = a_4, \\ \quad \cdots \\ x_n = a_n, \end{cases}$$

we obtain a sectional domain G_a in the x_1x_2 plane:

$$x_a x_a^T = x_1^2 + x_2^2 + \sum_{m=3}^{n} a_m^2 < 1.$$

Let $\Gamma_a : \xi_a \xi_a^T = \xi_1^2 + \xi_2^2 + \sum\limits_{m=3}^{n} a_m^2 = 1$ be the boundary of G_a, and its center be denoted by $O_a = (0, 0, a_3, a_4, ..., a_n)$.

For given continuous functions $\sum\limits_B{}' u'_B(\xi) e'_B$ $(\xi \in \partial\Omega), \phi''_B(\xi_a)$ $(\xi_a \in \Gamma_a)$ and the complex constants d''_{Ba}, we find a regular function $f(x) = \sum\limits_B{}' I'_B e'_B + \sum\limits_B{}'' I''_B e''_B$ in Ω, which is continuous in $\overline{\Omega}$ with the following pseudo-modified conditions:

$$\begin{cases} J_1 f(\xi) &= \sum\limits_B{}' u'_B(\xi) e'_B,\ \xi \in \partial\Omega, \\ \mathrm{Re} I''_B|_{\Gamma_a} &= \phi''_B(\xi_a) + h''_B(\xi_a),\ \xi_a \in \Gamma_a, \\ I''_B(O_a) &= d''_{Ba}, \end{cases}$$

where $h''_B(\xi_a) = h''_{Ba}$ $(\xi_a \in \Gamma_a)$ are all unknown real constants to be determined appropriately, and $\mathrm{Re} I''_B = \mathrm{Re}(F''_B + F''_{2B} e_2) = F''_B$. The above problem will be denoted by Problem D^*.

Theorem 1.5 *Suppose that* $\sum\limits_B{}' u'_B e'_B$ *is continuous on* $\partial\Omega$, *and for any fixed* $a_3, a_4, ..., a_n$, *the function* $\phi''_B(\xi_a)$ *is continuous on* Γ_a, *here*

$\xi = (\xi_1, \xi_2, a_3, ..., a_n)$. *Then there exists a unique solution of Problem* D^*. *Moreover the solution possesses the expressions* (1.5) – (1.10) *and satisfies*

$$\mathrm{Re}Q''_B(\xi_a) = -\mathrm{Re}[T_{12}R''_B(\xi_a)] + \phi''_B(\xi_a) + h''_B(\xi_a),\ \xi_a \in \Gamma_a, \tag{1.11}$$

$$Q''_B(O_a) = -T_{12}R''_B(O_a) + d''_{Ba}. \tag{1.12}$$

Proof Evidently, on the basis of the proof of Theorem 1.4, it is sufficient to add the following proof.

Firstly, we find the integral representation of the solution. Suppose that $f(x)$ is a solution of Problem D^*. From (1.7) and the boundary condition, we can derive

$$\mathrm{Re}[T_{12}R''_B(\xi_a) + Q''_B(\xi_a)] = \phi''_B(\xi_a) + h''_B(\xi_a),$$
$$T_{12}R''_B(O_a) + Q''_B(O_a) = I''_B(O_a) = d''_{Ba}.$$

Noting that

$$\partial_{\bar z_{12}} Q''_B(x) = 0,\ x \in \Omega,$$

it is clear that $Q''_B(x_a)$ satisfies the conditions

$$\begin{cases} \partial_{\bar z_{12}} Q''_B(x_a) = 0,\ x_a \in G_a, \\ \mathrm{Re}Q''_B(\xi_a) = -\mathrm{Re}[T_{12}R''_B(\xi_a)] + \phi''_B(\xi_a) + h''_B(\xi_a),\ \xi_a \in \Gamma_a, \\ Q''_B(O_a) = -T_{12}R''_B(O_a) + d''_{Ba}. \end{cases}$$

Since the modified Dirichlet problem for analytic functions has a unique solution [80]7), from (1.9), (1.11) and (1.12), we can find $Q''_B(x_a)$, $x \in G_a$, and then $Q''_B(x)$, $x \in \Omega$, because a is an arbitrary point.

That is to say, if $f(x)$ is a solution of Problem D^*, then the expressions (1.5) – (1.12) hold.

Next, we verify that the function $f(x)$ determined by the above expressions is a solution of Problem D^*. From (1.7) – (1.11), it follows that

$$\mathrm{Re}(I''_B)|_{\Gamma_a} = \mathrm{Re}[T_{12}R''_B(\xi_a)] + \mathrm{Re}[Q''_B(\xi_a)] = \phi''_B(\xi_a) + h''_B(\xi_a),\ \xi_a \in \Gamma_a,$$

and then $I''_B(O_a) = T_{12}R''_B(O_a) + Q''_B(O_a) = d''_{Ba}$. Therefore, the above function $f(x)$ is just a solution of Problem D^*.

Finally, we prove that the solution of Problem D^* is unique. Suppose that $f_1(x)$ and $f_2(x)$ are two solutions of Problem D^*, and denote $f_1(x)-$

$f_2(x)$ by $F(x)$. It is clear that $F(x)$ is regular in Ω and is a solution of the corresponding homogeneous equation of Problem D^* and $F^{(1)}|_{\partial\Omega} = \sum_B{}' u'_B(\xi)e'_B - \sum_B{}' u'_B(\xi)e'_B = 0$. For convenience, we shall adopt the same symbols for $f(x)$ as before, namely denote $F(x) = \sum_B{}' I'_B e'_B + \sum_B{}'' I''_B e''_B$. Since $F(x)$ is regular in Ω, thus it is harmonic in Ω, therefore for all B, $I'_B(x)$ are all harmonic in Ω. Since $\sum_B{}' I'_B e'_B|_{\partial\Omega} = F^{(1)}|_{\partial\Omega} = 0, I'_B|_{\partial\Omega} = 0$, again by using the uniqueness of the solution of the Dirichlet problem for harmonic functions in a ball (see [26]1)), we get $I'_B \equiv 0$ in Ω, thus $J_1F \equiv 0$, $I'_{\overline{mB}} \equiv 0$ in Ω. From the definition of $R''_B(x)$,

$$R''_B(x) = \frac{1}{2}\sum_{m=3}^{n} \delta'_{\overline{mB}} \overline{I}'_{\overline{mB}} x_m,$$

and then $R''_B \equiv 0$. Hence

$$I''_B = T_{12}R''_B(x) + Q''_B(x) = Q''_B(x). \tag{1.13}$$

Since $F(x)$ is a solution of the corresponding homogeneous equation of Problem D^*, from (1.9), (1.11) and (1.12), we derive

$$\begin{cases} \partial_{\overline{z}_{12}} Q''_B(x_a) = 0, \ x_a \in G_a, \\ \mathrm{Re}Q''_B(\xi_a) = h''_B(\xi_a), \ \xi_a \in \Gamma_a, \\ Q''_B(O_a) = 0, \ O_a \in G_a. \end{cases}$$

In addition, using the results about the existence and uniqueness of solutions of the modified Dirichlet problem for analytic functions (see [80]7)), we can obtain $Q''_B(x_a) \equiv 0$, $x_a \in G_a$. Hence $Q''_B(x) \equiv 0$, $x \in \Omega$. From (1.13), $I''_B \equiv 0$, $x \in \Omega$, and then $J_2F(x) \equiv 0$ in Ω. So $F(x) \equiv 0$, $x \in \Omega$, i.e. $f_1(x) = f_2(x)$, $x \in \Omega$. This shows the uniqueness of the solution of Problem D^*.

2 The Mixed Boundary Value Problem for Generalized Regular Functions in Real Clifford Analysis

In this section, we discuss the existence and uniqueness of solutions of the so-called mixed boundary value problem (Problem *P-R-H*) for generalized regular functions in real Clifford analysis; the material is derived from Huang Sha's paper [29]6).

Definition 2.1 We assume the linear elliptic system of second order equations

$$\Delta u = \sum_{m=1}^{n} d_m u_{x_m} + fu + g, \quad x \in \Omega, \tag{2.1}$$

where $\Omega(\in \mathbf{R}^n)$ is a bounded domain and $\Omega \in C^{2,\alpha}$ $(0 < \alpha < 1)$, $d_m = d_m(x) = d_m(x_1, ..., x_n) \in C^{0,\alpha}(\overline{\Omega})$, $d_m(x) \geq 0$, $x \in \overline{\Omega}$. The oblique derivative problem of equation (2.1) is to find a solution $u(x) = u(x_1, ..., x_n) \in C^2(\overline{\Omega})$ satisfying (2.1) and the boundary conditions

$$\begin{cases} \dfrac{\partial u}{\partial \nu} + \sigma(x)u(x) = \tau(x) + h, \ x \in \partial\Omega, \\ u(d) = u_0, \ d = (d_1, ..., d_n) \in \overline{\Omega}, \end{cases} \tag{2.2}$$

in which $\sigma(x), \tau(x) \in C^{1,\alpha}(\partial\Omega)$, $\sigma(x) \geq 0$, $x \in \partial\Omega$, h is an unknown real constant, u_0 is a real constant, ν is a vector on $x \in \partial\Omega$, $\cos(\nu, n_0) \geq 0$, n_0 is the outward normal vector on $x \in \partial\Omega$, and $\cos(\nu, n_0) \in C^{1,\alpha}(\partial\Omega)$. The above boundary problem is called Problem O.

Problem O is a non-regular oblique derivative problem. If the coefficients $\nu(x), \sigma(x)$ satisfy $\cos(\nu, n_0) \equiv 0$, $\sigma(x) \equiv 0$ on $\partial\Omega$, then Problem O is the Dirichlet problem. If $\nu(x), \sigma(x)$ satisfy $\cos(\nu, n_0) \equiv 1$, $\sigma(x) \equiv 0$ on $\partial\Omega$, then Problem O is the Neumann problem. If $\cos(\nu, n_0) \geq \delta > 0$, $\sigma(x) \geq 0$ on $\partial\Omega$, then Problem O is the regular oblique derivative problem. In [59], B. P. Ponejah proved the following lemma using the method of integral equations.

Lemma 2.1 *Problem O for equation (2.1) has a unique solution.*

Proof The existence and uniqueness of solutions for Problem O for equation (2.1) in the plane can be found in [80]4). Using a similar method in [80]4), we can also prove the uniqueness of the solution in Lemma 2.1. Using a priori estimates of solutions and the Leray-Schauder theorem [18], the existence of solutions in Lemma 2.1 can be proved.

For convenience, we order all $\omega_A(x)$ with numbers of the form 2^{n-1} in $\omega(x) = \sum\limits_A \omega_A(x)e_A$ according to the following method, and denote them by $\omega_1, \omega_2, \omega_2, ..., \omega_{2^{n-1}}$.

1) If none of the suffixes in $\omega_A = \omega_{h_1,...,h_r}$ is $h_i = n$, but there exists some suffix $k_j = n$ in $\omega_B = \omega_{k_1,...,k_s}$, then we arrange ω_A before ω_B.

2) If none of the suffixes in $\omega_A = \omega_{h_1,...,h_r}$, $\omega_B = \omega_{k_1,...,k_s}$ is n, then when $r < s$, we order ω_A before ω_B. When $\omega_A = \omega_{h_1,...,h_r}, \omega_C = \omega_{\alpha_1,...,\alpha_r}$,

and $h_1+\cdots+h_r<\alpha_1+\cdots+\alpha_r$, we order ω_A before ω_C. When $h_1+\cdots+h_r=\alpha_1+\cdots+\alpha_r$ and $(h_1,...,h_r)\neq(\alpha_1,...,\alpha_r)$, if the first unequal suffix is $h_i<\alpha_i$, we also order ω_A before ω_C.

3) If there exists some suffix in $\omega_A=\omega_{h_1,...,h_r}$ and $\omega_B=\omega_{k_1,...,k_s}$ is n, then $\omega_A=\omega_{h_1,...,h_{r-1},n}$, $\omega_B=\omega_{k_1,...,k_{s-1},n}$, we order $\omega_A=\omega_{h_1,...,h_{r-1}}$ and $\omega_B=\omega_{k_1,...,k_{s-1}}$ by using the method as in 2), and regard them as the order of $\omega_A=\omega_{h_1,...,h_r}$ and $\omega_B=\omega_{k_1,...,k_s}$.

We have ordered all suffixes with numbers in the form 2^{n-1} through 1), 2), 3), then we can denote them by $\omega_1,...,\omega_{2^{n-1}}$.

Definition 2.2 The oblique derivative problem for generalized regular functions in Ω is to find a solution $\omega(x)\in C^{1,\alpha}(\overline{\Omega})\bigcap C^2(\Omega)$for the elliptic system of first order equations

$$\bar{\partial}\omega=a\omega+b\bar{\omega}+l \tag{2.3}$$

satisfying the boundary conditions

$$\begin{cases}\dfrac{\partial\omega_k}{\partial\nu_k}+\sigma_k(x)\omega_k(x)=\tau_k(x)+h_k,\ x\in\partial\Omega,\\ \omega_k(d)=u_k,\ 1\le k\le 2^{n-1}\end{cases} \tag{2.4}$$

in which $\sigma_k(x),\tau_k(x)\in C^{1,\alpha}(\partial\Omega),\sigma_k(x)\ge 0\ on\ \partial\Omega$, h_k is an unknown real constant, u_k is a real constant, ν_k is the vector on $\partial\Omega$, n_0 is the outward normal vector on $\partial\Omega$; moreover, $\cos(\nu_k,n_0)\in C^{1,\alpha}(\partial\Omega)$. The above boundary value problem will be called Problem P.

Let $\omega(x)=\sum\limits_A\omega_A(x)e_A$ be a solution of Problem P. Then according to Property 3.3 and Property 3.4, Chapter I, we know that the following equalities for arbitrary index A are true:

$$\begin{aligned}\sum_{m=1,\overline{mB}=A}^{n}\delta_{\overline{mB}}\omega_{Bx_m}=&\sum_{C,\overline{C\underline{M}}=A}(a_C+b_C)\omega_{\underline{M}}\delta_{\overline{C\underline{M}}}\\&+\sum_{C,\overline{C\underline{\underline{M}}}}(a_C-b_C)\omega_{\underline{\underline{M}}}\delta_{\overline{C\underline{\underline{M}}}}+l_A,\end{aligned} \tag{2.5}$$

$$\begin{aligned}\Delta\omega_D=&\sum_{j,C}\delta_{\overline{jD}}\delta_{\overline{C\underline{M}}}(a_C+b_C)_{x_j}\omega_{\underline{M}}+\sum_{j,C}\delta_{\overline{jD}}\delta_{\overline{C\underline{M}}}(a_C+b_C)\omega_{\underline{M}x_j}\\&+\sum_{j,C}\delta_{\overline{jD}}\delta_{\overline{C\underline{\underline{M}}}}(a_C-b_C)_{x_j}\omega_{\underline{\underline{M}}}+\sum_{j,C}\delta_{\overline{jD}}\delta_{\overline{C\underline{\underline{M}}}}(a_C-b_C)\omega_{\underline{\underline{M}}x_j}\\&+\sum_{j=1}^{n}\delta_{\overline{jD}}l_{Ax_j},\end{aligned} \tag{2.6}$$

where $j = 1, 2, ..., n$, $\overline{jD} = A$, $\overline{C\underline{M}} = A$, $\overline{C\underline{\underline{M}}} = A$; $\underline{M}$, $\underline{\underline{M}}$ denote two kinds of indices of M respectively (see [29]6)).

Suppose that the equalities

$$(a_C + b_C)_{x_j} = 0,\ (a_C - b_C)_{x_j} = 0,\ \overline{jD} = A,\ \overline{CM} = A,\ C \neq j \quad (2.7)$$

in (2.6) are true. Especially, when $n = 3$, set $C = 23$, then according to the condition (2.7), we get $(a_{23} - b_{23})_{x_j} = 0$, $1 \leq j \leq 3$, namely $(a_{23} - b_{23})$ is a constant. If $C = 3$, by the condition (2.7), we get $(a_3 - b_3)_{x_j} = 0$, $j = 1, 2$, i.e. $(a_3 - b_3)$ only depends on x_3.

Noting the condition (2.7), the equality (2.6) can be written as

$$\begin{aligned}\Delta\omega_D = {} & \sum_j \delta_{\overline{jD}}\delta_{\overline{j\underline{M}}}(a_j + b_j)_{x_j}\omega_{\underline{M}} + \sum_{j,C} \delta_{\overline{jD}}\delta_{\overline{C\underline{M}}}(a_C + b_C)\omega_{\underline{M}x_j} \\ & + \sum_j \delta_{\overline{jD}}\delta_{\overline{j\underline{\underline{M}}}}(a_j - b_j)_{x_j}\omega_{\underline{\underline{M}}} + \sum_{j,C} \delta_{\overline{jD}}\delta_{\overline{C\underline{\underline{M}}}}(a_C - b_C)\omega_{\underline{\underline{M}}x_j} \\ & + \sum_{j=1}^{n} \delta_{\overline{jD}} l_{Ax_j},\end{aligned} \quad (2.8)$$

in which $\overline{C\underline{M}} = A$, $\overline{C\underline{\underline{M}}} = A$, $\overline{jD} = A$.

Suppose that D is the $\underline{A}$-type index. In the first term of the right side in (2.8), we see $\overline{j\underline{M}} = \overline{jD}$, $\underline{M} = D$, and then $\delta_{\overline{jD}}\delta_{\overline{j\underline{M}}} = 1$; and in the third term of the right side in (2.8), if the equality $\overline{j\underline{\underline{M}}} = \overline{j\underline{M}}$ holds, then $D = \underline{\underline{M}}$ is the $\underline{A}$-type index. This is a contradiction. Hence when D is the $\underline{A}$-type index, the third term of the equality (2.8) disappears. Thus the equality (2.8) can be written as

$$\begin{aligned}\Delta\omega_D = {} & \sum_{j=1}^{n} (a_j + b_j)_{x_j}\omega_D + \sum_{j,C} \delta_{\overline{jD}}\delta_{\overline{C\underline{M}}}(a_C + b_C)\omega_{\underline{M}x_j} \\ & + \sum_{j,C} \delta_{\overline{jD}}\delta_{\overline{C\underline{\underline{M}}}}(a_C - b_C)\omega_{\underline{\underline{M}}x_j} + \sum_{j=1}^{n} \delta_{\overline{jD}} l_{Ax_j},\end{aligned} \quad (2.9)$$

in which $\overline{C\underline{M}} = A, \overline{C\underline{\underline{M}}} = A, \overline{jD} = A$. Especially, when $n = 3, D = 1$, the equality (2.9) possesses the form

$$\begin{aligned}\Delta\omega_1 = {} & \sum_{j=1}^{3} (a_j + b_j)\omega_{1x_j} + \sum_{j=1}^{3} (a_j + b_j)_{x_j}\omega_1 + (a_1 - b_1)\omega_{2x_2} \\ & + (a_1 - b_1)\omega_{3x_3} - (a_2 - b_2)\omega_{2x_2} - (a_2 - b_2)\omega_{23x_3} \\ & - (a_3 - b_3)\omega_{3x_3} + (a_3 - b_3)\omega_{23x_2} - (a_{23} - b_{23})\omega_{23x_1} \\ & - (a_{23} - b_{23})\omega_{3x_2} + (a_{23} - b_{23})\omega_{2x_3} + \sum_{j=1}^{3} l_j x_j.\end{aligned} \quad (2.10)$$

Noting $\underline{M} = D = 1$, $\underline{\underline{M}} \neq D = 1$, it is clear that the third term of the equality (2.8): $\sum_{j=1}^{n}(a_j - b_j)_{x_j}\omega_{\underline{\underline{M}}}$ in (2.10) disappears. When $\underline{M} = D$, we have $C = \overline{\underline{M}A} = \overline{\underline{M}jD} = \overline{DjD} = j$, $\delta_{\overline{jD}}\delta_{\overline{C\underline{M}}} = \delta_{\overline{jD}}\delta_{\overline{CD}} = \delta_{\overline{jD}}\delta_{\overline{jD}} = 1$, and then the equality (2.9) can be rewritten as

$$\begin{aligned}\Delta\omega_D=&\sum_j(a_j+b_j)_{x_j}\omega_D+\sum_j(a_j+b_j)\omega_{Dx_j}+\sum_{j,C}\delta_{\overline{jD}}\delta_{\overline{C\underline{M}}}(a_C+b_C)\omega_{\underline{M}x_j}\\&+\sum_{j,C}\delta_{\overline{jD}}\delta_{\overline{C\underline{\underline{M}}}}(a_C-b_C)\omega_{\underline{\underline{M}}x_j}+\sum_j\delta_{\overline{jD}}l_{Ax_j},\end{aligned} \tag{2.11}$$

where $\underline{M} \neq D$, and A is the index satisfying $A = \overline{jD}$.

In addition, we first write equation (2.5) in the form

$$\begin{aligned}&\sum_{j=1}^{n}\delta_{\overline{j\underline{M}}}\omega_{\underline{M}x_j}+\sum_{j=1}^{n}\delta_{\overline{j\underline{\underline{M}}}}\omega_{\underline{\underline{M}}x_j}\\&=\sum_E(a_E+b_E)\omega_{\underline{M}}\delta_{\overline{E\underline{M}}}+\sum_E(a_E-b_E)\delta_{\overline{E\underline{\underline{M}}}}\omega_{\underline{\underline{M}}}+l_B,\end{aligned} \tag{2.12}$$

where $\overline{E\underline{M}} = B$, $\overline{B\underline{\underline{M}}} = B$, $\overline{j\underline{M}} = B$, and B runs all indexes. Moreover we use $\delta_{\overline{CD}}(a_C - b_C)$ to multiply every equation in (2.12), herein $\overline{CD} = B$, D is the fixed index, and sum according to the index B, we obtain

$$\begin{aligned}&\sum_{B,j}\delta_{\overline{CD}}\delta_{\overline{j\underline{M}}}(a_C-b_C)\omega_{\underline{M}x_j}+\sum_{B,j}\delta_{\overline{CD}}\delta_{\overline{j\underline{\underline{M}}}}(a_C-b_C)\omega_{\underline{\underline{M}}x_j}\\&=\sum_{B,E}(a_C-b_C)(a_E+b_E)\delta_{\overline{CD}}\delta_{\overline{E\underline{M}}}\omega_{\underline{M}}\\&\quad+\sum_{B,E}(a_C-b_C)(a_E-b_E)\delta_{\overline{CD}}\delta_{\overline{E\underline{\underline{M}}}}\omega_{\underline{\underline{M}}}+\sum_B(a_C-b_C)\delta_{\overline{CD}}l_B\end{aligned} \tag{2.13}$$

where $\overline{CD} = B, \overline{E\underline{M}} = B, \overline{E\underline{\underline{M}}} = B, \overline{jM} = B$.

According to Property 3.6 about the quasi-permutation in Section 3, Chapter I, $C = \overline{DB}$ and the arbitrariness of B, we know that C can run all indexes, so the second term in the left-hand side in (2.13) can also be written as

$$\begin{aligned}\sum_{B,j}\delta_{\overline{CD}}\delta_{\overline{j\underline{\underline{M}}}}(a_C-b_C)\omega_{\underline{\underline{M}}x_j}&=\sum_{C,j}\delta_{\overline{CD}}\delta_{\overline{j\underline{\underline{M}}}}(a_C-b_C)\omega_{\underline{\underline{M}}x_j}\\&=\sum_{j,C}\mu_{jDC}\delta_{\overline{jD}}\delta_{\overline{C\underline{\underline{M}}}}(a_C-b_C)\omega_{\underline{\underline{M}}x_j},\end{aligned} \tag{2.14}$$

where $\overline{j\underline{\underline{M}}} = B$, $\overline{CD} = B$.

Since D is given, we get $A = \overline{jD}\,(j = 1, 2, ..., n)$ and $C = \overline{MA} = \overline{Mj D}$. Suppose that the coefficients corresponding to C which do not conform to the condition $\overline{jD} = \overline{C\underline{\underline{M}}} = A$ satisfy

$$a_C - b_C = 0, \tag{2.15}$$

and the coefficients corresponding to $C = \overline{BD}$ with the condition $\mu_{jDC} = 1$ satisfy

$$a_C - b_C = 0; \tag{2.16}$$

then the equality (2.14) can also be written in the form

$$\sum_{\substack{B,j \\ (\overline{CD},\overline{j\underline{\underline{M}}}=B)}} \delta_{\overline{CD}}\delta_{\overline{j\underline{\underline{M}}}}(a_C - b_C)\omega_{\underline{\underline{M}}x_j} = - \sum_{\substack{j,C \\ (\overline{jD}=A,\overline{C\underline{\underline{M}}}=A)}} \delta_{\overline{jD}}\delta_{\overline{j\underline{\underline{M}}}}(a_C - b_C)\omega_{\underline{\underline{M}}x_j}. \tag{2.17}$$

In addition, from the equality (2.13), we have

$$\begin{aligned}
&\sum_{j,C} \delta_{\overline{jD}}\delta_{\overline{j\underline{\underline{M}}}}(a_C - b_C)\omega_{\underline{\underline{M}}x_j} \\
&= \sum_{\substack{j,B \\ (\overline{CD}=B,\overline{j\underline{M}}=B)}} \delta_{\overline{CD}}\delta_{\overline{j\underline{M}}}(a_C - b_C)\omega_{\underline{M}x_j} - \sum_{\substack{E,B \\ (\overline{CD}=B,\overline{E\underline{M}}=B)}} (a_C - b_C)(a_E + b_E)\delta_{\overline{CD}}\delta_{\overline{E\underline{M}}}\omega_{\underline{M}} \\
&\quad - \sum_{\substack{E,B \\ (\overline{CD}=B,\overline{E\underline{\underline{M}}}=B)}} (a_C - b_C)(a_E - b_E)\delta_{\overline{CD}}\delta_{\overline{E\underline{\underline{M}}}}\omega_{\underline{\underline{M}}} - \sum_{\substack{B \\ (\overline{CD}=B)}} (a_C - b_C)\delta_{\overline{CD}} l_B.
\end{aligned} \tag{2.18}$$

Substituting the equality (2.18) into the equality (2.11), we get

$$\begin{aligned}
\Delta\omega_D = &\sum_{\substack{j \\ (\overline{jD}=A)}} (a_j + b_j)_{x_j}\omega_D + \sum_{\substack{j \\ (\overline{jD}=A)}} (a_j + b_j)\omega_{Dx_j} \\
&+ \sum_{\substack{j,C \\ (\overline{jD}=A) \\ \overline{C\underline{M}}=A,\underline{M}\neq D}} \delta_{\overline{jD}}\delta_{\overline{C\underline{M}}}(a_C + b_C)\omega_{\underline{M}x_j} + \sum_{\substack{B,j \\ (\overline{j\underline{M}}=B,\overline{CD}=B)}} \delta_{\overline{CD}}\delta_{\overline{j\underline{M}}}(a_C - b_C)\omega_{\underline{M}x_j} \\
&- \sum_{\substack{B,E \\ (\overline{CD}=B,\overline{E\underline{M}}=B)}} (a_C - b_C)(a_E - b_E)\delta_{\overline{CD}}\delta_{\overline{E\underline{M}}}\omega_{\underline{M}} \\
&- \sum_{\substack{B,E \\ (\overline{CD}=B,\overline{E\underline{\underline{M}}}=B)}} (a_C - b_C)(a_E - b_E)\delta_{\overline{CD}}\delta_{\overline{E\underline{\underline{M}}}}\omega_{\underline{\underline{M}}} \\
&- \sum_{\substack{B \\ (\overline{CD}=B)}} (a_C - b_C)\delta_{\overline{CD}} l_B + \sum_{\substack{j \\ (\overline{jD}=A)}} \delta_{\overline{jD}} l_{Ax_j}.
\end{aligned} \tag{2.19}$$

In the fourth term of the equality (2.19), when $\underline{M} = D$, we have $\overline{CD} = B = \overline{j\underline{M}} = \overline{jD}$, and $C = j$, thus $\delta_{\overline{CD}} = \delta_{\overline{j\underline{M}}} = \delta_{\overline{jD}} = 1$. In the fifth term of the equality (2.19), when $\underline{M} = D$, in accordance with $\overline{CD} = \overline{E\underline{M}} = B$, we have $C = E$, and then $\delta_{\overline{CD}}\delta_{\overline{E\underline{M}}} = \delta_{\overline{CD}}\delta_{\overline{CD}} = 1$. So, the equality (2.19) can be written as

$$\begin{aligned}
&\Delta\omega_D = \\
&\left[\sum_{j,(\overline{jD}=A)}(a_j+b_j)_{x_j} - \sum_{C,(\overline{jD}=B)}(a_C^2-b_C^2)\right]\omega_D + \sum_{\substack{j\\ \overline{jD}=A,\overline{CD}=B}} 2a_C\omega_{Dx_j} \\
&+ \sum_{\substack{j,C\\ \overline{jD}=A,\overline{C\underline{M}}=A,\underline{M}\neq D}} \delta_{\overline{jD}}\delta_{\overline{C\underline{M}}}(a_C+b_C)\omega_{\underline{M}x_j} + \sum_{\substack{B,j\\ \overline{j\underline{M}}=B,\overline{CD}=B,\underline{M}\neq D}} \delta_{\overline{CD}}\delta_{\overline{j\underline{M}}}(a_C-b_C)\omega_{\underline{M}x_j} \\
&- \sum_{\substack{B,E\\ \overline{CD}=B,\overline{E\underline{M}}=B,\underline{M}\neq D}} (a_C-b_C)(a_E+b_E)\delta_{\overline{CD}}\delta_{\overline{E\underline{M}}}\omega_{\underline{M}} \\
&- \sum_{\substack{B,E\\ \overline{CD}=B,\overline{E\underline{\underline{M}}}=B}} (a_C-b_C)(a_E-b_E)\delta_{\overline{CD}}\delta_{\overline{E\underline{\underline{M}}}}\omega_{\underline{\underline{M}}} \\
&- \sum_{B,\ \overline{CD}=B} (a_C-b_C)\delta_{\overline{CD}}l_B + \sum_{j,\ \overline{jD}=A} \delta_{\overline{jD}}l_{Ax_j}.
\end{aligned} \tag{2.20}$$

In the sixth term of the right-hand side of equality (2.20), when $E \neq C$ and $\mu_{CDE} = -1$, by Property 3.6 in Section 3, Chapter I, we get $(a_C - b_C)(a_E-b_E)\delta_{\overline{CD}}\delta_{\overline{E\underline{\underline{M}}}} = -(a_C-b_C)(a_E-b_E)\delta_{\overline{ED}}\delta_{\overline{C\underline{\underline{M}}}} = -(a_E-b_E)(a_C-b_C)\delta_{\overline{ED}}\delta_{\overline{C\underline{\underline{M}}}}$, hence $\sum_{\substack{B,E\\ E\neq C,\mu_{CDE}=-1}} (a_C-b_C)(a_E-b_E) \times\delta_{\overline{CD}}\delta_{\overline{E\underline{\underline{M}}}}\omega_{\underline{\underline{M}}} = 0$;. While $E \neq C$, $\mu_{CDE} = -1$, suppose that the term corresponding to the sixth term of the right-hand side in (2.20) satisfies

$$(a_C - b_C)(a_E - b_E) = 0. \tag{2.21}$$

When $E = C$, we have $\overline{CD} = \overline{E\underline{\underline{M}}} = \overline{C\underline{\underline{M}}}$, so $D = \underline{\underline{M}}$. This contradicts that D is an $\underline{A}$-type index. Hence this condition does not hold, hence the equality (2.20) can also be written as

$$\begin{aligned}
\Delta\omega_D = &\left[\sum_{j,(\overline{jD}=A)}(a_j+b_j)_{x_j} - \sum_{C,(\overline{jD}=B)}(a_C^2-b_C^2)\right]\omega_D + \sum_j 2a_C\omega_{Dx_j} \\
&+ \sum_{\substack{j,C\\ \overline{jD}=A\\ \overline{C\underline{M}}=A,\underline{M}\neq D}} \delta_{\overline{jD}}\delta_{\overline{C\underline{M}}}(a_C+b_C)\omega_{\underline{M}x_j} \\
&+ \sum_{\substack{B,j\\ \overline{j\underline{M}}=B\\ \overline{CD}=B,\underline{M}\neq D}} \delta_{\overline{CD}}\delta_{\overline{j\underline{M}}}(a_C-b_C)\omega_{\underline{M}x_j} \\
&- \sum_{\substack{B,E\\ \overline{CD}=B,\overline{E\underline{M}}=B\\ \underline{M}\neq D,E\neq C}} (a_C-b_C)(a_E+b_E)\delta_{\overline{CD}}\delta_{\overline{E\underline{M}}}\omega_{\underline{M}}
\end{aligned}$$

$$-\sum_{B,(\overline{CD}=B)}(a_C-b_C)\delta_{\overline{CD}}l_B+\sum_{j,(\overline{jD}=A)}\delta_{\overline{jD}}l_{Ax_j}. \tag{2.22}$$

Let the third, fourth, and fifth terms in the right-hand side of (2.22) satisfy

$$a_C+b_C=0,\ a_C-b_C=0,\ (a_C-b_C)(a_E+b_E)=0, \tag{2.23}$$

respectively. Then equality (2.22) can be written in the form

$$\begin{aligned}\Delta\omega_D=&[\sum_{j,(\overline{jD}=A)}(a_j+b_j)_{x_j}-\sum_C(a_C^2-b_C^2)]\omega_D+\sum_{\substack{j\\ \overline{CD}=B,\overline{jD}=A}}2a_C\omega_{Dx_j}\\ &-\sum_{B,(\overline{CD}=B)}(a_C-b_C)\delta_{\overline{CD}}l_B+\sum_{j,(\overline{jD}=A)}\delta_{\overline{jD}}l_{Ax_j}.\end{aligned} \tag{2.24}$$

When $D=1$, we simply write the equality (2.24) as

$$\Delta\omega_1=\sum_{m=1}^{n}d_{m_1}\omega_{1x_m}+f_1\omega_1+g_1, \tag{2.25}$$

here $d_{m_1}=d_{m_1}(x)=d_{m_1}(x_1,...,x_n)$, $f_1=f_1(x)=f_1(x_1,...,x_n)$, $g_1=g_1(x)=g_1(x_1,...,x_n)$. Let

$$f_1(x)\geq 0. \tag{2.26}$$

Then according to Lemma 2.1, there exists a unique function $\omega_1(x)$ satisfying the boundary condition (2.4). After we get $\omega_1(x)$, we can consider that $\omega_1(x)$ in equation (2.5) is known as well as the known coefficients $a_A(x),b_A(x),l_A(x)$. Applying the same method concluding with the equality (2.25), we get the equality (2.24) when $D=2$, and write it simply as

$$\Delta\omega_2=\sum_{m=1}^{n}d_{m_2}(x)\omega_{2x_m}+f_2\omega_2+g_2, \tag{2.27}$$

where $d_{m_2}=d_{m_2}(x)$, $f_2=f_2(x)$, $g_2=g_2(x,\omega_1(x))$. Set $f_2(x)\geq 0$; on the basis of Lemma 2.1, there exists a unique $\omega_2(x)$ satisfying the boundary condition (2.4). After getting $\omega_2(x)$, we can regard $\omega_1(x)$, $\omega_2(x)$ as the known functions. Using the above method, we can get

$$\Delta\omega_3=\sum_{m=1}^{n}d_{m_3}\omega_{3x_m}+f_3\omega_3+g_3, \tag{2.28}$$

in which $d_{m3} = d_{m3}(x)$, $f_3 = f_3(x)$, $g_3 = g_3(x, \omega_1(x), \omega_2(x))$. In accordance with the above steps and the order of $\omega_1, ..., \omega_{2^{n-1}}$, we proceed until $\omega_{2^{n-2}}$. At last, we get the unique $\omega_1, ..., \omega_{2^{n-2}}$ satisfying

$$\Delta\omega_k = \sum_{m=1}^{n} d_{m_k}\omega_{kx_m} + f_k\omega_k + g_k, \tag{2.29}$$

where

$$d_{m_k} = d_{m_k}(x),\ f_k = f_k(x),\ g_k = g_k(x, \omega_1(x),\ \omega_2(x), ..., \omega_{k-1}(x)),$$
$$f_k(x) \geq 0,\ 1 \leq k \leq 2^{n-2}.$$

For simplicity, denote by $\overline{U}$ the conditions (2.7), (2.15), (2.16), (2.17), (2.21), (2.23),... and $f_k(x) \geq 0\, (1 \leq k \leq 2^{n-2})$ of system (2.3).

From the above discussion, by means of Lemma 2.1, if system (2.3) satisfies the condition $\overline{U}$, then there exists a unique solution ω_k, , $1 \leq k \leq 2^{n-2}$.

In order to further discuss ω_k $(2^{n-2}+1 \leq k \leq 2^{n-1})$, we need to study the system (2.5). Suppose that the suffix A of l_A in the equality (2.5) has been arranged according to the above method, and l_A has been written $l_1, ..., l_{2^{n-2}}, l_{2^{n-2}+1}, ..., l_{2^{n-1}}$. We may only discuss the system of equations corresponding to $l_{2^{n-2}+1}, l_{2^{n-2}+2}, ..., l_{2^{n-1}}$ in the equality (2.5), namely

$$\begin{aligned}\sum_{\substack{m=1\\ \overline{mB}=A}}^{n} \delta_{\overline{mB}}\omega_B x_m = &\sum_{C,(\overline{C\underline{M}}=A)} (a_C + b_C)\omega_{\underline{M}}\delta_{\overline{C\underline{M}}} \\ &+ \sum_{C,(\overline{C\underline{\underline{M}}}=A)} (a_C - b_C)\omega_{\underline{\underline{M}}}\delta_{\overline{C\underline{\underline{M}}}} + l_A,\end{aligned} \tag{2.30}$$

where $l_A = l_k$, $2^{n-2} + 1 \leq k \leq 2^{n-1}$, $\omega(x) = \sum_A \omega_A(x)e_A$, $x \in \Omega$. The following assumption is called the condition $\overline{V}$. Set $n = 2m$, and denote $x_{2k-1} + x_{2k}i = z_k$, $k = 1, ..., m$, i is the imaginary unit, and $\omega_{2k-1} + i\omega_{2k} = \overline{w_k}$, $k = 1, ..., 2^{n-2}$. Let $\Omega = G_1 \times \cdots \times G_m$ be a multiply circular cylinder about complex variables $z_1, ..., z_m$. Then we regard $\omega_1, ..., \omega_{2^{n-2}}$ as the known functions, by using the result in [80]4), the elliptic system of first order equations: (2.30) about $\omega_{2^{n-2}+1}, ..., \omega_{2^{n-1}}$ can be written as

$$\frac{\partial\overline{w}_k}{\partial\overline{z}_k} = f_{kl}(z_1, ..., z_m, \overline{w}_{2^{n-3}+1}, \overline{w}_{2^{n-3}+2}, ..., \overline{w}_{2^{n-2}}), \tag{2.31}$$

in which $k = 2^{n-3} + 1, ..., 2^{n-2}$, $l = 1, ..., m$.

Denote (see [43])

$$N_1^{\perp}=\left\{f \left| \begin{array}{l} f \text{ is a Hölder continuous function defined on characteristic} \\ \text{manifold } \partial G_1\times\cdots\times\partial G_m, \text{ whose real index is } \beta, \text{ and} \\ \displaystyle\int_{\partial G_1}\cdots\int_{\partial G_m}\frac{f(\zeta_1,...,\zeta_m)}{\zeta_1^{J_1}...\zeta_m^{J_m}}d\zeta_1...d\zeta_m=0, \\ J_l>-k_l,\ l=1,...,m \end{array}\right.\right\}.$$

If (i) f_{kl} is continuous with respect to $z=(z_1,...,z_m)\in\overline{\Omega}$, $\overline{\omega}=(\overline{\omega}_{2^{n-3}+1},\overline{\omega}_{2^{n-3}+2},...,\overline{\omega}_{2^{n-2}})\in B_\theta$, here $B_\theta=\{\omega||\overline{\omega}_j|<\theta,\ j=2^{n-3}+1,2^{n-3}+2,...,2^{n-2}\}$, $\theta>0$, moreover, f_{kl} is holomorphic about $\overline{\omega}\in B_\theta$, and has continuous mixed partial derivatives until $m-1$ order for different $\overline{z}_j$ $(j\neq l)$, namely $\dfrac{\partial^\lambda f_{kl}}{\partial\overline{z}_{i_1}...\partial\overline{z}_{i_\lambda}}$ is continuous about $\overline{\omega}\in B_\theta, z\in\overline{\Omega}$, here $\lambda\le m-1$, $1\le i_1<\cdots<i_\lambda\le m$, $i_k\neq l$, $k=1,...,\lambda$.

(ii) the system (2.31) is completely integrable, that is

$$\frac{\partial f_{kl}}{\partial\overline{z}_j}+\sum_{p=2^{n-3}+1}^{2^{n-2}}\frac{\partial f_{kl}}{\partial\overline{\omega}_p}f_{p_j}=\frac{\partial f_{kj}}{\partial\overline{z}_l}+\sum_{p=2^{n-3}+1}^{2^{n-2}}\frac{\partial f_{kj}}{\partial\overline{\omega}_p}f_{pl},$$

$$k=2^{n-3}+1,...,2^{n-2},\ 1\le j,\ l\le m.$$

(iii) the set

$$M_\theta=\left\{\overline{\omega} \left| \begin{array}{l} \overline{\omega} \text{ are several complex variable functions defined on } \overline{\Omega}, \\ \text{with continuous mixed partial derivatives up to order } m \\ \text{for different } \overline{z}_j, \text{ and satisfy} \\ |\overline{\omega}_j|<\theta,\ |\dfrac{\partial^\lambda\overline{\omega}_j}{\partial\overline{z}_{i_1}...\partial\overline{z}_{i_\lambda}}|\le\theta,\ j=2^{n-3}+1,...,2^{n-2}, \\ 1\le i_1<\cdots<i_\lambda\le m,\ 1\le\lambda\le m,\ z\in\overline{\Omega} \end{array}\right.\right\}$$

is defined. When $\overline{\omega}\in M_\theta$, the composite function f_{kl} and its continuous mixed partial derivatives up to order m for different $\overline{z}_j\,(j\neq l)$ are uniformly bounded, we denote its bound by K_θ. Moreover, for arbitrary $\overline{\omega},\tilde{\overline{\omega}}\in M_\theta$, the composite function f_{kl} and its mixed partial derivatives satisfy the Lipschitz condition, that is

$$\left|\frac{\partial^\lambda f_{kl}(z_1,...,z_m,\overline{\omega}_{2^{n-3}+1},...,\overline{\omega}_{2^{n-2}})}{\partial\overline{z}_{i_1}...\partial\overline{z}_{i_\lambda}}\right.$$
$$\left.-\frac{\partial^\lambda f_{kl}(z_1,...,z_m,\tilde{\overline{\omega}}_{2^{n-3}+1},...,\tilde{\overline{\omega}}_{2^{n-2}})}{\partial\overline{z}_{i_1}...\partial\overline{z}_{i_\lambda}}\right|$$

$$\leq L_\theta \max_{\substack{2^{n-3}+1\leq\varepsilon\leq 2^{n-2}\\ 1\leq j_1<\cdots<j_\alpha\leq m}} \sup_{\Omega} \left|\frac{\partial^\alpha(\overline{\omega}_E-\tilde{\overline{\omega}}_\varepsilon)}{\partial\overline{z}_{j_1}...\partial\overline{z}_{j_\alpha}}\right|,$$

$$1\leq i_1<\cdots<i_\lambda\leq m, i_p\neq l.$$

(iv) the real functions $\psi_j(z_1,...,z_m)$ on $\partial G_1\times\cdots\times\partial G_m$ satisfy the Hölder condition, namely $\psi_j(z_1,...,z_m)\in C^\beta(\partial G_1\times\cdots\times\partial G_m)$. In addition, we assume an unknown real function h_j on $N_1^\perp$. The above conditions are called the condition $\overline{V}$.

Definition 2.3 For the solution $\omega = \sum_{k=1}^{2^{n-1}}\omega_k(x)e_k$ of Problem P, if $\overline{\omega}_{2^{n-3}+1},...,\overline{\omega}_{2^{n-2}}(\overline{\omega}_k = \omega_{2k-1}+i\omega_{2k})$ satisfy generalized Riemann-Hilbert boundary condition on $\partial G_1\times\cdots\times\partial G_m$ (see [43]):

$$\begin{aligned}&\mathrm{Re}[z_1^{-k_1},...,z_m^{-k_m},\overline{\omega}_j(z_1,...,z_m)] = \psi_j(z_1,...,z_m)+h_j,\\ &j=2^{n-3}+1,...,2^{n-2}, z=(z_1,...,z_m)\in\partial G_1\times\cdots\times\partial G_m,\end{aligned}\tag{2.32}$$

then the problem for generalized regular functions is called the mixed boundary problem, which will be denoted by Problem *P-R-H*.

On the basis of the result in [43], under the condition $\overline{V}$, when $K_j<0\,(j=1,...,m)$, and $K_\theta, L_\theta, \sum_{j=2^{n-3}+1}^{2^{n-2}} C_\beta(\psi_j)$ are small enough, there exists a unique solution $(\overline{\omega}_{2^{n-3}+1},...,\overline{\omega}_{2^{n-2}})$ for the modified problem (2.31), (2.32), so $\omega_{2^{n-2}+1},...,\omega_{2^{n-1}}$ are uniquely determined.

From the above discussion, we get the existence and uniqueness of the solution of Problem *P-R-H* for generalized regular functions in real Clifford analysis.

Theorem 2.2 *Under the condition $\overline{U},\overline{V}$, when $K_j<0$ $(j=1,2,...,m)$, and $K_\theta, L_\theta, \sum_{j=2^{n-3}+1}^{2^{n-2}} C_\beta(\psi_j)$ are small enough, there exists a unique solution $\omega(x)=\sum_A\omega_A(x)e_A=\sum_{k=1}^{2^{n-1}}\omega_k(x)e_k\,(x\in\Omega)$ of Problem $P-R-H$ for generalized regular functions, where $\omega_1(x),...,\omega_{2^{n-2}}(x)$ satisfy equation (2.3) and the boundary condition (2.4) of Problem P. Denote $\overline{\omega}_k=\omega_{2k-1}+i\omega_{2k}\,(k=2^{n-3}+1,\ ...,2^{n-2})$, then $\overline{\omega}_{2^{n-3}+1},...,\ \overline{\omega}_{2^{n-2}}$ satisfy equation (2.31), and the corresponding functions $\omega_{2^{n-2}+1}$, $\omega_{2^{n-2}+2}$, ..., $\omega_{2^{n-1}}$ satisfy equation (2.3) and the generalized $R-H$ boundary condition (2.32).*

3 A Nonlinear Boundary Value Problem With Haseman Shift for Regular Functions in Real Clifford Analysis

This section deals with the nonlinear boundary value problem with Haseman shift $d(t)$ in real Clifford analysis, whose boundary condition is as follows:

$$\begin{aligned} &a(t)\Phi^+(t) + b(t)\Phi^+(d(t)) + c(t)\Phi^-(t) \\ &= g(t) \cdot f(x, \Phi^+(t), \Phi^-(t), \Phi^+(d(t)), \Phi^-(d(t))). \end{aligned} \tag{3.1}$$

We shall prove the existence of solutions for the problem (3.1) by using the Schauder fixed point theorem (see [29]1)). It is easy to see that when $a(t) = g(t) \equiv 0$, $b(t) \equiv 1$, the problem (3.1) becomes the Haseman problem

$$\Phi^+(d(t)) = G(t)\Phi^-(t). \tag{3.2}$$

The problem (3.2) was first solved by C. Haseman [22]. In general, all boundary value conditions for holomorphic functions can be expressed as the pasting condition of the unknown functions, hence the boundary value problem can be regarded as the conformal pasting problem [87] in function theory. But the method of conformal pasting cannot be used to handle all problems of multiple elements. In 1974, A. M. Hekolaeshuk [23] gave an example, i.e. for the boundary value problem

$$a(t)\Phi^+(t) + b(t)\Phi^-(d(t)) + c(t)\Phi^-(t) = g(t), \tag{3.3}$$

the method of conformal pasting cannot be eliminated the shift $d(t)$. For the problem (3.1) discussed in this section, we choose the linear case of (3.3) as its example.

Firstly, we reduce the problem to the integral equation problem, and then use the fixed point theorem to prove the existence of solutions for the problem.

Assume a connected open set $\Omega \in \mathbf{R^n}$, whose boundary $\partial\Omega$ is a smooth, oriented, compact Liapunov surface (see Section 2, Chapter 1). Suppose that $a(t), b(t), c(t), d(t), g(t)$ are given on $\partial\Omega$, and $d(t)$ is a homeomorphic mapping, which maps $\partial\Omega$ onto $\partial\Omega$. Denote $\Omega^+ = \Omega$, $\Omega^- = R^n \backslash \overline{\Omega}$, $\overline{\Omega} = \Omega \bigcup \partial\Omega$; we shall find a regular function $\Phi(x)$ in Ω^+, which is continuous on $\Omega^\pm \bigcup \partial\Omega$, and satisfies $\Phi^-(\infty) = 0$ and the nonlinear boundary condition (3.1) with Haseman shift. The above problem is called Problem SR. Set

$$\Phi(x) = \frac{1}{\omega_n} \int_{\partial\Omega} E(x,t)m(t)\varphi(t)ds_t, \tag{3.4}$$

where $\omega_n = 2\pi^{\frac{n}{2}}/\Gamma(\frac{n}{2})$ is the surface area of a unit ball in $\mathbf{R}^n$, $E(x,t) = \frac{\overline{t-x}}{|t-x|^n}$, $m(t) = \sum\limits_{j=1}^{n} e_j \cos(m, e_j)$ is the outward normal unit vector on $\partial\Omega$, and ds_t is the area element, and $\varphi(t)$ is the unknown Hölder continuous function on $\partial\Omega$. According to the Plemelj formula (2.24) in Section 2, Chapter I,

$$\Phi^+(x) = \frac{\varphi(x)}{2} + P\varphi(x),\ x \in \partial\Omega, \tag{3.5}$$

$$\Phi^-(x) = -\frac{\varphi(x)}{2} + P\varphi(x),\ x \in \partial\Omega, \tag{3.6}$$

in which the operator

$$P\varphi(x) = \frac{1}{\omega_n} \int_{\partial\Omega} E(x,t)m(t)\varphi(t)dS_t,\ x \in \partial\Omega.$$

In addition

$$\Phi^+(d(t)) = \frac{\varphi_1(x)}{2} + P_1\varphi(x),\ \ x \in \partial\Omega, \tag{3.7}$$

$$\Phi^-(d(t)) = -\frac{\varphi_1(x)}{2} + P_1\varphi(x),\ \ x \in \partial\Omega, \tag{3.8}$$

where $\varphi_1(x) = \varphi(d(x))$, and

$$\begin{aligned} P_1\varphi(x) &= P\varphi(d(x)) \\ &= \frac{1}{\omega_n} \int_{\partial\Omega} E(d(x),t)m(t)\varphi(t)dS_t. \end{aligned} \tag{3.9}$$

Substituting (3.5) – (3.8) into (3.1), we get

$$a(\frac{\varphi}{2} + P\varphi) + b(\frac{\varphi_1}{2} + P_1\varphi) + c(-\frac{\varphi}{2} + P\varphi) = g \cdot f. \tag{3.10}$$

Introducing the operator

$$F\varphi = (a+c)(-\frac{\varphi}{2} + P\varphi) + b(\frac{\varphi_1}{2} + P_1\varphi) + (1+a)\varphi - gf,$$

the equation (3.10) becomes

$$\varphi = F\varphi. \tag{3.11}$$

Thus the problem SR is reduced to solving the integral equation (3.11).

Denote by $H(\partial\Omega, \beta)$ the set of the above Hölder continuous functions with the Hölder index $\beta\,(0 < \beta < 1)$. For arbitrary $\varphi \in H(\partial\Omega, \beta)$, the norm of φ is defined as

$$\| \varphi \|_\beta = C(\varphi, \partial\Omega) + H(\varphi, \partial\Omega, \beta),$$

where $C(\varphi, \partial\Omega) = \max\limits_{t\in\partial\Omega} |\varphi(t)|$, $H(\varphi, \partial\Omega, \beta) = \sup\limits_{t_1\neq t_2,\, t_1,t_2\in\partial\Omega} \dfrac{|\varphi(t_1)-\varphi(t_2)|}{|t_1-t_2|^\beta}$.
It is evident that $H(\partial\Omega, \beta)$ is a Banach space; moreover we easily verify

$$\| f+g \|_\beta \le \| f \|_\beta + \| g \|_\beta, \quad \| f\cdot g \|_\beta \le 2^{n-1}\| f \|_\beta \| g \|_\beta, \tag{3.12}$$

where $f, g \in H(\partial\Omega, \beta)$.

Theorem 3.1 *Suppose that the operator $\theta : \theta\varphi = \frac{\varphi}{2} - P\varphi$ and the function $\varphi(t) \in H(\partial\Omega, \beta)$ are given; then there exists a constant J_1 independent of φ, such that*

$$\| \theta\varphi \|_\beta \le J_1 \| \varphi \|_\beta. \tag{3.13}$$

Proof On the basis of Theorem 2.7, Chapter I, we know

$$\frac{1}{\omega_n}\int_{\partial\Omega} E(x,t)m(t)ds_t = \frac{1}{2}, \quad x \in \partial\Omega,$$

and then

$$|(\theta\varphi)(x)| \le \frac{1}{\omega_n} H(\varphi, \partial\Omega, \beta)\int_{\partial\Omega} \frac{ds_t}{|t-x|^{n-1-\beta}} = M_1 H(\varphi, \partial\Omega, \beta), \tag{3.14}$$

in which M_1 is a constant independent of φ.

In order to consider $H(\theta\varphi, \partial\Omega, \beta)$, we choose arbitrary $x, \hat{x} \in \partial\Omega$, and denote $\delta = |x - \hat{x}|$. Firstly, suppose $6\delta < d$ (d is the constant about a Liapunov surface in Section 2, Chapter I); we can make a sphere with the center at x and radius 3δ. The inner part of this sphere is denoted $\partial\Omega_1$ and the remaining part is denoted $\partial\Omega_2$, thus we have

$$\begin{aligned} |(\theta\varphi)(x) - (\theta\varphi)(\hat{x})| &\le \frac{1}{\omega_n}\left|\int_{\partial\Omega_1} E(x,t)m(t)(\varphi(x)-\varphi(t))ds_t\right| \\ &+ \frac{1}{\omega_n}\left|\int_{\partial\Omega_1} E(\hat{x},t)m(t)(\varphi(\hat{x})-\varphi(t))ds_t\right| \\ &+ \frac{1}{\omega_n}\left|\int_{\partial\Omega_2} E(x,t)m(t)(\varphi(x)-\varphi(t))ds_t - \int_{\partial\Omega_2} E(\hat{x},t)m(t)(\varphi(\hat{x})-\varphi(t))ds_t\right| \\ &= L_1 + L_2 + L_3. \end{aligned}$$

For x, we use the result about $N_0 \in \partial\Omega$ in Section 2, Chapter I, and denote by π_1 the projection field of $\partial\Omega_1$ on the tangent plane of x; then

$$L_1 \leq M_2 H(\varphi, \partial\Omega, \beta) \int_0^{3\delta} \frac{\rho_0^{n-2}}{\rho_0^{n-1-\beta}} d\rho_0 = M_3 H(\varphi, \partial\Omega, \beta)|x - \hat{x}|^\beta,$$

where M_2, M_3 are constants independent of x, $\hat{x}$. In the following we shall denote by M_i the constant having this property. Similarly, $L_2 \leq M_4 H(\varphi, \partial\Omega, \beta)|x - \hat{x}|^\beta$. Next, we estimate L_3:

$$\begin{aligned} L_3 &\leq \frac{1}{\omega_n} \Big| \int_{\partial\Omega_2} (E(x,t) - E(\hat{x},t)) m(t)(\varphi(x) - \varphi(t)) ds_t \Big| \\ &+ \frac{1}{\omega_n} \Big| \int_{\partial\Omega_2} E(\hat{x},t) m(t)(\varphi(x) - \varphi(\hat{x})) ds_t \Big| = O_1 + O_2. \end{aligned}$$

By using Hile's lemma (see Section 2, Chapter 1), we get

$$\begin{aligned} |E(x,t) - E(\hat{x},t)| &= \left| \frac{\overline{t-x}}{|t-x|^n} - \frac{\overline{t-\hat{x}}}{|t-\hat{x}|^n} \right| \\ &\leq \sum_{k=0}^{n-2} \left| \frac{t-x}{t-\hat{x}} \right|^{-(k+1)} |t - \hat{x}|^{-n} |x - \hat{x}|. \end{aligned}$$

For arbitrary $t \in \partial\Omega_2$, we have $|t - \hat{x}| \geq 2\delta$, and then

$$\frac{1}{2} \leq \left| \frac{t-x}{t-\hat{x}} \right| \leq 2.$$

Thus $O_1 \leq M_5 H(\varphi, \partial\Omega, \beta)|x - \hat{x}|^\beta$. Noting that $\varphi \in H(\varphi, \partial\Omega)$, it is easy to see that $O_2 \leq M_6 H(\varphi, \partial\Omega, \beta)|x - \hat{x}|^\beta$. Hence

$$L_3 \leq M_7 H(\varphi, \partial\Omega, \beta)|x - \hat{x}|^\beta.$$

From the above discussion, when $6|x - \hat{x}| < d$, we have

$$|(\theta\varphi)(x) - (\theta\varphi)(\hat{x})| \leq M_8 H(\varphi, \partial\Omega, \beta)|x - \hat{x}|^\beta. \tag{3.15}$$

On the basis of the results in [53], we obtain the above estimation for $6|x - \hat{x}| \geq d$. Moreover, according to (3.14),(3.15), there exists a positive constant J_1, such that $\| \theta\varphi \|_\beta \leq J_1 \| \varphi \|_\beta$. This completes the proof.

Taking into account

$$P\varphi = \frac{\varphi}{2} - \theta\varphi,$$

we get

$$\| P\varphi \|_\beta \leq \frac{1}{2} \| \varphi \|_\beta + \| \theta\varphi \|_\beta \leq (\frac{1}{2} + J_1) \| \varphi \|_\beta. \tag{3.16}$$

Similarly, it is easy to prove the following corollary.

Corollary 3.2 *For arbitrary $\varphi \in H(\partial\Omega, \beta)$, there exists a constant J_2 independent of φ, such that*

$$\| \frac{\varphi}{2} + P\varphi \|_\beta \le J_2 \| \varphi \|_\beta. \tag{3.17}$$

Theorem 3.3 *Let the shift $d = d(x)\,(x \in \partial\Omega)$ satisfy the Lipschitz condition on $\partial\Omega$. Then for arbitrary $x, \hat{x} \in \partial\Omega$, we have*

$$|d(x) - d(\hat{x})| \le J_3 |x - \hat{x}|. \tag{3.18}$$

We introduce the operator

$$G\varphi = \frac{\varphi_1}{2} + P_1\varphi = \frac{\varphi(d(x))}{2} + P\varphi(d(x)),$$

then for arbitrary $\varphi \in H(\partial\Omega, \beta)$, there exists a constant J_6 independent of φ, such that

$$\| G\varphi \|_\beta \le J_6 \| \varphi \|_\beta. \tag{3.19}$$

Proof According to (3.16), we get

$$C(G\varphi, \partial\Omega) \le \frac{\| \varphi \|_\beta}{2} + (\frac{1}{2} + J_1) \| \varphi \|_\beta = (1 + J_1) \| \varphi \|_\beta. \tag{3.20}$$

Similar to the proof of (3.15), we have

$$|P\varphi(d(x)) - P\varphi(d(\hat{x}))| \le J_4 \| \varphi \|_\beta |d(x) - d(\hat{x})|^\beta \le J_5 \| \varphi \|_\beta |x - \hat{x}|^\beta. \tag{3.21}$$

From (3.20), (3.21), it follows that the inequality $\| G\varphi \|_\beta \le J_6 \| \varphi \|_\beta$ holds.

Corollary 3.4 *Under the same condition as in Theorem 3.3, the following inequality holds:*

$$\| \frac{-\varphi_1}{2} + P_1\varphi \|_\beta \le J_7 \| \varphi \|_\beta. \tag{3.22}$$

Theorem 3.5 *Suppose that the shift $d = d(x)$ in Problem SR satisfies the condition (3.18) and $a(t), b(t), c(t), g(t) \in H(\partial\Omega, \beta)$. Then, if the function $f(t, \Phi^{(1)}, \Phi^{(2)}, \Phi^{(3)}, \Phi^{(4)})$ is Hölder continuous for the arbitrary fixed Clifford numbers $\Phi^{(1)}, \Phi^{(2)}, \Phi^{(3)}, \Phi^{(4)}$ about fixed $t \in \partial\Omega$*

and satisfies the Lipschitz condition for the arbitrary fixed $t \in \partial\Omega$ *about* $\Phi^{(1)}, \Phi^{(2)}, \Phi^{(3)}, \Phi^{(4)}$, *namely*

$$\begin{aligned}&|f(t_1, \Phi_1^{(1)}, \Phi_1^{(2)}, \Phi_1^{(3)}, \Phi_1^{(4)}) - f(t_2, \Phi_2^{(1)}, \Phi_2^{(2)}, \Phi_2^{(3)}, \Phi_2^{(4)})|\\ &\le J_8|t_1 - t_2|^\beta + J_9|\Phi_1^{(1)} - \Phi_2^{(1)}| + \cdots + J_{12}|\Phi_1^{(4)} - \Phi_2^{(4)}|,\end{aligned} \tag{3.23}$$

where J_i, $i = 8, ..., 12$ *are positive constants independent of* t_i, $\Phi_j^{(i)}$ $(i = 1, 2, 3, 4,\ j = 1, 2)$, $f(0, 0, 0, 0, 0) = 0$; *and if* $\| a + c \|_\beta < \varepsilon < 1$, $\| b \|_\beta < \varepsilon < 1$, $\| 1 + a \|_\beta < \varepsilon < 1$, $0 < \mu = \varepsilon \cdot 2^{n-1}(J_1 + J_6 + 1) < 1$, *and* $\| g \|_\beta < \delta$; *then when* $0 < \delta \le \dfrac{M(1-\mu)}{2^{n-1}(J_{17} + J_{18}M)}$, *Problem SR is solvable, where* M *is the given positive number* ($\| \varphi \|_\beta \le M$), J_{17}, J_{18} *are the positive numbers dependent on* $J_i, i = 1, 2, 6, 7, ..., 12$.

Proof Denote by

$$T = \{\varphi | \varphi \in H(\partial\Omega,\ \beta), \| \varphi \|_\beta \le M\}$$

the subset of the continuous function space $C(\partial\Omega)$. According to (3.11), we have

$$\begin{aligned}||F\varphi||_\beta \le\, & 2^{n-1}\|a+c\|_\beta\|\theta\varphi\|_\beta + 2^{n-1}||b||_\beta||G\varphi||_\beta + 2^{n-1}||1+a||_\beta||\varphi||_\beta\\ &+ 2^{n-1}||g||_\beta \cdot ||f(t, \frac{\varphi}{2} + P\varphi, -\frac{\varphi}{2} + P\varphi, \frac{\varphi_1}{2} + P_1\varphi, -\frac{\varphi_1}{2} + P_1\varphi)||_\beta.\end{aligned}$$

From Theorems 3.1, 3.3, Corollaries 3.2, 3.4 and the condition (3.23), it follows that

$$C(f, \partial\Omega) \le J_{13} + J_{14}||\varphi||_\beta. \tag{3.24}$$

Moreover using (3.23) we have

$$\begin{aligned}&\Big|f\left(t_1, \frac{\varphi(t_1)}{2} + P\,\varphi(t_1), \frac{-\varphi(t_1)}{2} + P\varphi(t_1), \frac{\varphi_1(t_1)}{2} + P_1\varphi(t_1)\,,\right.\\ &\left.-\frac{\varphi_1(t_1)}{2} + P_1\varphi(t_1)\right) - f\left(t_2, \frac{\varphi(t_2)}{2} + P\varphi(t_2)\,, \frac{-\varphi(t_2)}{2} + P\varphi(t_2),\right.\\ &\left.\frac{\varphi_1(t_2)}{2} + P_1\varphi(t_2), \frac{-\varphi(t_2)}{2} + P_1\varphi(t_2)\right)\Big|\\ &\le (J_{15} + J_{16}||\varphi||_\beta)|t_1 - t_2|^\beta\ (J_{15} = J_8).\end{aligned} \tag{3.25}$$

In accordance with (3.24),(3.25), we obtain

$$||f||_\beta \le J_{17} + J_{18}||\varphi||_\beta, \tag{3.26}$$

hence when $\varphi \in T$, applying the condition in this theorem, the inequality

$$\begin{aligned} ||F\varphi||_\beta &\le \mu||\varphi||_\beta + 2^{n-1}\delta(J_{17} + J_{18}||\varphi||_\beta) \\ &< \mu M + \delta 2^{n-1}(J_{17} + J_{18}M) \le M \end{aligned}$$

is concluded. This shows that F maps the set T into itself.

In the following, we shall prove that F is a continuous mapping. Choose arbitrary $\varphi^{(n)}(x) \in T$, such that $\{\varphi^{(n)}(x)\}$ uniformly converges to $\varphi(x), x \in \partial\Omega$. It is clear that for arbitrary given number $\varepsilon > 0$, when n is large enough, $||\varphi^{(n)} - \varphi||_\beta$ may be small enough. Now we consider $P\varphi^{(n)}(x) - P\varphi(x)$. Let $6\delta < d$, $\delta > 0$. Then we can make a sphere with the center at x and radius 3δ. The inner part of the sphere is denoted by $\partial\Omega_1$, and the rest part is denoted $\partial\Omega_2$. Thus we have

$$\begin{aligned} &|P\varphi^{(n)}(x) - P\varphi(x)| \le \frac{1}{\omega_n}\left|\int_{\partial\Omega} E(x,t)m(t)[\varphi^{(n)}(t) - \varphi(t)]dS_t\right| \\ = &\frac{1}{\omega_n}|\int_{\partial\Omega} E(x,t)m(t)[\varphi^{(n)}(t) - \varphi^{(n)}(x) + \varphi(x) - \varphi(t) + \varphi^{(n)}(x) \\ &-\varphi(x)]dS_t| \le \frac{1}{\omega_n}|\int_{\partial\Omega} E(x,t)m(t)[(\varphi^{(n)}(t) - \varphi^{(n)}(x)) \\ &+(\varphi(x) - \varphi(t))]dS_t| + \frac{1}{\omega_n}\left|\int_{\partial\Omega} E(x,t)m(t)\left(\varphi^{(n)}(x) - \varphi(x)\right)dS_t\right| \\ \le &\frac{1}{\omega_n}|\int_{\partial\Omega_1} E(x,t)m(t)[(\varphi^{(n)}(t) - \varphi^{(n)}(x)) + (\varphi(x) - \varphi(t))]dS_t| \\ &+\frac{1}{\omega_n}|\int_{\partial\Omega_2} E(x,t)m(t)[(\varphi^{(n)}(t) - \varphi^{(n)}(x)) + (\varphi(x) - \varphi(t))]dS_t| \\ &+\frac{1}{2}||\varphi^{(n)} - \varphi||_\beta = L_4 + L_5 + \frac{||\varphi^{(n)} - \varphi||_\beta}{2} \end{aligned}$$

where

$$\begin{aligned} L_4 &= \frac{1}{\omega_n}|\int_{\partial\Omega_1} E(x,t)m(t)[(\varphi^{(n)}(t) - \varphi^{(n)}(x)) + (\varphi(x) - \varphi(t))dS_t| \\ &\le J_{19}\int_0^{3\delta} \frac{1}{\rho_0^{n-1-\beta}}\rho_0^{n-2}d\rho_0 = J_{19}\int_0^{3\delta} \rho_0^{\beta-1}d\rho_0 = J_{20}\delta^\beta, \end{aligned}$$

and

$$\begin{aligned} L_5 &= \frac{1}{\omega_n}|\int_{\partial\Omega_2} E(x,t)m(t)[(\varphi^{(n)}(t) - \varphi(t)) - (\varphi^{(n)}(x) - \varphi(x))dS_t| \\ &\le J_{21}||\varphi^{(n)} - \varphi||_\beta, \end{aligned}$$

hence

$$|P\varphi^{(n)}(x) - P\varphi(x)| \le J_{20}\delta^\beta + J_{22}||\varphi^{(n)} - \varphi||_\beta.$$

We choose a sufficiently small positive number δ, such that $J_{20}\delta^{\beta} < \varepsilon/2$, and then choose a sufficiently large positive integer n, such that $J_{22}||\varphi^{(n)} - \varphi||_{\beta} < \frac{\varepsilon}{2}$. Thus for the arbitrary $x \in \partial\Omega$, we have

$$|P\varphi^{(n)}(x) - P\varphi(x)| < \varepsilon. \tag{3.27}$$

Similarly, when n is large enough, for arbitrary $x \in \partial\Omega$, we can derive

$$|\varphi_1^{(n)}(x) - \varphi_1(x)| < \varepsilon, \tag{3.28}$$

$$|P_1\varphi^{(n)}(x) - P_1\varphi(x)| < \varepsilon. \tag{3.29}$$

Taking into account (3.11), (3.23) and (3.27) – (3.29), we can choose a sufficiently large positive integer n, such that

$$|F\varphi^{(n)}(x) - F\varphi(x)| < \varepsilon, \text{ for arbitrary } x \in \partial\Omega.$$

This shows that F is a continuous mapping, which maps T into itself. By means of the Ascoli-Arzela theorem, we know that T is a compact set in the continuous function space $C(\partial\Omega)$. Hence the continuous mapping F maps the closed convex set T in $C(\partial\Omega)$ onto itself, and $F(T)$ is also a compact set in $C(\partial\Omega)$. By the Schauder fixed point theorem, there exists a function $\varphi_0 \in H(\partial\Omega, \beta)$ satisfying the integral equation (3.11). This shows that Problem SR is solvable.

Theorem 3.6 *If $f \equiv 1$ in Theorem 3.5, then Problem SR has a unique solution.*

In fact, for arbitrary $\varphi_1, \varphi_2 \in H(\partial\Omega, \beta)$, by using the similar method as before, we can obtain

$$||F\varphi_1 - F\varphi_2||_{\beta} < \mu||\varphi_1 - \varphi_2||_{\beta}.$$

Taking account of the condition $0 < \mu < 1$, we know that $F\varphi$ (when $f \equiv 1$) is a contracting mapping from the Banach space $H(\partial\Omega, \beta)$ into itself, hence there exists a unique fixed point $\varphi_0(x)$ of the functional equation $\varphi_0 = F\varphi_0$, i.e. Problem SR has a unique solution

$$\Phi(x) = \frac{1}{\omega_n}\int_{\partial\Omega} E(x,t)m(t)\varphi_0(t)dS_t, \ \Phi^{-}(\infty) = 0.$$

In 1991, Sha Huang discussed the boundary value problem with conjugate value

$$a(t)\Phi^{+}(t) + b(t)\overline{\Phi^{+}}(t) + c(t)\Phi^{-}(t) + d(t)\overline{\Phi^{-}}(t)$$
$$= g(t), \ t \in \Omega$$

for regular functions in real Clifford analysis (see [29]3)). Similarly, we can discuss the nonlinear boundary value problem with shift and conjugate value for regular functions in real Clifford analysis.

4 The Dirichlet Problem of Hyperbolic Harmonic Functions in Real Clifford Analysis

One of the generalized forms of a Cauchy-Riemann system in high dimensional space is the following system of equations:

$$\begin{cases} x_n(\dfrac{\partial u_1}{\partial x_1} - \dfrac{\partial u_2}{\partial x_2} - \cdots - \dfrac{\partial u_n}{\partial x_n}) + (n-1)u_n = 0, \\ \dfrac{\partial u_i}{\partial x_k} = \dfrac{\partial u_k}{\partial x_i}, \ i, k = 2, ..., n, \\ \dfrac{\partial u_1}{\partial x_k} = -\dfrac{\partial u_k}{\partial x_1}, \ k = 2, ..., n. \end{cases} \qquad (H_n)$$

The system (H_n) appeared in a remark of H. Hasse paper [21] in 1949, but to our knowledge has not been treated so far. In 1992, H. Leutwiler established the relation between solutions for system (H_n) and classical holomorphic functions [41]. In this section, on the basis of [41], we study the Schwarz integral representation for hyperbolic harmonic functions and the existence of solutions for a kind of boundary value problems for hyperbolic harmonic functions for a high dimension ball in real Clifford analysis. We also discuss hyperbolic harmonic functions in real Clifford analysis and the relation with solutions of system (H_n). The material comes from Sha Huang's paper [29]7).

4.1 The Relation Between Solutions for System (H_n) and Holomorphic Functions

Setting $x = (x_1, x_2, ..., x_n) \in \mathbf{R}^n$, we denote $l(x) = [\sum\limits_{k=2}^{n} x_k^2]^{\frac{1}{2}}$, $I(x) = \sum\limits_{k=2}^{n} x_k e_k / l(x)$. In the following, we shall introduce a kind of mapping from $\mathbf{R}^n$ to $\mathcal{A}_n(\mathbf{R})$. For any complex variable function $f(z) = u(x, y) + iv(x, y)$, we consider its corresponding function $\tilde{f} = \tilde{f}(x_1, x_2, ..., x_n) = u(x_1, l(x)) + I(x)v(x_1, l(x))$. In [41], H. Leutwiler gave the following result.

Theorem 4.1 *Let* $\Omega \subset (\mathbf{R}^2)^+ = \{z \,|\, z = (x, y) \in \mathbf{R}^2 \,, \, y > 0\}$ *be an* *open set, and* $f(z) = u + iv$ *be holomorphic in* Ω*. Then*

$$\tilde{f}(x) = \tilde{f}(x_1, x_2, ..., x_n) = u(x_1, l(x)) + I(x)v(x_1, l(x)) \tag{4.1}$$

is a solution of system (H_n) *in* $\widetilde{\Omega} = \{x = \sum_{k=1}^{n} x_k e_k \in \mathbf{R}^n | x_1 + il(x) \in \Omega\}$*. Moreover, if we denote* $\tilde{f} = \sum_{k=1}^{n} u_k e_k$ *in (4.1), then* $u_i x_k = u_k x_i$*,* $i, k = 2, ..., n$*.*

Proof Since $f(z)$ is holomorphic, we have

$$\begin{cases} u'_x = v'_y, \\ u'_y = -v'_x, \end{cases} \tag{4.2}$$

$$\tilde{f} = u(x_1, l(x)) + (\frac{x_2 e_2}{l(x)} + \frac{x_3 e_3}{l(x)} + \cdots + \frac{x_n e_n}{l(x)})v(x_1, l(x)),$$

and then

$$u_1(x) = u(x_1, l(x)),$$
$$u_k(x) = \frac{x_k}{l(x)} v(x_1, l(x)), \ k = 2, 3, ..., n.$$

Hence

$$\begin{aligned} \frac{\partial u_1}{\partial x_1} &= u'_x(x_1, l(x)), \\ \frac{\partial u_k}{\partial x_k} &= (\frac{x_k}{l(x)})'_{x_k} v(x_1, l(x)) + \frac{x_k}{l(x)}[v(x_1, l(x))]_{x_k} \\ &= \frac{l(x) - x_k l'_{x_k}(x)}{[l(x)]^2} v(x_1, l(x)) + \frac{x_k}{l(x)} v'_y(x_1, l(x)) l'_{x_k}(x) \\ &= \frac{l(x)v(x_1, l(x)) - x_k^2[(l(x))^{-1} v(x_1, l(x)) + v'_y(x_1, l(x))]}{[l(x)]^2}. \end{aligned}$$

Substitute the above equality into the first equality of system (H_n); it is obvious that the first equality holds. After a similar computation, the other equalities are all true.

Sha Huang gave the corresponding results about the above functions in [29]7).

Theorem 4.2 *Suppose we have complex constants* $a = a_1 + ib$*,* $c = c_1 + id$ *and complex variable number* $z = x_1 + iy$*. Then*

1) $\tilde{i} = I(x)$*,* $\tilde{z} = x = x_1 + \sum_{k=2}^{n} x_k e_k$*.*

2) $\tilde{a} = a_1 + I(x)b$, *specially*, $\tilde{a} = a$, $\widetilde{a+z} = \tilde{a} + \tilde{z}$, *when* a *is a real number.*

3) $\widetilde{(\frac{1}{z})} = \frac{1}{\tilde{z}}$ $\widetilde{(\frac{a}{z})} = \frac{\tilde{a}}{\tilde{z}}$.

4) $\widetilde{(\frac{1}{a})} = \tilde{\bar{a}}|a|^{-2}$, *here*, $|a| = \sqrt{a_1^2 + b^2}$, $\widetilde{(\frac{z}{a})} = \tilde{z}\widetilde{(\frac{1}{a})}$.

5) $\widetilde{(\frac{z+a}{c-z})} = \frac{\widetilde{z+a}}{\widetilde{c-z}}$.

We can verify by direct computation that all above terms are true. Here, the proof is omitted.

Theorem 4.3 *Suppose that* $\tilde{f} = \sum_{k=1}^{n} u_k e_k$ *is defined in the ball* $\widetilde{B} \subset (\mathbf{R}^n)^+$, *which does not intersect with the real axis in* $\mathbf{R}^n$, *moreover* $\tilde{f}$ *is a solution for system* (H_n) *satisfying* $u_i x_k = u_k x_i$, $i, k = 2, ..., n$. *Then there exists a holomorphic function* $f(z) = u + iv$ *defined in a circular disk* B, *such that* $\tilde{f}(x) = u(x_1, \sqrt{\sum_{k=2}^n x_k^2}) + I(x)v(x_1, \sqrt{\sum_{k=2}^n x_k^2})$, *where* $x \in \widetilde{B} = \left\{x = \sum_{k=1}^n x_k e_k \in \mathbf{R^n} | (x_1 + i\sqrt{\sum_{k=2}^n x_k^2}) \in B\right\}$.

4.2 The Integral Representation of Hyperbolic Harmonic Functions in Real Clifford Analysis

The components $u_1, ..., u_{n-1}$ of (twice continuously differentiable) solution $(u_1, ..., u_n)$ of (H_n) satisfy the hyperbolic version of the Laplace equation i.e. the hyperbolic Laplace equation in mathematics and physics is

$$x_n \Delta u - (n-1)\frac{\partial u}{\partial x_n} = 0, \tag{4.3}$$

where $u : \mathbf{R}^n \to \mathbf{R}$ is a real-valued function with n variables.

Definition 4.1 The twice continuously differentiable solution $u(x)$ of equation (4.3) is called the real hyperbolic harmonic function of n variables.

In [41], H. Leutwiler introduced the definition of hyperbolic harmonic function in real Clifford analysis.

Definition 4.2 Let $\tilde{f} = \sum_{k=1}^{n} u_k(x) e_k : \mathbf{R}^n \to \mathbf{R}^n$ possess twice continuously differentiable derivatives, the components $u_1, u_2, ..., u_{n-1}$ be hyperbolic harmonic, and u_n satisfy the equation

$$x_n^2 \Delta u - (n-1)x_n \frac{\partial u}{\partial x_n} + (n-1)u = 0. \tag{4.4}$$

Then $\tilde{f}$ is called a hyperbolic harmonic function.

Theorem 4.4 *Let $\tilde{f} = u_1 + u_2e_2 + \cdots + u_ne_n$ be a twice continuously differentiable function. Then $\tilde{f}$ is a solution of system (H_n) if and only if $\tilde{f}$ and the functions*

$$x\tilde{f}e_k + e_k\tilde{f}x,\ k = 1, ..., n-1$$

are hyperbolic harmonic in the above sense.

Proof The Clifford numbers $\omega_k = \dfrac{1}{2}(x\tilde{f}e_k + e_k\tilde{f}x)$ $(k = 2, ..., n-1)$ are vectors, whose components ω_{ki} $(i = 1, ..., n)$ are given by

$$\begin{cases} \omega_{k1} = -x_1u_k - x_ku_1, \quad k = 2, ..., n-1, \\ \omega_{ki} = -x_iu_k + x_ku_i, \quad i = 2, ..., n,\ k = 2, ..., n-1,\ i \neq k, \\ \omega_{kk} = x_1u_1 - x_2u_2 - \cdots - x_nu_n, \quad k = 2, ..., n-1. \end{cases}$$

In case $k = 1$, i.e. $\omega_1 = \dfrac{1}{2}(x\tilde{f} + \tilde{f}x)$, we have

$$\begin{cases} \omega_{11} = x_1u_1 - x_2u_2 - \cdots - x_nu_n, \\ \omega_{1i} = x_iu_1 + x_1u_i, \quad i = 2, ..., n. \end{cases}$$

It is easy to verify that $\tilde{f}$ and ω_k $(k = 1, ..., n-1)$ are hyperbolic harmonic if and only if $\tilde{f}$ satisfies system (H_n).

Theorem 4.5 *Suppose that the ball $\widetilde{B}$ with the radius $R > 0$, $\widetilde{B} \subset (\mathbf{R}^n)^+$, (or $\widetilde{B} \subset (\mathbf{R}^n)^-$), $\tilde{f} = \sum\limits_{k=1}^{n} u_k(x)e_x$ is a hyperbolic harmonic function in $\widetilde{B}$ and continuous on the boundary of $\widetilde{B}$, and denote by ω_{ki} $(1 \le k \le n-1, 1 \le i \le n)$ the components of $\omega_k(x) = (x\tilde{f}e_k + e_k\tilde{f}x)/2$. Let the following four conditions hold : i) $\omega_{ki} = 0\,(2 \le k \le n-1,\ 2 \le i \le n)$; ii) The other ω_{k1} $(2 \le k \le n-1), \omega_{1i}\,(1 \le i \le n-1)$ are real-valued hyperbolic harmonic functions; iii) $x_iu_n = x_nu_i\,(2 \le i \le n-1)$; iv) ω_{n1} satisfies (4.4). Then, $\tilde{f}(x)$ possesses the integral representation:*

$$\tilde{f}(x) = \frac{1}{2\pi}\int_0^{2\pi} \Phi(t)\frac{\tilde{t} + x - 2\tilde{a}}{\tilde{t} - x}d\varphi + I(x)v(a), \tag{4.5}$$

where $x \in \widetilde{B}$, t belongs to the boundary of a circular disk B with the center at $a \in (\mathbf{R}^2)^+$ (see Theorem 4.1) and radius R, $(t-a) = Re^{i\varphi}$, $f(z) = u + iv$ is analytic in B, and $\mathrm{Re}f(x) = \Phi(t)$.

Proof According to the conditions i), ii), we see that $\omega_{ki}\,(1 \le k \le n-1, 1 \le i \le n-1)$ are real-valued hyperbolic harmonic. From the conditions i), iv), it follows that $\omega_{kn}(x)\,(1 \le k \le n-1)$ satisfies (4.4). Hence $\omega_k(x)\,(1 \le k \le n-1)$ are hyperbolic harmonic functions. Moreover by Theorem 4.4, we know that $\tilde{f}$ is a solution of system (H_n). It is clear that from the formula (1.7) in [41], when $2 \le k \le n-1,\ 2 \le i \le n$, we have

$$\begin{aligned}
\omega_k &= \frac{1}{2}(x\tilde{f}e_k + e_k\tilde{f}x) = x\mathrm{Re}[\tilde{f}e_k] + e_k\mathrm{Re}[\tilde{f}x] - (x, e_k)\overline{\tilde{f}} \\
&= (x_1 + \sum_{i=2}^{n} x_i e_i)(-u_k) + e_k[u_1 x_1 - \sum_{i=2}^{n} u_i x_i] - x_k(u_1 - \sum_{i=2}^{n} u_i e_i) \\
&= -x_1 u_k + \sum_{i=2}^{n}(-1)x_i u_k e_i + [u_1 x_1 - \sum_{i=2}^{n} u_i x_i]e_k - x_k u_1 + \sum_{i=2}^{n} x_k u_i e_i \\
&= -(x_1 + u_k + x_k u_1) + \sum_{\substack{i=2 \\ (i\neq k)}}^{n} (x_k u_i - x_i u_k)e_i + [u_1 x_1 - \sum_{i=2}^{n} u_i x_i]e_k,
\end{aligned}$$

and its components are

$$\omega_{ki} = -x_i u_k + x_k u_i,\ 2 \le k \le n-1,\ 2 \le i \le n, i \ne k.$$

In addition, by the conditions i), iii), we have $x_i u_k = x_k u_i\ (i, k = 2, ..., n)$ when $i = k$, hence the above equality is true. By using Theorem 4.3, there exists $f(z) = u + iv$, which is analytic in the circular disc B : $|z-a| < R, B \subset (\mathbf{R}^2)^+$, such that

$$\tilde{f}(x) = u\left(x_1, \sqrt{\sum_{k=2}^{n} x_k^2}\right) + I(x)v\left(x_1, \sqrt{\sum_{k=2}^{n} x_k^2}\right),$$

where $x \in \tilde{B} = \{x = \sum_{k=1}^{n} x_k e_k \in \mathbf{R}^n | (x_1 + i\sqrt{\sum_{k=2}^{n} x_k^2}) \in B\}$, and in the following we denote $\tilde{f} = u(x) + I(x)v(x)$, $[\tilde{f}(x)]_1 = u(x)$. Finally, according to the Schwarz formula of the holomorphic function:

$$f(z) = \frac{1}{2\pi}\int_0^{2\pi} \Phi(t)\frac{t+z-2a}{t-z}d\varphi + iv(a)(z \in B) \tag{4.6}$$

in which $(t-a) = Re^{i\varphi}, \Phi(t) = \mathrm{Re}f(t)$, and using Theorem 4.2 and (4.6), we obtain

$$\tilde{f}(x) = \frac{1}{2\pi}\int_0^{2\pi} \Phi(t)\frac{\tilde{t}+x-2\tilde{a}}{\tilde{t}-x}d\varphi + I(x)v(a),$$

where $x \in \tilde{B}, t \in B$.

4.3 The Existence and Integral Representation of Solutions for a Kind of Boundary Value Problem

Let $\Phi(t)$ be a continuous function defined on the boundary $\dot{B}$ $(|t-a| = R)$ of the disk $B : |z-a| < R$ in $(\mathbf{R}^2)^+$, and $\widetilde{B} = \{x|x = \sum_{k=1}^n x_k e_k \in \mathbf{R}^n, x_1 + i\sqrt{\sum_{k=2}^n x_k^2} \in B\}$ and $t = t_1 + ih \in \dot{B}$. Denote $h = \sqrt{\sum_{k=2}^n t_k^2}$, $\underline{t} = t_1 + \sum_{k=2}^n t_k e_k$, and $\Phi(t) = \Phi(t_1, h) = \Phi(t_1, \sqrt{\sum_{k=2}^n t_k^2}) = \Phi(\underline{t})$. In the following, we shall discuss the hyperbolic harmonic function $\tilde{f} : \mathbf{R}^n \to \mathbf{R}^n$ in $\widetilde{B}$, and find a solution of the problem of $Re\tilde{f}(t_1) = \Phi(\underline{t})$; here $Re\tilde{f}(x) = u(x_1, l(x))$, u is a real number. This problem is called Problem D.

Theorem 4.6 *Problem D for hyperbolic harmonic functions in a high dimension ball $\widetilde{B}$ is solvable.*

Proof On the basis of the existence of solutions of the Dirichlet problem for holomorphic functions, we see that there exists a holomorphic function $f(z) = u + iv$, such that when $z \to t$, $\mathrm{Re}f(z) \to \Phi(t)$, where $t = t_1 + ih \in \dot{B}$, $z = x_1 + iy \in B$. By Theorem 4.1, we know that

$$\tilde{f}(x) = u\left(x_1, \sqrt{\sum_{k=2}^n x_k^2}\right) + I(x)v\left(x_1, \sqrt{\sum_{k=2}^n x_k^2}\right)$$

is a solution of system (H_n) in $\widetilde{B}$. If we denote the above function as $\tilde{f}(x) = \sum_{k=1}^n u_k v_k$, then $u_i x_k = u_k x_i$, $i, k = 2, ..., n$. From Theorem 4.4, it is clear that $\tilde{f}(x)$ is hyperbolic harmonic in $\widetilde{B}$. Denote $x = x_1 + \sum_{k=2}^n x_k e_k \in \widetilde{B}$, and when $x \to \underline{t} = t_1 + \sum_{k=2}^n t_k e_k$, we have $z = x_1 + iy = x_1 + i\sqrt{\sum_{k=2}^n x_k^2} \to t = t_1 + ih = t_1 + i\sqrt{\sum_{k=2}^n t_k^2}$, thus $\mathrm{Re}f(z) \to \Phi(t)$, i.e. $u(x_1, y) = u(z) \to \Phi(t) = \Phi(t_1, h)$. Again because $u(x_1, y) = u(x_1, \sqrt{\sum_{k=2}^n x_k^2}) = u(x)$, $\Phi(t_1, h) = \Phi(\underline{t}\,)$, we have $Re\tilde{f}(x) = u(x) \to \Phi(\underline{t})(x \to \underline{t})$, namely $Re\tilde{f}(\underline{t}) = \Phi(\underline{t})$. This shows that $\tilde{f}(x)$ is a solution of Problem D.

Theorem 4.7 *The solution $\tilde{f}(x)$ as in Theorem 4.6 possesses the integral representation (4.5).*

Proof In fact, the hyperbolic harmonic function $\tilde{f}(x)$ in Theorem 4.6 is a solution of Problem D, hence it is also a solution of system (H_n) satisfying (4.7). According to the proof of Theorem 4.5, we know that $\tilde{f}(x)$ possesses the integral representation (4.5).

CHAPTER III

NONLINEAR BOUNDARY VALUE PROBLEMS FOR GENERALIZED BIREGULAR FUNCTIONS IN REAL CLIFFORD ANALYSIS

This chapter deals with boundary value problems for the functions in real Clifford analysis. In the first section we consider boundary value problems for biregular functions. In the second section we consider boundary value problems for generalized biregular functions. In the last section we consider boundary value problems for biregular function vectors.

1 A Nonlinear Boundary Value Problem for Biregular Functions in Real Clifford Analysis

The regular function in Clifford analysis is similar to the holomorphic function in complex analysis. The biregular function discussed in this section is the regular function with two variables. In Chapter II, we have given some boundary value problems for regular functions. Similarly to holomorphic functions of several complex variables, Cauchy type integral formulas, the Hartogs theorem and the Cousin problem of biregular functions were obtained (see [7], [39]), and the Plemeli formula for several complex variables was considered (see [88]1)).

In this section, we give the Plemelj formula for biregular functions and prove the existence of solutions for a nonlinear boundary value problem, especially showing that the linear boundary value problem ($f \equiv 1$) has a unique solution [29]2).

1.1 Cauchy principal value of Cauchy type integrals

Denote by $\Omega = \Omega_1 \times \Omega_2$ an open connected set in the Euclidean space $\mathbf{R}^m \times \mathbf{R}^k$, $1 \le m \le n, 1 \le k \le n$, and by

$$F_\Omega^{(r)} = \left\{ f \,\middle|\, \begin{array}{l} f:\Omega \to \mathcal{A}_n(R), f(x,y) = \sum\limits_A f_A(x,y)e_A, \\ f_A(x,y) \in C^r(\Omega),\ x \in \mathbf{R}^m,\ y \in \mathbf{R}^k, \end{array} \right\}$$

the set of C^r functions in Ω with values in $\mathcal{A}_n(R)$. We introduce the right and left Dirac operators, i.e.

$$\overline{\partial}_x f = \sum_{i=1}^m \sum_A e_i e_A \frac{\partial f_A}{\partial x_i},\ f\overline{\partial}_y = \sum_{j=1}^k \sum_A e_A e_j \frac{\partial f_A}{\partial y_j},$$

respectively. A function f is called biregular if and only if

$$\begin{cases} \overline{\partial}_x f = 0, \\ f\overline{\partial}_y = 0. \end{cases}$$

Let the boundaries $\partial\Omega_1, \partial\Omega_2$ of Ω_1, Ω_2 be differentiable, oriented, compact Liapunov surfaces (see Section 2, Chapter I). Now, we consider the Cauchy principal value of a Cauchy type integral.

A function $\lambda(\mu,\nu) = \sum\limits_A \lambda_A(\mu,\nu)e_A, (\mu,\nu) \in \partial\Omega_1 \times \partial\Omega_2$ is said to be Hölder continuous on $\partial\Omega_1 \times \partial\Omega_2$, if and only if $|\lambda(\mu_1,\nu_1) - \lambda(\mu_2,\nu_2)| \leq G|(\mu_1,\nu_1) - (\mu_2,\nu_2)|^\beta$, where G, $\beta\,(0 < \beta < 1)$ are positive constants and $|(\mu_1,\nu_1) - (\mu_2,\nu_2)| = [|\mu_1 - \mu_2|^2 + |\nu_1 - \nu_2|^2]^{1/2}$. Denote by $H(\partial\Omega_1 \times \partial\Omega_2, \beta)$ the set of Hölder continuous functions on $\partial\Omega_1 \times \partial\Omega_2$, and define the norm in $H(\partial\Omega_1 \times \partial\Omega_2, \beta)$ as $||f||_\beta = C(f, \partial\Omega_1 \times \partial\Omega_2) + H(f, \partial\Omega_1 \times \partial\Omega_2, \beta)$, in which

$$H(f,\partial\Omega_1 \times \partial\Omega_2,\beta) = \sup_{(\mu_i,\nu_i)\in\partial\Omega_1\times\partial\Omega_2,(\mu_1,\nu_1)\neq(\mu_2,\nu_2)} \frac{|f(\mu_1,\nu_1) - f(\mu_2,\nu_2)|}{|(\mu_1,\nu_1) - (\mu_2,\nu_2)|^\beta},$$

$$C(f,\partial\Omega_1 \times \partial\Omega_2) = \max_{(\mu,\nu)\in\partial\Omega_1\times\partial\Omega_2} |f(\mu,\nu)|,\ f \in H(\partial\Omega_1 \times \partial\Omega_2, \beta).$$

It is easy to prove that $H(\partial\Omega_1 \times \partial\Omega_2, \beta)$ is a Banach space and

$$\| f+g \|_\beta \leq \| f \|_\beta + \| g \|_\beta,\ \| fg \|_\beta \leq J_1 \| f \|_\beta \| g \|_\beta. \tag{1.1}$$

Theorem 1.1 *Let* $x\overline{\in}\partial\Omega_1$, $y\overline{\in}\partial\Omega_2, \varphi(\mu,\nu) \in H(\partial\Omega_1 \times \partial\Omega_2, \beta)$. *Then the Cauchy type integral*

$$\Phi(x,y) = \lambda \int_{\partial\Omega_1\times\partial\Omega_2} E_m(x,\mu) d\sigma_\mu \varphi(\mu,\nu) d\sigma_\nu E_k(y,\nu) \tag{1.2}$$

is a biregular function and $\Phi(x,\infty)=\Phi(\infty,y)=\Phi(\infty,\infty)=0$, *where* $E_l(t_1,t_2)=\bar{t}_2-\bar{t}_1/|t_2-t_1|^l$ $(l=m,k)$ *and* $\lambda=1/\omega_m\omega_k$, ω_m,ω_k *are the area of a unit sphere in* $\mathbf{R}^m,\mathbf{R}^k$ *respectively. In this chapter,* $E_l(t_1,t_2)$ *and* λ *possess the same meaning.*

Proof It is clear that $\overline{\partial}_x E_m(x,\mu)=E_k(y,\nu)\overline{\partial}_y=0$ and $\lim\limits_{x\to\infty}E_m(x,\mu)$ $=\lim\limits_{y\to\infty}E_k(y,\nu)=0$ (see [80]7)), hence the result is obvious.

If the point $(t_1,t_2)\in\partial\Omega_1\times\partial\Omega_2$ and $O_1:|x-t_1|<\delta$, $O_2:|y-t_2|<\delta$, then $O((t_1,t_2),\delta)=O_1\times O_2$ is called the δ-neighborhood of (t_1,t_2), and denote $\lambda_\delta=(\partial\Omega_1\times\partial\Omega_2)\cap[O((t_1,t_2),\delta)]$.

Definition 1.1 The integral

$$\Phi(t_1,t_2)=\lambda\int_{\partial\Omega_1\times\partial\Omega_2}E_m(t_1,\mu)d\sigma_\mu\varphi(\mu,\nu)d\sigma_\nu E_k(t_2,\nu),\ (t_1,t_2)\in\partial\Omega_1\times\partial\Omega_2 \tag{1.3}$$

is called a singular integral on $\partial\Omega_1\times\partial\Omega_2$ [6].

Definition 1.2 If $\lim\limits_{\delta\to 0}\Phi_\delta(t_1,t_2)=I$, where

$$\Phi_\delta(t_1,t_2)=\lambda\int_{\partial\Omega_1\times\partial\Omega_2-\lambda_\delta}E_m(t_1,\mu)d\sigma_\mu\varphi(\mu,\nu)d\sigma_\nu E_k(t_2,\nu),$$

then I is called the Cauchy principal value of a singular integral and written as $I=\Phi(t_1,t_2)$.

Theorem 1.2 *If* $\varphi(\mu,\nu)\in H(\partial\Omega_1\times\partial\Omega_2,\beta)$, *then there exists the Cauchy principal value of singular integrals and*

$$\Phi(t_1,t_2)=-\frac{1}{4}\varphi(t_1,t_2)+\chi(t_1,t_2)+\frac{1}{4}(P_1\varphi+P_2\varphi), \tag{1.4}$$

where

$$\begin{aligned}
\chi(t_1,t_2)&=\lambda\int_{\partial\Omega_1\times\partial\Omega_2}E_m(t_1,\mu)d\sigma_\mu\psi(\mu,\nu)d\sigma_\nu E_k(t_2,\nu),\\
\psi(\mu,\nu)&=\varphi(\mu,\nu)-\varphi(\mu,t_2)-\varphi(t_1,\nu)+\varphi(t_1,t_2),\\
\chi(x,y)&=\lambda\int_{\partial\Omega_1\times\partial\Omega_2}E_m(x,\mu)d\sigma_\mu\psi(\mu,\nu)d\sigma_\nu E_k(y,\nu),
\end{aligned}$$

and

$$\begin{aligned}
P_1\varphi&=\frac{2}{\omega_m}\int_{\partial\Omega_1}E_m(t_1,\mu)d\sigma_\mu\varphi(\mu,t_2),\\
P_2\varphi&=\frac{2}{\omega_k}\int_{\partial\Omega_2}\varphi(t_1,\nu)d\sigma_\nu E_k(t_2,\nu)
\end{aligned}$$

are singular integral operators, where $\Phi_i\,(i=1,2,3)$ *are as stated in Definition 1.1.*

Proof Denote $\Phi_\delta = \overline{I}_1 + \overline{I}_2 + \overline{I}_3 + \overline{I}_4$, where

$$\overline{I}_1 = \lambda \int_{\partial\Omega_1\times\partial\Omega_2-\lambda_\delta} E_m(t_1,\mu)d\sigma_\mu\varphi(t_1,t_2)d\sigma_\nu E_k(t_2,\nu),$$
$$\overline{I}_2 = \lambda \int_{\partial\Omega_1\times\partial\Omega_2-\lambda_\delta} E_m(t_1,\mu)d\sigma_\mu\psi(\mu,\nu)d\sigma_\nu E_k(t_2,\nu),$$
$$\overline{I}_3 = \lambda \int_{\partial\Omega_1\times\partial\Omega_2-\lambda_\delta} E_m(t_1,\mu)d\sigma_\mu[\varphi(\mu,t_2)-\varphi(t_1,t_2)]d\sigma_\nu E_k(t_2,\nu),$$
$$\overline{I}_4 = \lambda \int_{\partial\Omega_1\times\partial\Omega_2-\lambda_\delta} E_m(t_1,\mu)d\sigma_\mu[\varphi(t_1,\nu)-\varphi(t_1,t_2)]d\sigma_\nu E_k(t_2,\nu).$$

By using [33],

$$\frac{1}{\omega_m}\int_{\partial\Omega_1} E_m(t_1,\mu)d\sigma_\mu = \frac{1}{2},\ \frac{1}{\omega_k}\int_{\partial\Omega_2} d\sigma_\nu E_k(t_2,\nu) = \frac{1}{2}, \tag{1.5}$$

we have $\overline{I}_1 \longrightarrow \frac{1}{4}\varphi(t_1,t_2)$ as $\delta\to 0$. Since $\varphi\in H(\partial\Omega_1\times\partial\Omega_2,\beta)$, $|\psi(\mu,\nu)|\le A_0|\mu-t_1|^\beta$, $|\psi(\mu,\nu)|\le A_1|\nu-t_2|^\beta$, where $A_0=A_1=2\|\,\varphi\,\|_\beta$, from [88]1), the inequality

$$|\psi(\mu,\nu)|\le A_2|\mu-t_1|^{\frac{\beta}{2}}|\nu-t_2|^{\frac{\beta}{2}},\ A_2=2\|\,\varphi\,\|_\beta \tag{1.6}$$

is derived.

By (2.11) in Chapter I and (1.6), we have

$$|E_m(t_1,\mu)d\sigma_\mu\psi(\mu,\nu)d\sigma_\nu E_k(t_2,\nu)|\le 4A_2(\rho_{01}^{\frac{\beta}{2}-1}d\rho_{01})(\rho_{02}^{\frac{\beta}{2}-1}d\rho_{02}), \tag{1.7}$$

where $\rho_{0i}\,(i=1,2)$ are ρ_0 in (2.11) of Chapter I for $t_i(i=1,2)$, which correspond to N_0 in Section 2, Chapter I. Therefore, $\overline{I}_2\to\chi(t_1,t_2)$ as $\delta\to 0$. By (1.5) and $|E_m(t_1,\mu)d\sigma_\mu[\varphi(\mu,t_2)-\varphi(t_1,t_2)]|\le A_2\rho_{01}^{\beta-1}d\rho_{01}$, there exists a limit of $\overline{I}_3$ as $\delta\to 0$. Similarly, we can obtain the limit of $\overline{I}_4$ as $\delta\to 0$. From [33], when $\delta\to 0$, we can get

$$\overline{I}_3\to\frac{1}{4}(P_1\varphi-\varphi),\ \overline{I}_4\to\frac{1}{4}(P_2\varphi-\varphi).$$

Consequently (1.4) is proved.

1.2 The Plemelj formula of Cauchy type integrals

Theorem 1.3 *Let* $\varphi(\mu,\nu) \in H(\partial\Omega_1 \times \partial\Omega_2, \beta), (x,y)\overline{\in}\partial\Omega_1 \times \partial\Omega_2$, *then*

$$\lim_{(x,y)\to(t_1,t_2)} \chi(x,y) = \chi(t_1,t_2),\ (t_1,t_2) \in \partial\Omega_1 \times \partial\Omega_2.$$

Proof Denote

$$\begin{aligned}
&\chi(x,y) - \chi(t_1,t_2)\\
= &\ \lambda \int_{\partial\Omega_1\times\partial\Omega_2} [E_m(x,\mu) - E_m(t_1,\mu)]d\sigma_\mu\psi(\mu,\nu)d\sigma_\nu E_k(t_2,\nu)\\
&+\lambda \int_{\partial\Omega_1\times\partial\Omega_2} [E_m(x,\mu)-E_m(t_1,\mu)]d\sigma_\mu\psi(\mu,\nu)d\sigma_\nu[E_k(y,\nu)-E_k(t_2,\nu)]\\
&+\lambda \int_{\partial\Omega_1\times\partial\Omega_2} E_m(t_1,\mu)d\sigma_\mu\psi(\mu,\nu)d\sigma_\nu[E_k(y,\nu)-E_k(t_2,\nu)]\\
= &\ L_1(\partial\Omega_1\times\partial\Omega_2)+L_2(\partial\Omega_1\times\partial\Omega_2)+L_3(\partial\Omega_1\times\partial\Omega_2),
\end{aligned}$$

and suppose that $6\delta < d_i, i = 1,2$, $\delta > 0$, $O((t_1,t_2),\delta)$ is the δ neighborhood of (t_1,t_2) with the center at point (t_1,t_2) and radius δ. Let $\partial\Omega_{i1} = (\partial\Omega_i)\cap[O((t_1,t_2),\delta)]$ $(i = 1,2)$, $\partial\Omega_{i2} = (\partial\Omega_i) - (\partial\Omega_{i1})$ $(i = 1,2)$; we have

$$\begin{aligned}
L_j(\partial\Omega_1 \times \partial\Omega_2) = &\ L_j(\partial\Omega_{11} \times \partial\Omega_{21})+L_j(\partial\Omega_{11} \times \partial\Omega_{22})\\
&+L_j(\partial\Omega_{12} \times \partial\Omega_{21})+L_j(\partial\Omega_{12} \times \partial\Omega_{22}),\ j=1,2,3.
\end{aligned}$$

On the basis of the Hile inequality (2.20) in Chapter I, we get

$$\begin{aligned}
&|E_m(x,\mu) - E_m(t_1,\mu)| = \left|\frac{\overline{\mu}-\overline{x}}{|\mu-x|^m} - \frac{\overline{\mu}-\overline{t}_1}{|\mu-t_1|^m}\right|\\
&= \left|\frac{\mu-x}{|\mu-x|^m} - \frac{\mu-t_1}{|\mu-t_1|^m}\right|\\
&\le \frac{\sum\limits_{i=0}^{m-2}|\mu-x|^{(m-2)-i}|\mu-t_1|^i}{|\mu-x|^{m-1}|\mu-t_1|^{m-1}}|(\mu-x)-(\mu-t_1)|\\
&= \sum_{i=0}^{m-2}\left|\frac{\mu-t_1}{\mu-x}\right|^i\left|\frac{x-t_1}{\mu-x}\right||\mu-t_1|^{-m+1}.
\end{aligned}$$

Similarly we estimate $|E_k(\nu,y) - E_k(\nu,t_2)|$, and

$$\begin{cases}
|E_m(x,\mu) - E_m(t_1,\mu)| \le \sum\limits_{i=0}^{m-2}|\dfrac{\mu-t_1}{\mu-x}|^i\,|\dfrac{x-t_1}{\mu-x}||\mu-t_1|^{-m+1},\\
|E_k(\nu,y) - E_k(\nu,t_2)| \le \sum\limits_{j=0}^{k-2}|\dfrac{\nu-t_2}{\nu-y}|^j\,|\dfrac{y-t_2}{\nu-y}||\nu-t_2|^{-k+1}
\end{cases} \tag{1.8}$$

are obtained.

It is easy to see that we only need to prove the result when x, y limits t_1, t_2 along the line that are not in the tangent plane of $\partial\Omega_1$ at t_1, $\partial\Omega$ at t_2 respectively. If we take an angle between the direction of $x \to t_1$ and the tangent plane of $\partial\Omega_1$ at t_1, which is greater than $2\beta_0$, and similarly take an angle between the direction of $y \to t_2$ and the tangent plane of $\partial\Omega_2$ at t_2, which is greater than $2\beta_0$, then

$$|\frac{\mu - t_1}{\mu - x}| \le M, |\frac{x - t_1}{\mu - x}| \le M, |\frac{\nu - t_2}{\nu - y}| \le M, |\frac{y - t_2}{\nu - y}| \le M, M = M(\beta_0). \tag{1.9}$$

By (1.6) – (1.9), we have

$$|L_1(\partial\Omega_{11} \times \partial\Omega_{21})| \le A_3\delta^\beta, \ |L_1(\partial\Omega_{12} \times \partial\Omega_{21})| \le A_4\delta^{\frac{\beta}{2}},$$

and $|L_1(\partial\Omega_{11} \times \partial\Omega_{22})| \le A_5\delta^{\frac{\beta}{2}}$. Now we consider $L_1(\partial\Omega_{12} \times \partial\Omega_{22})$, and first rewrite (1.8) in the form

$$\begin{cases} |E_m(x,\mu) - E_m(t_1,\mu)| \le \displaystyle\sum_{i=0}^{m-2} |\frac{\mu - t_1}{\mu - x}|^{i+1} \frac{|x - t_1|}{|\mu - t_1|^m}, \\ |E_k(\nu, y) - E_k(\nu, t_2)| \le \displaystyle\sum_{j=0}^{k-2} |\frac{\nu - t_2}{\nu - y}|^{j+1} \frac{|y - t_2|}{|\nu - t_2|^k}. \end{cases} \tag{1.10}$$

Next from (1.6) – (1.10), we can get $|L_1(\partial\Omega_{12} \times \partial\Omega_{22})| \le A_6|x - t_1|$. This shows that $L_1(\partial\Omega_1 \times \partial\Omega_2) \to 0$ as $(x, y) \to (t_1, t_2)$. Similarly, we have $L_2(\partial\Omega_1 \times \partial\Omega_2) \to 0$ and $L_3(\partial\Omega_1 \times \partial\Omega_2) \to 0$ as $(x, y) \to (t_1, t_2)$. Thus $\lim\limits_{(x,y)\to(t_1,t_2)} \chi(x, y) = \chi(t_1, t_2)$, $(t_1, t_2) \in \partial\Omega_1 \times \partial\Omega_2$.

Set $\Omega_i^+ = \Omega_i$, $(i = 1, 2)$, $\Omega_1^- = R^m\backslash\overline{\Omega}_1$, $\Omega_2^- = R^k\backslash\overline{\Omega}_2$, and denote $x(\in \Omega_1^\pm) \to t_1 \in \partial\Omega_1$ by $x \to t_1^\pm$. Moreover denote $y(\in \Omega_2^\pm) \to t_2 \in \partial\Omega_2$ by $y \to t_2^\pm$, and $\Phi(x, y) \to \Phi^{\pm\pm}(t_1, t_2)$ by $(x, y) \to (t_1^\pm, t_2^\pm)$; we have the following Sohotskiĭ–Plemelj formula (see [88]1))

Theorem 1.4 *Suppose $\varphi(\mu, \nu) \in H(\partial\Omega_1 \times \partial\Omega_2, \beta)$. Then*

$$\begin{cases} \Phi^{++}(t_1, t_2) = \frac{1}{4}[\varphi(t_1, t_2) + P_1\varphi + P_2\varphi + P_3\varphi], \\ \Phi^{+-}(t_1, t_2) = \frac{1}{4}[-\varphi(t_1, t_2) - P_1\varphi + P_2\varphi + P_3\varphi], \\ \Phi^{-+}(t_1, t_2) = \frac{1}{4}[-\varphi(t_1, t_2) + P_1\varphi - P_2\varphi + P_3\varphi], \\ \Phi^{--}(t_1, t_2) = \frac{1}{4}[\varphi(t_1, t_2) - P_1\varphi - P_2\varphi + P_3\varphi], \end{cases} \tag{1.11}$$

in which $(t_1, t_2) \in \partial\Omega_1 \times \partial\Omega_2$, *and* $P_i\,(i = 1, 2)$ *are as in Theorem 1.2,* $P_3 = 4\Phi(t_1, t_2)$.

Proof We first write (1.2) as

$$\begin{aligned}\Phi(x,y) &= \chi(x,y) + \lambda \int_{\partial\Omega_1\times\partial\Omega_2} E_m(x,\mu)d\sigma_\mu \varphi(t_1,t_2)d\sigma_\nu E_k(y,\nu)\\ &+ \lambda \int_{\partial\Omega_1\times\partial\Omega_2} E_m(x,\mu)d\sigma_\mu[\varphi(\mu,t_2)-\varphi(t_1,t_2)]d\sigma_\nu E_k(y,\nu)\\ &+ \lambda \int_{\partial\Omega_1\times\partial\Omega_2} E_m(x,\mu)d\sigma_\mu[\varphi(t_1,\nu)-\varphi(t_1,t_2)]d\sigma_\nu E_k(y,\nu).\end{aligned}$$

From Theorems 1.2 and 1.3, $\chi^{\pm\pm}(t_1,t_2) = \chi^{\pm\mp}(t_1,t_2) = \chi(t_1,t_2) = \frac{1}{4}[\varphi - P_1\varphi - P_2\varphi + P_3\varphi]$ is derived. Next by [33], (1.11) can be verified.

Corollary 1.5 *If* $\varphi(\mu,\nu) \in H(\partial\Omega_1 \times \partial\Omega_2, \beta)$, $(t_1,t_2) \in \partial\Omega_1 \times \partial\Omega_2$, *then*

$$\begin{cases}\Phi^{++}(t_1,t_2)-\Phi^{+-}(t_1,t_2)-\Phi^{-+}(t_1,t_2)+\Phi^{--}(t_1,t_2) = \varphi(t_1,t_2),\\ \Phi^{++}(t_1,t_2)-\Phi^{+-}(t_1,t_2)+\Phi^{-+}(t_1,t_2)-\Phi^{--}(t_1,t_2) = P_1\varphi,\\ \Phi^{++}(t_1,t_2)+\Phi^{+-}(t_1,t_2)-\Phi^{-+}(t_1,t_2)-\Phi^{--}(t_1,t_2) = P_2\varphi,\\ \Phi^{++}(t_1,t_2)+\Phi^{+-}(t_1,t_2)+\Phi^{-+}(t_1,t_2)+\Phi^{--}(t_1,t_2) = P_3\varphi.\end{cases} \tag{1.12}$$

1.3 Existence of solutions for Problem R

We assume that $A(t_1,t_2)$, $B(t_1,t_2)$, $C(t_1,t_2)$, $D(t_1,t_2)$, $g(t_1,t_2)$, $(t_1,t_2) \in \partial\Omega_1 \times \partial\Omega_2$ and $f(t_1,t_2,\Phi^{(1)},\Phi^{(2)},\Phi^{(3)},\Phi^{(4)})$ is a function on $(\partial\Omega_1\times\partial\Omega_2)\times\mathcal{A}_n(R)\times\mathcal{A}_n(R)\times\mathcal{A}_n(R)\times\mathcal{A}_n(R)$. We identify a sectionally biregular function $\Phi(x,y)$, in $\Omega_1^+\times\Omega_2^+, \Omega_1^+\times\Omega_2^-, \Omega_1^-\times\Omega_2^+, \Omega_1^-\times\Omega_2^-$, which is continuous in $(\Omega_1^+\times\Omega_2^+)\cup(\partial\Omega_1\times\partial\Omega_2)$, $(\Omega_1^+\times\Omega_2^-)\cup(\partial\Omega_1\times\partial\Omega_2)$, $(\Omega_1^-\times\Omega_2^+)\cup(\partial\Omega_1\times\partial\Omega_2)$, $(\Omega_1^-\times\Omega_2^-)\cup(\partial\Omega_1\times\partial\Omega_2)$ and $\Phi^{+-}(x,\infty) = \Phi^{-+}(\infty,y) = \Phi^{--}(\infty,\infty) = 0$, and satisfies the nonlinear boundary condition

$$\begin{aligned}&A\Phi^{++}(t_1,t_2)+B\Phi^{+-}(t_1,t_2)+C\Phi^{-+}(t_1,t_2)+D\Phi^{--}(t_1,t_2)\\ &= g(t_1,t_2)f[t_1,t_2,\Phi^{++}(t_1,t_2),\Phi^{+-}(t_1,t_2),\Phi^{-+}(t_1,t_2),\Phi^{--}(t_1,t_2)].\end{aligned} \tag{1.13}$$

The above boundary value problem is called Problem R.

Noting (1.2), (1.11) and (1.13), we can obtain

$$F\varphi = \varphi, \tag{1.14}$$

where

$$\begin{aligned}F\varphi = \ & (A+B)(\varphi+P_1\varphi+P_2\varphi+P_3\varphi)\\ & +(C+D)(-\varphi+P_1\varphi-P_2\varphi+P_3\varphi)\\ & +(B+D)(2\varphi-2P_1\varphi)+(1-4B)\varphi-4gf.\end{aligned}$$

Therefore, Problem R is transformed into the singular integral equation (1.14). From (3.17) and (3.22) of Chapter II, we get

Lemma 1.6 *Let $\varphi(t_1,t_2)\in H(\partial\Omega_1\times\partial\Omega_2,\beta)$. Then*

$$\| 2\varphi\pm 2P_i\varphi \|_\beta \le J_2\| \varphi \|_\beta,\ \| 2P_i\varphi \|_\beta \le J_2\| \varphi \|_\beta,\ i=1,2, \tag{1.15}$$

where J_2 is a positive constant.

Theorem 1.7 *If $\varphi(t_1,t_2)\in H(\partial\Omega_1\times\partial\Omega_2,\beta)$, then*

$$\| P_2\varphi\pm P_3\varphi \|_\beta \le J_3\| \varphi \|_\beta, \tag{1.16}$$

in which J_3 is a positive constant which is independent of φ.

Proof From (1.4), it follows that

$$P_2\varphi-P_3\varphi=\varphi-P_1\varphi-4\chi(t_1,t_2). \tag{1.17}$$

Moreover, from Lemma 1.6 we only need to prove $\|\chi(t_1,t_2)\|_\beta\le J_4\|\varphi\|_\beta$. We first use (1.6), (1.7) to get

$$|\chi(t_1,t_2)|\le B_1\| \varphi \|_\beta, \tag{1.18}$$

then we rewrite $\psi(\mu,\nu)$ as $\psi_0(t_1,t_2)$, thus

$$\chi(t_1,t_2)=\lambda\int_{\partial\Omega_1\times\partial\Omega_2}E_m(t_1,\mu)d\sigma_\mu\psi_0(t_1,t_2)d\sigma_\nu E_k(t_2,\nu).$$

Now we consider $H(\chi,\partial\Omega_1\times\partial\Omega_2,\beta)$ and write $\delta=|(t_1,t_2)-(t_1',t_2')|=(\delta_1^2+\delta_2^2)^{\frac{1}{2}}$ for any $(t_1,t_2),(t_1',t_2')\in\partial\Omega_1\times\partial\Omega_2$ and denote by $\rho_1,\rho_2,\rho_1',\rho_2'$ the projections of $|\mu-t_1|,|\nu-t_2|,|\mu-t_1'|,|\nu-t_2'|$ on the tangent plane respectively. Moreover we construct spheres $O_i(t_i,3\delta_i)$ with the center at t_i and radius $3\delta_i$, where $6\delta_i<d_i,\delta_i<1,i=1,2$. Denote by $\partial\Omega_{i1},\partial\Omega_{i2}$, the part of $\partial\Omega_i$ lying inside the sphere O_i and its surplus part $(i=1,2)$ respectively, and set $R(\partial\Omega_1\times\partial\Omega_2)=\chi(t_1,t_2)-\chi(t_1',t_2')=\overline{\chi}(\partial\Omega_1\times\partial\Omega_2)-\overline{\overline{\chi}}(\partial\Omega_1\times\partial\Omega_2)$. Thus

$$\begin{aligned}&R(\partial\Omega_1\times\partial\Omega_2)\\ =\ &R(\partial\Omega_{11}\times\partial\Omega_{21})+R(\partial\Omega_{11}\times\partial\Omega_{22})\\ &+R(\partial\Omega_{12}\times\partial\Omega_{21})+R(\partial\Omega_{12}\times\partial\Omega_{22}).\end{aligned} \tag{1.19}$$

Firstly, we consider $R(\partial\Omega_{11}\times\partial\Omega_{21})$. By (1.6) and (1.7), we have

$$\begin{aligned}&|R(\partial\Omega_{11}\times\partial\Omega_{21})|\\ \le\;& |\bar{\chi}(\partial\Omega_{11}\times\partial\Omega_{21})|+|\bar{\bar{\chi}}(\partial\Omega_{11}\times\partial\Omega_{21})|\\ \le\;& B_2\|\,\varphi\,\|_\beta|(t_1,t_2)-(t_1',t_2')|^\beta,\end{aligned}\tag{1.20}$$

and

$$\begin{aligned}&|\psi_0(t_1,t_2)|\le 2\|\,\varphi\,\|_\beta|\mu-t_1|^\beta,\\ &|\psi_0(t_1,t_2)|\le 2\|\,\varphi\,\|_\beta|\nu-t_2|^\beta,\\ &|\psi_0(t_1',t_2')|\le 2\|\,\varphi\,\|_\beta|\mu-t_1'|^\beta,\\ &|\psi_0(t_1',t_2')|\le 2\|\,\varphi\,\|_\beta|\nu-t_2'|^\beta.\end{aligned}\tag{1.21}$$

Noting that $|\nu-t_2'|\ge 2\delta_2,\ |\nu-t_2|\ge 3\delta_2>0$ on $\partial\Omega_{22}$, we have

$$|R(\partial\Omega_{11}\times\partial\Omega_{22})|\le B_3\|\,\varphi\,\|_\beta|(t_1,t_2)-(t_1',t_2')|^\beta.\tag{1.22}$$

Similarly, we can discuss the case of $R(\partial\Omega_{12}\times\partial\Omega_{21})$. Moreover we write $R(\partial\Omega_{12}\times\partial\Omega_{22})$ as

$$\begin{aligned}&R(\partial\Omega_{12}\times\partial\Omega_{22})\\ =\;&\lambda\int_{\partial\Omega_{12}\times\partial\Omega_{22}}[E_m(t_1,\mu)-E_m(t_1',\mu)]d\sigma_\mu\psi_0(t_1,t_2)d\sigma_\nu E_k(t_2,\nu)\\ &+\lambda\int_{\partial\Omega_{12}\times\partial\Omega_{22}}E_m(t_1',\mu)d\sigma_\mu\psi_0(t_1,t_2)d\sigma_\nu[E_k(t_2,\nu)-E_k(t_2',\nu)]\\ &+\lambda\int_{\partial\Omega_{12}\times\partial\Omega_{22}}E_m(t_1',\mu)d\sigma_\mu[\psi_0(t_1,t_2)-\psi_0(t_1',t_2')]d\sigma_\nu E_k(t_2',\nu)\\ =\;&S_1+S_2+S_3.\end{aligned}$$

For S_2, by the Hile lemma, it is easy to see that

$$|E_k(t_2,\mu)-E_k(t_2',\mu)|\le\sum_{j=0}^{k-2}\left|\frac{\nu-t_2}{\nu-t_2'}\right|^{j+1}\frac{|t_2'-t_2|}{|\nu-t_2|^k}.\tag{1.23}$$

Next from (1.6), (1.7) and (1.23) and noting that $|\mu-t|\ge 3\delta_1>0$, $|\mu-t'|\ge 2\delta_1>0$ on $\partial\Omega_{12}$, $|\nu-t_2|\ge 3\delta_2>0$, $|\nu-t_2'|\ge 2\delta_2>0$ on $\partial\Omega_{22}$, we get $|S_2|\le C_1\|\,\varphi\,\|_\beta|(t_1,t_2)-(t_1',t_2')|^\beta$. Furthermore, we can similarly discuss the case of S_1. As for S_3, by (1.5), it is clear that

$$\begin{aligned}S_3=\;&\frac{1}{2\omega_m}\int_{\partial\Omega_{12}}E_m(t_1',\mu)d\sigma_\mu[\varphi(\mu,t_2')-\varphi(\mu,t_2)]\\ &+\frac{1}{2\omega_k}\int_{\partial\Omega_{22}}[\varphi(t_1',\nu)-\varphi(t_1,\nu)]d\sigma_\nu E_k(t_2',\nu)+\frac{\varphi(t_1,t_2)-\varphi(t_1',t_2')}{4}.\end{aligned}$$

Again by (1.6) and (1.7),$|S_3| \le C_2\| \varphi \|_\beta |(t_1,t_2)-(t_1',t_2')|^\beta$ is obtained. The above discussion shows that

$$|R(\partial\Omega_{12} \times \partial\Omega_{22})| \le C_3\| \varphi \|_\beta |(t_1,t_2)-(t_1',t_2')|^\beta. \tag{1.24}$$

Secondly by (1.19), (1.20), (1.22) and (1.24) and when $6\delta_i < d_i, d_i < 1, \delta_i > 0, i = 1,2$, we have

$$|R(\partial\Omega_1 \times \partial\Omega_2)| \le C_4\| \varphi \|_\beta |(t_1,t_2)-(t_1',t_2')|^\beta. \tag{1.25}$$

It is easily seen that (1.25) holds for any $(t_1,t_2),(t_1',t_2') \in \partial\Omega_1 \times \partial\Omega_2$.

Finally by (1.18) and (1.25), we can obtain $\| \chi(t_1,t_2) \|_\beta \le J_4\| \varphi \|_\beta$. Hence (1.16) is derived.

Corollary 1.8 *Let $\varphi(t_1,t_2) \in H(\partial\Omega_1 \times \partial\Omega_2, \beta)$. Then*

$$\| \Phi^{++}(t_1,t_2) \|_\beta \le J_5\| \varphi \|_\beta, \ \| \Phi^{+-}(t_1,t_2) \|_\beta \le J_5\| \varphi \|_\beta,$$

$$\| \Phi^{-+}(t_1,t_2) \|_\beta \le J_5\| \varphi \|_\beta, \ \| \Phi^{--}(t_1,t_2) \|_\beta \le J_5\| \varphi \|_\beta,$$

$$\| \varphi + P_1\varphi + P_2\varphi + P_3\varphi \|_\beta \le J_5\| \varphi \|_\beta,$$

$$\| -\varphi - P_1\varphi + P_2\varphi + P_3\varphi \|_\beta \le J_5\| \varphi \|_\beta,$$

$$\| -\varphi + P_1\varphi - P_2\varphi + P_3\varphi \|_\beta \le J_5\| \varphi \|_\beta,$$

$$\| \varphi - P_1\varphi - P_2\varphi + P_3\varphi \|_\beta \le J_5\| \varphi \|_\beta.$$

Theorem 1.9 *Suppose the functions $A(x,y), B(x,y), C(x,y), D(x,y), g(x,y) \in H(\partial\Omega_1 \times \partial\Omega_2, \beta)$. Then the function $f(t_1,t_2,\Phi^{(1)}, \Phi^{(2)}, \Phi^{(3)}, \Phi^{(4)})$ is a Hölder continuous function for $(t_1,t_2) \in \partial\Omega_1 \times \partial\Omega_2$ and satisfies the Lipschitz condition for $\Phi^1, \Phi^2, \Phi^3, \Phi^4$ and any (t_1,t_2), namely*

$$|f(t_{11},t_{21},\Phi_1^{(1)},\Phi_1^{(2)},\Phi_1^{(3)},\Phi_1^{(4)}) - f(t_{12},t_{22},\Phi_2^{(1)},\Phi_2^{(2)},\Phi_2^{(3)},\Phi_2^{(4)})|$$
$$\le J_6|(t_{11},t_{21})-(t_{12},t_{22})|^\beta + J_7|\Phi_1^{(1)}-\Phi_2^{(1)}| + \cdots + J_{10}|\Phi_1^{(4)}-\Phi_2^{(4)}|, \tag{1.26}$$

where $J_i\,(i = 6,...,10)$ are positive constants independent of $t_{1j}, t_{2j}, \Phi_j^{(1)}, ..., \Phi_j^{(4)}, j = 1,2$. Next let $f(0,0,0,0,0,0) = 0$, $\| A+B \|_\beta < \varepsilon$, $\| C+D \|_\beta < \varepsilon$, $\| D+B \|_\beta < \varepsilon$, $\| 1-4B \|_\beta < \varepsilon$, $0 < \varepsilon < 1, 0 < \mu = \varepsilon J_1(2J_5+J_2+1) < 1$, $\| g \|_\beta < \delta$. If

$$0 < \delta < \frac{M(1-\mu)}{4J_1(J_{13}+J_{14}M)},$$

then Problem R has at least one solution as in the form (1.2), in which $M(\| \varphi \|_\beta < M), J_{13}, J_{14}$ *are positive constants, such that* $\|f\|_\beta \le J_{13} + J_{14}\|\varphi\|_\beta$.

Proof First of all, let $T = \{\varphi | \varphi \in H(\partial\Omega_1 \times \partial\Omega_2, \beta), \| \varphi \|_\beta \le M\}$ be denoted a subset of $C(\partial\Omega_1 \times \partial\Omega_2)$. By condition (1.26) and Theorem 1.2, it is not difficult to see that $C(f, \partial\Omega_1 \times \partial\Omega_2) \le J_{11} + J_{12}\| \varphi \|_\beta$. Similarly, we have $\| f \|_\beta \le J_{13} + J_{14}\| \varphi \|_\beta$. Hence, by (1.1) and (1.14), $\| F\varphi \|_\beta \le M$ is derived. This shows that the operator F is a mapping from T into itself.

Now we prove that the operator F is a continuous mapping. In fact, suppose that the sequence of functions $\{\varphi_n\}$ uniformly converges to a function $\varphi(t_1, t_2), (t_1, t_2) \in \partial\Omega_1 \times \partial\Omega_2$, where $\varphi_n \in T$. For any $\varepsilon > 0$, by [88]1), if n is large enough, then

$$|P_1\varphi_n - P_1\varphi| < \varepsilon, \; |P_2\varphi_n - P_2\varphi| < \varepsilon. \tag{1.27}$$

Moreover we consider $(P_3\varphi_n - P_3\varphi)$ and write $\psi_n(\mu, \nu) = \varphi_n(\mu, \nu) - \varphi_n(t_1, \nu) - \varphi_n(\mu, t_2) + \varphi_n(t_1, t_2)$. By the result in [88]1), it is easy to see that

$$|\psi_n(\mu, \nu)| \le A_8|\mu - t_2|^{\frac{\beta}{2}}|\nu - t_2|^{\frac{\beta}{2}}. \tag{1.28}$$

Set $P_3\varphi_n(t_1, t_2) - P_3\varphi(t_1, t_2) = I_1(\partial\Omega_1 \times \partial\Omega_2) + \cdots + I_5(\partial\Omega_1 \times \partial\Omega_2)$, where

$$I_1(\partial\Omega_1 \times \partial\Omega_2) = 4\lambda \int_{\partial\Omega_1 \times \partial\Omega_2} E_m(t_1, \mu) d\sigma_\mu \psi_n(\mu, \nu) d\sigma_\nu E_k(t_2, \nu),$$

$$I_2(\partial\Omega_1 \times \partial\Omega_2) = -4\lambda \int_{\partial\Omega_1 \times \partial\Omega_2} E_m(t_1, \mu) d\sigma_\mu \psi(\mu, \nu) d\sigma_\nu E_k(t_2, \nu),$$

$$I_3(\partial\Omega_1 \times \partial\Omega_2) = 4\lambda \int_{\partial\Omega_1 \times \partial\Omega_2} E_m(t_1, \mu) d\sigma_\mu [\varphi_n(t_1, \nu) - \varphi(t_1, \nu)] d\sigma_\nu E_k(t_2, \nu),$$

$$I_4(\partial\Omega_1 \times \partial\Omega_2) = 4\lambda \int_{\partial\Omega_1 \times \partial\Omega_2} E_m(t_1, \mu) d\sigma_\mu [\varphi_n(\mu, t_2) - \varphi(\mu, t_2)] d\sigma_\nu E_k(t_2, \nu),$$

$$I_5(\partial\Omega_1 \times \partial\Omega_2) = 4\lambda \int_{\partial\Omega_1 \times \partial\Omega_2} E_m(t_1, \mu) d\sigma_\mu [\varphi(t_1, t_2) - \varphi_n(t_1, t_2)] d\sigma_\nu E_k(t_2, \nu).$$

Assume that $6\sigma < d_i, i = 1, 2, \sigma > 0$. Denote by $O((t_1, t_2), 3\sigma)$ the 3σ neighborhood of (t_1, t_2) with the center at the point $(t_1, t_2) \in \partial\Omega_1 \times \partial\Omega_2$ and radius 3σ, and assume that $\partial\Omega_{i1}, \partial\Omega_{i2}$ are as stated before. Consequently, $I_j(\partial\Omega_1 \times \partial\Omega_2) = I_j(\partial\Omega_{11} \times \partial\Omega_{21}) + I_j(\partial\Omega_{11} \times \partial\Omega_{22}) + I_j(\partial\Omega_{12} \times \partial\Omega_{21}) + I_j(\partial\Omega_{12} \times \partial\Omega_{22}), 1 \le j \le 5$. From (1.6), (1.7) and (1.28), it follows that

$$|I_1(\partial\Omega_{11} \times \partial\Omega_{21})| \le J_{15}\sigma^\beta \le J_{16}\sigma^{\frac{\beta}{2}},$$

and

$$|I_1(\partial\Omega_{12}\times\partial\Omega_{21})|\le J_{17}\int_{\partial\Omega_{12}}\rho_1^{\frac{\beta}{2}-1}d\rho_1\int_0^{3\sigma}\rho_2^{\frac{\beta}{2}-1}d\rho_2\le J_{18}\sigma^{\frac{\beta}{2}}.$$

Similarly, we consider $I_1(\partial\Omega_{11}\times\partial\Omega_{22}), I_2(\partial\Omega_{11}\times\partial\Omega_{21}), I_2(\partial\Omega_{12}\times\partial\Omega_{21}), I_2(\partial\Omega_{11}\times\partial\Omega_{22})$ and write

$$\begin{aligned}&I_1(\partial\Omega_{12}\times\partial\Omega_{22})+I_2(\partial\Omega_{12}\times\partial\Omega_{22})\\ &=4\lambda\int_{\partial\Omega_1\times\partial\Omega_2}E_m(t_1,\mu)d\sigma_\mu W(\mu,\nu)d\sigma_\nu E_k(t_2,\nu),\end{aligned}$$

where

$$\begin{aligned}W(\mu,\nu)=&\{[\varphi_n(\mu,\nu)-\varphi(\mu,\nu)]-[\varphi_n(t_1,\nu)-\varphi(t_1,\nu)]\}\\ &+\{[\varphi_n(t_1,t_2)-\varphi(t_1,t_2)]-[\varphi_n(\mu,t_2)-\varphi(\mu,t_2)]\}.\end{aligned}$$

In accordance with [10], it is clear that

$$|W(\mu,\nu)|\le 2\|\varphi_n-\varphi\|_\beta|\mu-t_1|^{\frac{\beta}{2}}|\nu-t_2|^{\frac{\beta}{2}},$$

and then by (1.7),

$$I_1(\partial\Omega_{12}\times\partial\Omega_{22})+I_2(\partial\Omega_{12}\times\partial\Omega_{22})\le J_{19}\|\varphi_n-\varphi\|_\beta$$

is concluded. Again by (1.5) and (1.27), we have

$$|I_3(\partial\Omega_1\times\partial\Omega_2)|<\varepsilon, |I_4(\partial\Omega_1\times\partial\Omega_2)|<\varepsilon,$$

and

$$|I_5(\partial\Omega_1\times\partial\Omega_2)|=|\varphi_n-\varphi|.$$

The above discussion shows that

$$|P_3\varphi_n-P_3\varphi|\le J_{20}(\varepsilon+\sigma^{\frac{\beta}{2}}+\|\varphi_n-\varphi\|_\beta).$$

Thus for any $\varepsilon>0$, we first choose a sufficiently small number σ and next select a sufficiently large positive integer n; we can then obtain

$$|P_3\varphi_n-P_3\varphi|<\bar{G}\varepsilon, \tag{1.29}$$

in which $\bar{G}$ is a positive constant.

Finally, by (1.14), (1.27), (1.29) and condition (1.26), we can choose n large enough such that $|F\varphi_n-F\varphi|<W\varepsilon$, for any $(t_1,t_2)\in\partial\Omega_1\times\partial\Omega_2$, here W is a positive constant. This proves that the operator F is a

continuous mapping from T into itself. According to the Ascoli-Arzela theorem, it is evident that T is a compact set in the space $C(\partial\Omega_1\times\partial\Omega_2)$. Thus the continuous operator F maps the convex closed subset T of $C(\partial\Omega_1\times\partial\Omega_2)$ onto a compact subset $F(T)$ of T. On the basis of the Schauder fixed point principle, there exists a function $\varphi\in H(\partial\Omega_1\times\partial\Omega_2,\beta)$ satisfying the integral equation (1.14). This completes the proof.

Theorem 1.10 *When $f\equiv 1$ in Theorem 1.9, Problem R has a unique solution.*

Proof By the principle of contracting mapping, it is not difficult to verify that the statement in this theorem is true.

2 Nonlinear Boundary Value Problems of Generalized Bireguler Functions With Haseman Shift in Real Clifford Analysis

In complex analysis the generalized regulae function is a generalization of a holomorphic function that has formed an important mathematical branch and has found important applications in fluid mechanics and elasticity mechanics (see [3],[77],[81],[83] and so on). Similarly to holomorphic functions, some results about generalized regular functions have also been obtained in Clifford analysis. In this section, we introduce the integral expression and the Plemej formula of generalized biregular functions in Clifford analysis. In addition, we also discuss the solution of a kind of nonlinear boundary value problem for generalized biregular functions with shift in Clifford analysis (see [29]1)8), [64]2)).

2.1 Formulation of the boundary value problem

Let F^r_Ω be the class of Hölder continuous functions on $\Omega=\Omega_1\times\Omega_2$ and the Dirac operator are as stated before. The so-called generalized regular function f is the function f on Ω satisfying the system:

$$\begin{cases} \overline{\partial}_x f=F_1, \\ f\overline{\partial}_y=F_2. \end{cases} \tag{2.1}$$

We suppose that Ω_1^+, Ω_2^+ are the unit sphere of R^{m+1} and R^{k+1} respectively, and $W^{\pm\pm}(t_1,t_2)$ are the limit values of the function W, when $(x,y)\to(t_1^{\pm,\pm},t_2^{\pm,\pm})$. The nonlinear boundary value problem is formulated as follows.

Problem R^∇ We find a generalized biregular function $W(x,y)(\in F^{(r)}_\Omega)$

in $\Omega_1^+ \times \Omega_2^+$, $\Omega_1^+ \times \Omega_2^-$, $\Omega_1^- \times \Omega_2^+$, $\Omega_1^- \times \Omega_2^-$, such that it is continuous in $(\Omega_1^\pm \times \Omega_2^+) \cup (\partial\Omega_1 \times \partial\Omega_2)$, $(\Omega_1^\pm \times \Omega_2^-) \cup (\partial\Omega_1 \times \partial\Omega_2)$, and satisfies $W(\infty,\infty) = W(x,\infty) = W(\infty,y) = 0$ and the nonlinear boundary condition with shift:

$$\begin{aligned}
&A(t_1,t_2)W^{++}(\alpha(t_1),t_2) + B(t_1,t_2)W^{+-}(\alpha(t_1),t_2))\\
&\quad + C(t_1,t_2)W^{-+}(t_1,t_2) + D(t_1,t_2)W^{--}(t_1,t_2)\\
&= g(t_1,t_2)f(t_1,t_2,W^{++}(t_1,t_2),W^{+-}(t_1,t_2),W^{-+}(t_1,t_2),W^{--}(t_1,t_2)),
\end{aligned} \tag{2.2}$$

where $\alpha(t_1): \partial\Omega_1 \to \partial\Omega_1$ is a Haseman shift (see [54]1)), $A, B, C, D \in H(\partial\Omega_1 \times \partial\Omega_2, \beta)$ and f are all known functions.

2.2 Plemelj formula of generalized biregular functions

We are given the operator

$$T_1F(x,y) = \frac{-1}{\omega_m}\int_{\Omega_1^+} E_m(x,u)F(u,y)du - \frac{1}{\omega_m}\int_{\Omega_1^+} G_m(x,u)F(\frac{1}{u},y)du,$$

$$T_2F(x,y) = \frac{-1}{\omega_k}\int_{\Omega_2^+} F(x,v)E_k(y,v)dv - \frac{1}{\omega_k}\int_{\Omega_2^+} F(x,\frac{1}{v})G_k(y,v)dv,$$

where $G_l(t_1,t_2) = (1/\bar{t}_2 - \bar{t}_1)/(|1/t_2 - t_1)||t_2|)^l$, $l = m, k$.

Lemma 2.1 (see [19]) *If $F(x,y) \in L^{p,m}(R^m)$ for every fixed $y \in R^k$, and the norm $|F|_{p,m} = |F,\Omega_1^+|_p + |F^{(m)},\Omega_1^+|_p$ is independent of y, $p > m$, then for every $y \in \mathbf{R}^k$, we have*

1. $|T_1F| \le M(m,p)|F|_{p,m}$, $x \in \mathbf{R}^m$.
2. *For any $x_1, x_2 \in \mathbf{R}^m$, we have*

$$|T_1F(x_1,y) - T_1F(x_2,y)| \le M(m,p)|F|_{p,m}|x_1 - x_2|^\alpha,\ \alpha = \frac{p-m}{p}.$$

3. *For $|x| \ge 3$ we have* $|T_1F| \le M(m,p)|F|_{p,m}|x|^{((m/p)-m+1)}$.
4. $\bar{\partial}_x(T_1F) = F$, $x \in \mathbf{R}^m$.

In all cases $M(m,p)$ is a positive constant dependent on m, p and independent of x, y.

Lemme 2.2 *Let $F(x,y) \in L^{q,k}(R^k)$ for every fixed $x \in \mathbf{R}^m$ and let the norm $|F|_{q,k} = |F,\Omega_2^+|_q + |F^{(k)},\Omega_2^+|_q$ be independent of x, $q > k$. Then for every fixed $x \in \mathbf{R}^m$, we have*

1. $|T_2F| \leq M(k,q)|F|_{q,k}$, $y \in \mathbf{R}^k$.

2. *For any* $y_1, y_2 \in \mathbf{R}^k$, *we have*

$$|T_2F(x,y_1) - T_2F(x,y_2)| \leq M(k,q)|F|_{q,k}|y_1 - y_2|^{\mu}, \mu = \frac{q-k}{q}.$$

3. *For* $|y| \geq 3$, *we have* $|T_2F| \leq M(k,q)|F|_{q,k}|y|^{((k/q)-k+1)}$.

4. $(T_2F)\overline{\partial}_y = F$, $y \in \mathbf{R}^k$.

In all cases $M(k,q)$ *is a positive constant dependent on* k, q *and independent of* x, y.

Corollary 2.3 *Under the conditions of Lemmas 2.1 and 2.2, and if* $q = pk/m$ *and* $m \geq 2$, $k \geq 2$, *then*

1. *For every fixed* $y \in \mathbf{R}^k$, *we have* $T_1F \in C^{\alpha}(\mathbf{R}^m)$, *and* $T_1F(\infty, y) = 0$, *here* $\alpha = 1 - m/p = 1 - k/q$.

2. *For every fixed* $x \in \mathbf{R}^m$, *we have* $T_2F \in C^{\alpha}(R^k)$, *and* $T_2F(x, \infty) = 0$, *here* α *is as stated above.*

Proof According to Lemmas 2.1 and 2.2, we have $\mu = (q-k)/q = 1 - k/q = 1 - m/p = \alpha$ and $(m/p) - m + 1 < 1 - m + 1 \leq 0$, $(k/q) - k + 1 \leq 0$, then the result can be obtained.

Theorem 2.4 *Let* $k \geq 2$, $m \geq 2$, $F(x,y) \in C^{\alpha}(R^k)$, $0 < \alpha < 1$, *for every fixed* $x \in \mathbf{R}^m$, $y_1, y_2 \in \mathbf{R}^k$, *and* $|F(x,y_1) - F(x,y_2)| \leq M_1|y_1 - y_2|^{\alpha}$, *here* M_1 *is independent of* x, $p > m, \alpha = 1 - m/p$. *Then for any fixed* $x \in \mathbf{R}^m$, *we have* $T_1F \in C^{\alpha}(R^k)$ *and the Hölder constant is independent of* x.

Proof Let p' satisfy $(1/p) + (1/p') = 1$, note $p > m$; then we have $1 < p' < m/(m-1)$, $m - 1 < (m-1)p' < m$, and

$$|T_1F(x,y_1) - T_1F(x,y_2)| \leq \frac{M_1}{\omega_m}[\int_{\Omega_1^+} |u - x|^{-(m-1)p'}|du|]^{\frac{1}{p'}} \left| |y_1 - y_2|^{\alpha}, \Omega_1^+ \right|_p$$

$$+ \frac{M_1}{\omega_m}[\int_{\Omega_1^+} |\frac{1}{u} - x|^{-(m-1)p'}|u|^{-mp'}|du|]^{\frac{1}{p'}} \left| |y_1 - y_2|^{\alpha}, \Omega_1^+ \right|_p \leq M_2|y_1 - y_2|^{\alpha}.$$

Theorem 2.5 *Let* $k \geq 2$, $m \geq 2$, $F(x,y) \in C^{\alpha}(\mathbf{R}^m)$, $0 < \alpha < 1$, *for every fixed* $y \in \mathbf{R}^k$ *and any* $x_1, x_2 \in \mathbf{R}^m$, $|F(x_1,y) - F(x_2,y)| \leq M_1|x_1 - x_2|^{\alpha}$, *here* M_1 *is independent of* y, $q > k$, $\alpha = 1 - k/q$. *Then for every fixed* $y \in \mathbf{R}^k$, *we have* $T_2F \in C^{\alpha}(\mathbf{R}^m)$ *and the Hölder constant is independent of* y.

The proof of this theorem is similar to that in Theorem 2.4, so we omit it.

Theorem 2.6 *Suppose that Ω_i $(i = 1, 2)$ are two unit balls in $\mathbf{R}^m$, $\mathbf{R}^k$ respectively, and $F(x, y)$ satisfies the same conditions as in Theorems 2.4 and 2.5, and $p > m$, $q > k$, $\alpha = 1 - m/p = 1 - k/q$, $q = p(k)/m$. Then*

$$T_i F(x, y) \in C^\alpha(R^m, R^k),\ i = 1, 2.$$

Proof In accordance with the conditions in this theorem, when $x \in \mathbf{R}^m$, we see $F(x, y) \in L^{q,k}(\mathbf{R}^k)$, and when $y \in \mathbf{R}^k$, we have $F(x, y) \in L^{p,m}(\mathbf{R}^m)$. On the basis of Lemmas 2.1, 2.2 and Theorems 2.4, 2.5, for any $(x_1, y_1), (x_2, y_2) \in \mathbf{R}^m \times \mathbf{R}^k$ we get

$$|T_i F(x_1, y_1) - T_i F(x_2, y_2)|$$

$$\leq |T_i F(x_1, y_1) - T_i F(x_2, y_1) + T_i F(x_2, y_1) - T_i F(x_2, y_2)|$$

$$\leq M_3 |x_1 - x_2|^\alpha + M_4 |y_1 - y_2|^\alpha \leq M_5 |(x_1, y_1) - (x_2, y_2)|^\alpha,\ i = 1, 2.$$

Theorem 2.7 *Let F_1, F_2 in (2.2) satisfy $F_1, F_2 \in \mathbf{C}^1(\Omega)$, where $\Omega = \Omega_1 \times \Omega_2$, $F_1 \overline{\partial}_y = \overline{\partial}_x F_2$ and the condition on F is as in Theorem 2.6, moreover $F_1 \overline{\partial}_y$ also satisfy the condition on F as in Theorem 2.6. Then the generalized biregular function $W(x, y)$, i.e. a solution of equation (2.1) in $\Omega_1^\pm \times \Omega_2^+$, $\Omega_1^\pm \times \Omega_2^-$, has the integral expression*

$$W(x, y) = T_1 F_1 + T_2[F_2 - (T_1 F_1)\overline{\partial}_y] + \Phi(x, y), \tag{2.3}$$

where $\Phi(x, y)$ is a biregular function in $\Omega_1^\pm \times \Omega_2^+$, $\Omega_1^\pm \times \Omega_2^-$.

Proof When $(x, y) \in \Omega_1^\pm \times \Omega_2^+$, $\Omega_1^\pm \times \Omega_2^-$, by means of Lemmas 2.1 and 2.2, we know $\overline{\partial}_x (T_1 F_i) = F_i$, $(T_2 F_i)\overline{\partial}_y = F_i$ $(i = 1, 2)$, then according to the condition $F_1 \overline{\partial}_y = \overline{\partial}_x F_2$, we obtain

$$\overline{\partial}_x \{T_1 F_1 + T_2[F_2 - (T_1 F_1)\overline{\partial}_y]\} = \overline{\partial}_x T_1 F_1 + \overline{\partial}_x T_2 F_2 - \overline{\partial}_x T_2[(T_1 F_1)\overline{\partial}_y]$$

$$= F_1 + T_2(\overline{\partial}_x F_2) - \overline{\partial}_x T_2[T_1(F_1 \overline{\partial}_y)] = F_1 + T_2(\overline{\partial}_x F_2) - T_2[\overline{\partial}_x (T_1(F_1 \overline{\partial}_y))]$$

$$= F_1 + T_2(\overline{\partial}_x F_2) - T_2(F_1 \overline{\partial}_y) = F_1 + T_2(\overline{\partial}_x F_2) - T_2(\overline{\partial}_x F_2) = F_1.$$

Moreover,

$$\{T_1 F_1 + T_2[F_2 - (T_1 F_1)\overline{\partial}_y]\}\overline{\partial}_y = (T_1 F_1)\overline{\partial}_y + F_2 - (T_1 F_1)\overline{\partial}_y = F_2,$$

that is to say $T_1 F_1 + T_2[F_2 - (T_1 F_1)\overline{\partial}_y]$ and $W(x, y)$ are both generalized biregular functions, so

$$\{T_1 F_1 + T_2[F_2 - (T_1 F_1)\overline{\partial}_y]\} - W(x, y)$$

is a biregular function, hence (2.3) is derived. This completes the proof.

Now we introduce the operator $T_3(F_1\overline{\partial}_y) = T_2[(T_1F_1)\overline{\partial}_y]$.

Theorem 2.8 *Under the conditions described in Theorem 2.7, if* $F_1(x, \infty) = F_2(\infty, y) = 0$, *then we have*

$$\overline{\partial}_x[(T_3(F_1\overline{\partial}_y)] = T_2(F_1\overline{\partial}_y),\ [T_3(F_1\overline{\partial}_y)]\overline{\partial}_y = T_1(F_1\overline{\partial}_y),$$
$$T_3(F_1\overline{\partial}_y) \in C^\alpha(R^m, R^k),$$
$$T_3(F_1\overline{\partial}_y)(\infty, y) = T_3(F_1\overline{\partial}_y)(x, \infty) = T_3(F_1\overline{\partial}_y)(\infty, \infty) = 0.$$

Proof According to the proof of Theorem 2.7, we know that the first term is true. Using Theorem 2.6, we see that the second term is valid. According to Corollary 2.3, the third term is also true. Thus Theorem 2.8 is proved.

Theorem 2.9 *Under the conditions described in Theorem 2.4, we obtain the Plemelj formula of generalized biregular functions as follows:*

$$\begin{cases} W^{++}(t_1,t_2)=T_1F_1+T_2F_2-T_3(F_1\overline{\partial}_y)+\frac{1}{4}[\varphi+P_1\varphi+P_2\varphi+P_3\varphi], \\ W^{+-}(t_1,t_2)=T_1F_1+T_2F_2-T_3(F_1\overline{\partial}_y)+\frac{1}{4}[-\varphi-P_1\varphi+P_2\varphi+P_3\varphi], \\ W^{-+}(t_1,t_2)=T_1F_1+T_2F_2-T_3(F_1\overline{\partial}_y)+\frac{1}{4}[-\varphi+P_1\varphi-P_2\varphi+P_3\varphi], \\ W^{--}(t_1,t_2)=T_1F_1+T_2F_2-T_3(F_1\overline{\partial}_y)+\frac{1}{4}[\varphi-P_1\varphi-P_2\varphi+P_3\varphi], \end{cases} \tag{2.4}$$

for any (t_1, t_2) *on* $\partial\Omega_1 \times \partial\Omega_2$, *moreover,* $W(\infty, y) = W(x, \infty) = W(\infty, \infty) = 0$.

Proof Similarly to [29]2) (see formula (1.11) of Chapter III), we can use the Cauchy type integral to express $\Phi(x, y)$ in (2.3). Moreover noting the result in [29]2), Theorem 2.7, Corollary 2.3, we see that (2.4) is valid. According to $F_1(x, \infty) = F_2(\infty, y) = 0$, and Theorems 2.7, 2.8, we obtain $T_1F_1(x, \infty) = T_2F_2(\infty, y) = 0$ and $W(x, \infty) = W(\infty, y) = W(\infty, \infty) = 0$.

Corollary 2.10 *Let* $\alpha(t_1)$ *be a Haseman shift on* $\partial\Omega_1 \to \partial\Omega_1$, *and the conditions in Theorem 2.8 be satisfied. Then for any* $(t_1, t_2) \in \partial\Omega_1 \times \partial\Omega_2$, *we have*

$$\begin{cases} W^{++}(\alpha(t_1),t_2) = \{T_1F_1+T_2F_2-T_3(F_1\overline{\partial}_y)\}(\alpha(t_1),t_2) \\ \qquad\qquad +\dfrac{1}{4}[\varphi_1+q_1\varphi+q_2\varphi+q_3\varphi], \\ W^{+-}(\alpha(t_1),t_2) = \{T_1F_1+T_2F_2-T_3(F_1\overline{\partial}_y)\}(\alpha(t_1),t_2) \\ \qquad\qquad +\dfrac{1}{4}[-\varphi_1-q_1\varphi+q_2\varphi+q_3\varphi], \end{cases} \tag{2.5}$$

where $\varphi_1(t_1,t_2)=\varphi(\alpha(t_1),t_2)$, $q_i\varphi(t_1,t_2)=P_i\varphi(\alpha(t_1),t_2)$, $i=1,2,3$, $P_i\,(i=1,2,3)$ *are as stated in Section 1.*

2.3 The existence of the problem R^∇

The boundary condition (2.2) of Problem R^∇ can be reduced to the singular integral equation

$$L\varphi=\varphi, \tag{2.6}$$

in which L is a singular integral operator in the space $H(\partial\Omega_1\times\partial\Omega_2,\beta)$:

$$\begin{aligned} L\varphi &= (A+B)(\varphi_1+q_1\varphi+q_2\varphi+q_3\varphi)-2B(\varphi_1+q_1\varphi) \\ &\quad +(C+D)[-\varphi+P_1\varphi-P_2\varphi+P_3\varphi]+(2D+1))\varphi-2DP_1\varphi-4gf \\ &\quad +4(A+B)[T_1F_1+T_2F_2-T_3(F_1\overline{\partial}_y)](\alpha(t_1),t_2) \\ &\quad +4(C+D)[T_1F_1+T_2F_2-T_3(F_1\overline{\partial}_y)](t_1,t_2), \end{aligned} \tag{2.7}$$

where φ_1, P_i, $q_i\,(i=1,2,3)$ are as stated in [29]1).

Lemma 2.11 *Let* $\Omega_i\,(i=1,2)$ *be unit balls, and let* f *satisfy the following condition on* $\partial\Omega_i\,(i=1,2)$*:*

$$\begin{aligned} &|f(t_{11},t_{21},W_1^{(1)},W_1^{(2)},W_1^{(3)},W_1^{(4)})-f(t_{12},t_{22},W_2^{(1)},W_2^{(2)},W_2^{(3)},W_2^{(4)})| \\ &\le J_1|(t_{11},t_{21})-(t_{12},t_{22})|^\beta+J_2|W_1^{(1)}-W_2^{(1)}|+\cdots+J_5|W_1^{(4)}-W_2^{(4)}|, \end{aligned} \tag{2.8}$$

in which $J_i\,(i=1,2,3,4,5)$ *are positive constants independent of* $W_j^{(m)}$, *for* t_{1j}, t_{2j} $(j=1,2,\ m=1,2,3,4)$,

$$|(t_{11},t_{21})-(t_{12},t_{22})|^\beta=\left(\sqrt{(t_{11}-t_{12})^2+(t_{21}-t_{22})^2}\right)^\beta,$$

and $f(0,0,0,0,0,0)=0$. *Then there exist the constants* J_6, J_7 *such that*

$$\|f\|_\beta\le J_6+J_7\|\varphi\|_\beta.$$

Proof According to (2.8) and the bounded property of $\Omega = \Omega_1 \times \Omega_2$, it is easy to see that

$$C(f, \partial\Omega_1 \times \partial\Omega_2) = \max_{(t_1,t_2)\in\partial\Omega_1\times\partial\Omega_2} |f|$$

$$= \max_{(t_1,t_2)\in\partial\Omega_1\times\partial\Omega_2} |f(t_1, t_2, W^{++}(t_1, t_2), ..., W^{--}(t_1, t_2))$$

$$-f(0, 0, 0, 0, 0, 0)| \leq J_8 + J_9\|\varphi\|_\beta,$$

where J_8, J_9 are positive constants. Moreover we have

$$|f(t_{11}, t_{21}, W^{++}(t_{11}, t_{21}), W^{+-}(t_{11}, t_{21}), W^{-+}(t_{11}, t_{21}), W^{--}(t_{11}, t_{21})$$

$$-f(t_{12}, t_{22}, W^{++}(t_{12}, t_{22}), W^{+-}(t_{12}, t_{22}), W^{-+}(t_{12}, t_{22}), W^{--}(t_{12}, t_{22}))|$$

$$\leq (J_{10} + J_{11}\|\varphi\|_\beta)|(t_{11}, t_{21}) - (t_{12}, t_{22})|^\beta,$$

thus $\|f\|_\beta \leq J_6 + J_7\|\varphi\|_\beta$.

Lemma 2.12 *Suppose that Ω_i $(i = 1, 2)$, $\alpha(t)$ are as stated before, and f satisfies the conditions in Lemma 2.11, moreover*

1) F_i $(i = 1, 2)$ satisfy the condition of F in Lemma 2.1, 2.2, $F_1\overline{\partial}_y = \overline{\partial}_x F_2$ and $F_1(x, \infty) = F_2(\infty, y) = 0$.

2) T_1F_1, T_2F_2, $T_3(F_1\overline{\partial}_y)$, A, B, C, D, $g \in H(\partial\Omega_1 \times \partial\Omega_2, \beta)$, $\alpha < \beta < 1, \alpha = 1 - m/p$, f are as stated as in Lemma 2.11.

3) Set $\gamma = J_{12}[J_{13}(\|A + B\|_\beta + \|C + D\|_\beta + \|B\|_\beta + 2\|D\|_\beta) + \|2D + 1\|_\beta] < 1$, where J_{12}, J_{13} are both positive constants, J_{12} is similar to J_1 in [29]1), J_{13} is similar to the maximum one among J_2, J_3, J_4 in [29]1), A, B, C, D, g are coefficients of the boundary condition. In addition we give $\delta > 0$ such that $\delta < M(1 - \gamma)/4(1/J_{13} + J_{12}(J_6 + J_7M))$, where M is a positive constant such that $\|\varphi\|_\beta \leq M$, $\|g\|_\beta < \delta$, $\|T_1F_1 + T_2F_2 - T_3(F_1\overline{\partial}_y)\|_\beta < \delta$; herein g, F_1, F_2 are all known functions satisfying the requirement conditions. Then, we have the following results:

1. The operator L maps the subspace of $C(\partial\Omega_1 \times \partial\Omega_2)$, $T = \{\varphi| \varphi \in H(\partial\Omega_1 \times \partial\Omega_2, \beta), \|\varphi\|_\beta \leq M\}$ into itself.

2. The operator L is a continuous mapping on T.

Proof 1. According to the definition of the operator L and a property of the norm $\| * \|_\beta$, we can get

$$\|L\varphi\|_\beta \le J_{12}\|A+B\|_\beta\|\varphi_1+q_1\varphi+q_2\varphi+q_3\varphi\|_\beta+J_{12}\|B\|_\beta\|2\varphi_1$$
$$+2q_1\varphi\|_\beta+J_{12}\|C+D\|_\beta\|\varphi_1+P_1\varphi-P_2\varphi+P_3\varphi\|_\beta$$
$$+J_{12}\|2D+1\|_\beta\|\varphi\|_\beta+J_{12}\|2D\|_\beta\|P_1\varphi\|_\beta$$
$$+4J_{12}\|A+B+C+D\|_\beta\|T_1F_1+T_2F_2-T_3(F_1\overline{\partial}_y)\|_\beta+4J_{12}\|f\|_\beta\|g\|_\beta.$$

From the conditions 3) $J_{12}J_{13}(\|A+B\|_\beta+\|C+D\|_\beta)\le\gamma<1$, we can obtain

$$\|A+B+C+D\|_\beta\le\|A+B\|_\beta+\|C+D\|_\beta<\frac{1}{J_{12}J_{13}}.$$

Similarly

$$\|L\varphi\|_\beta\le J_{12}J_{13}\|\varphi\|_\beta[\|A+B\|_\beta+\|C+D\|_\beta+\|B\|_\beta$$
$$+2\|D\|_\beta]+J_{12}\|2D+1\|_\beta\|\varphi\|_\beta+4J_{12}\frac{1}{J_{12}J_{13}}\delta$$
$$+4J_{12}[J_6+J_7\|\varphi\|_\beta]\delta\le M\gamma+\delta[\tfrac{4}{J_{13}}+4J_{12}(J_6+J_7\|\varphi\|_\beta)]$$
$$\le M\gamma+M(1-\gamma)=M,$$

which shows that the operator L maps T into itself.

2. We arbitrarily choose $\varphi_n\in T$, $n=1,2,\ldots$, such that $\{\varphi_n\}$ uniformly converge to φ on $\partial\Omega_1\times\partial\Omega_2$. For any $\varepsilon>0$, when n is large enough, $\|\varphi_n-\varphi\|_\beta$ can be small enough. According to [4], when n is large enough, for any $(t_1,t_2)\in\partial\Omega_1\times\partial\Omega_2$, we have $|P_1\varphi_n-P_1\varphi|<\varepsilon$, $|P_2\varphi_n-P_2\varphi|<\varepsilon$. Similarly, we can obtain $|P_3\varphi_n-P_3\varphi|<\varepsilon$, $|q_i\varphi_n-q_i\varphi|<\varepsilon\,(i=1,2,3)$, and $|L\varphi_n-L\varphi|<\overline{W}\varepsilon$, where $\overline{W}$ is a positive constant, thus we prove that L is a continuous mapping in T.

Theorem 2.13 *Under the same conditions as in Lemma 2.12, there exists $\varphi_0\in T$ such that $L\varphi_0=\varphi_0$. Moreover, Problem R^∇ is solvable, and the solution can be expressed as*

$$W(x,y)=T_1F_1+T_2[F_2-(T_1F_1)\overline{\partial}_y]+\Phi(x,y), \tag{2.9}$$

in which

$$\Phi(x,y)=\lambda\int_{\partial\Omega_1\times\partial\Omega_2}E_m(x,\mu)d\sigma_\mu\varphi_0(\mu,\nu)d\sigma_\nu E_k(y,\nu),$$

here $E_m(x,\mu)$, $E_k(y,\nu)$ are the same as Theorem 1.1, Section 1.

Proof According to Lemma 2.12, we see that L is a mapping from T to itself and it is continuous. By the Schauder fixed-point theorem, we know that there exists $\varphi_0 \in T$ such that $L\varphi_0 = \varphi_0$. Substituting φ_0 into the above formula, we get $\Phi(x,y)$, moreover substituting $\Phi(x,y)$ into the formula (2.3), the function $W(x,y)$ is obtained, which is just a solution of Problem R^∇.

Similarly to Theorem 2.13 (see [29]3)) we can also discuss a kind of nonlinear boundary value problem for generalized biregular functions with the shift and with conjugate value

$$\begin{aligned}
&A_1(t_1,t_2)W^{++}(t_1,t_2) + A_2(t_1,t_2)\overline{W^{++}(d(t_1,t_2))}\\
&+B_1(t_1,t_2)W^{+-}(t_1,t_2) + B_2(t_1,t_2)\overline{W^{+-}(d(t_1,t_2))}\\
&+C_1(t_1,t_2)W^{-+}(t_1,t_2) + C_2(t_1,t_2)\overline{W^{-+}(d(t_1,t_2))}\\
&+D_1(t_1,t_2)W^{--}(t_1,t_2) + D_2(t_1,t_2)\overline{W^{--}(t_1,t_2)(d(t_1,t_2))}\\
&= g(t_1,t_2)f(t_1,t_2,W^{++}(t_1,t_2),W^{+-}(t_1,t_2),W^{-+}(t_1,t_2),W^{--}(t_1,t_2),\\
&W^{++}(d(t_1,t_2)),W^{+-}(d(t_1,t_2)),W^{-+}(d(t_1,t_2)),W^{--}d((t_1,t_2)))
\end{aligned}$$

where $d(t_1,t_2) : \partial\Omega_1 \times \partial\Omega_2 \to \partial\Omega_1 \times \partial\Omega_2$ is a Haseman shift.

3 A Nonlinear Boundary Value Problem for Biregular Function Vectors in Real Clifford Analysis

On the basis of the results of the above two sections and enlightened by the vector value analysis [25], in this section the existence and uniqueness of solutions of a boundary value problem for biregular function vectors will be discussed (see [9],[30]).

3.1 Formulation of the boundary value problem

Definition 3.1 Let $F_i(x,y) \in F_\Omega^{(r)}$ $(i = 1,\dots,p)$ be functions with values in the real Clifford algebra $\mathcal{A}_n(R)$. We call $F(x,y) = (F_1(x,y), \dots, F_p(x,y))$ a function vector, and call $F_1,\dots,F_p$ the components of F. For $F(x,y)=(F_1(x,y),\dots,F_p(x,y))$, $G(x,y) = (G_1(x,y), \dots, G_p(x,y))$, the addition operation of two function vectors is defined by $F+G = (F_1+G_1,\dots,F_p+G_p)$, and the multiplication operation of function vectors is defined by $F \otimes G = (F_1G_1,\dots,F_pG_p)$. Moreover we define the multiplication of a function vector and a function as $\varphi F = (\varphi F_1,\dots,\varphi F_p)$, $F\psi = (F_1\psi,\dots,F_p\psi)$, where $\varphi(x,y)$, $\psi(x,y)$ are functions. We define the norm of the function vector as $|F(x,y)| = (\sum_{i=1}^{p}|F_i(x,y)|^2)^{1/2}$. It is easy to see that

$$|F+G| \le |F| + |G|,\ |F \otimes G| \le J_0|F||G|, \tag{3.1}$$

where J_0 is a positive constant.

Definition 3.2 We call a function vector $F(x,y) = (F_1(x,y), \ldots, F_p(x,y))$ biregular in Ω, that is to say that $F_i(x,y)$ $(i = 1,\ldots,p)$ are biregular functions in Ω, and $\Omega = \Omega_1 \times \Omega_2$ is stated as in section 1.

We call a function vector $F(u,v) = (F_1(u,v),\ldots,F_p(u,v))$, $(u,v) \in \partial\Omega_1 \times \partial\Omega_2$ Hölder continuous in the characteristic manifold $\partial\Omega_1 \times \partial\Omega_2$, if it satisfies

$$\begin{aligned}|F(u_1,v_1) - F(u_2,v_2)| &= (\sum_{i=1}^{p}|F_i(u_1,v_1) - F_i(u_2,v_2)|^2)^{1/2}\\ &\le B|(u_1,v_1) - (u_2,v_2)|^{\beta},\end{aligned}$$

in which $0 < \beta < 1, |(u_1,v_1) - (u_2,v_2)| = (|u_1-u_2|^2 + |v_1-v_2|^2)^{1/2}$, and B is a positive constant independent of (u_i,v_i) $(i = 1,2)$.

Theorem 3.1 *The number $\beta\,(0 < \beta < 1)$ is the index of a Hölder continuous function vector $F(u,v)$ in $\partial\Omega_1 \times \partial\Omega_2$, if and only if the number β is the index of every Hölder continuous component $f_i(u,v)$ $(i = 1,\ldots,p)$.*

It is easy to verify by the definition, so we omit it.

Noting that the Hölder continuous function vector set with the index $\beta(0 < \beta < 1)$ in $\partial\Omega_1 \times \partial\Omega_2$ is denoted by $H_p(\partial\Omega_1 \times \partial\Omega_2, \beta)$, we take $F(u,v) \in H_p(\partial\Omega_1 \times \partial\Omega_2, \beta)$, and define the norm of $F(u,v)$ as $||F||_\beta = C_p(F, \partial\Omega_1 \times \partial\Omega_2) + H_p(F, \partial\Omega_1 \times \partial\Omega_2, \beta)$, where

$$\begin{aligned}C_p(F, \partial\Omega_1 \times \partial\Omega_2) &= \max_{(u,v)\in\partial\Omega_1\times\partial\Omega_2} |F(u,v)|\\ &= \max_{(u,v)\in\partial\Omega_1\times\partial\Omega_2} (\sum_{i=1}^{p}|F_i(u,v)|^2)^{1/2},\end{aligned}$$

$$\begin{aligned}H_p(F, \partial\Omega_1 \times \partial\Omega_2, \beta) &= \sup_{\substack{(u_i,v_i)\in\partial\Omega_1\times\partial\Omega_2\\ (u_1,v_1)\ne(u_2,v_2)}} \frac{|F(u_1,v_1) - F(u_2,v_2)|}{|(u_1,v_1) - (u_2,v_2)|^{\beta}}\\ &= \sup_{\substack{(u_i,v_i)\in\partial\Omega_1\times\partial\Omega_2\\ (u_1,v_1)\ne(u_2,v_2)}} \frac{(\sum_{i=1}^{p}|F_i(u_1,v_1) - F_i(u_2,v_2)|^2)^{1/2}}{|(u_1,v_1) - (u_2,v_2)|^{\beta}}.\end{aligned}$$

It is easy to see that $H_p(\partial\Omega_1 \times \partial\Omega_2, \beta)$ is a Banach space, and we can verify that the following inequalities hold:

$$||F+G||_\beta \leq ||F||_\beta + ||G||_\beta,\ ||F \otimes G||_\beta \leq J_0||F||_\beta||G||_\beta, \tag{3.2}$$

where J_0 is the same as before, $F, G \in H_p(\partial\Omega_1 \times \partial\Omega_2, \beta)$.

Problem SR We are given a connected open set $\Omega = \Omega_1 \times \Omega_2 \subset \mathbf{R}^m \times \mathbf{R}^k$, whose boundaries $\partial\Omega_i (i = 1, 2)$ are all smooth, oriented, compact Liapunov surfaces. Let $A(t_1, t_2), B(t_1, t_2), C(t_1, t_2), D(t_1, t_2), G(t_1, t_2)$ and $F_*(t_1, t_2, \Phi^{(1)}, \Phi^{(2)}, \Phi^{(3)}, \Phi^{(4)})$ be given function vectors in $\partial\Omega_1 \times \partial\Omega_2$ and $\partial\Omega_1 \times \partial\Omega_2 \times \mathcal{A}_n(R) \times \mathcal{A}_n(R) \times \mathcal{A}_n(R) \times \mathcal{A}_n(R)$ respectively. We find a biregular function vector $\Phi(x, y)$ in $\Omega_1^+ \times \Omega_2^+, \Omega_1^+ \times \Omega_2^-, \Omega_1^- \times \Omega_2^+, \Omega_1^- \times \Omega_2^-$, such that it is continuous in $(\Omega_1^\pm \times \Omega_2^+) \cup (\partial\Omega_1 \times \partial\Omega_2)$, $(\Omega_1^\pm \times \Omega_2^-) \cup (\partial\Omega_1 \times \partial\Omega_2)$, namely its every component is continuous, $\Phi(x, \infty) = \Phi(\infty, y) = \Phi(\infty, \infty) = 0$ and satisfies the nonlinear boundary condition

$$\begin{aligned}&A(t_1,t_2)\otimes\Phi^{++}(t_1,t_2)+B(t_1,t_2)\otimes\Phi^{+-}(t_1,t_2)+C(t_1,t_2)\\ &\otimes\Phi^{-+}(t_1,t_2)+D(t_1,t_2)\otimes\Phi^{--}(t_1,t_2)=G(t_1,t_2)\otimes F_*[t_1,t_2,\\ &\qquad\Phi^{++}(t_1,t_2),\Phi^{+-}(t_1,t_2),\Phi^{-+}(t_1,t_2),\Phi^{--}(t_1,t_2)],\end{aligned}$$

where A, B, C, D, G, F_* are all known function vectors, and $\Phi(t_1, t_2)$ is an unknown function vector as

$$\begin{aligned}\Phi^{++}(t_1,t_2) &= [\Phi_1^{++}(t_1,t_2),\ldots,\Phi_p^{++}(t_1,t_2)],\\ \Phi^{+-}(t_1,t_2) &= [\Phi_1^{+-}(t_1,t_2),\ldots,\Phi_p^{+-}(t_1,t_2)],\\ \Phi^{-+}(t_1,t_2) &= [\Phi_1^{-+}(t_1,t_2),\ldots,\Phi_p^{-+}(t_1,t_2)],\\ \Phi^{--}(t_1,t_2) &= [\Phi_1^{--}(t_1,t_2),\ldots,\Phi_p^{--}(t_1,t_2)].\end{aligned}$$

Here we simply write $F \otimes G$ as FG, so the above boundary condition can be briefly written as

$$\begin{aligned}&A(t_1,t_2)\Phi^{++}(t_1,t_2)+B(t_1,t_2)\Phi^{+-}(t_1,t_2)+C(t_1,t_2)\Phi^{-+}(t_1,t_2)\\ &+D(t_1,t_2)\Phi^{--}(t_1,t_2)=G(t_1,t_2)F_*(t_1,t_2,\Phi^{++},\Phi^{+-},\Phi^{-+},\Phi^{--}).\end{aligned} \tag{3.3}$$

3.2 The Plemelj formula of biregular function vectors with Cauchy type integrals

We can obtain the following theorem about biregular function vectors by the above lemma.

Theorem 3.2 *Let $x \notin \partial\Omega_1$, $y \notin \partial\Omega_2$ and the function vector $\varphi(u,v) \in H_p(\partial\Omega_1 \times \partial\Omega_2, \beta)$. Then the Cauchy type integral*

$$\Phi(x,y) = \lambda \int_{\partial\Omega_1 \times \partial\Omega_2} E_m(x,u) d\sigma_u \varphi(u,v) d\sigma_v E_k(y,v) \tag{3.4}$$

is a biregular function vector satisfying the condition $\Phi(\infty, y) = \Phi(x, \infty) = \Phi(\infty,\infty) = 0$, where $E_m(x,u) = (\bar{u} - \bar{x})/|u-x|^m$, $E_k(y,v) = (\bar{v} - \bar{y})/|v-y|^k$ and λ is as same as in Section 1.

Proof It is clear that

$$\begin{aligned}
\Phi(x,y) &= \lambda \int_{\partial\Omega_1 \times \partial\Omega_2} E_m(x,u) d\sigma_u \varphi(u,v) d\sigma_v E_k(y,v) \\
&= [\lambda \int_{\partial\Omega_1 \times \partial\Omega_2} E_m(x,u) d\sigma_u \varphi_1(u,v) d\sigma_v E_k(y,v), \\
&\quad \ldots, \lambda \int_{\partial\Omega_1 \times \partial\Omega_2} E_m(x,u) d\sigma_u \varphi_p(u,v) d\sigma_v E_k(y,v)] \\
&= [\Phi_1(x,y), \ldots, \Phi_p(x,y)].
\end{aligned}$$

On the basis of Theorem 3.1, we see that the function vector $\varphi(u,v) \in H_p(\partial\Omega_1 \times \partial\Omega_2, \beta)$, hence its every component $\varphi_i(u,v) \in H(\partial\Omega_1 \times \partial\Omega_2, \beta)\,(i = 1, \ldots, p)$. By (1.2), we know that $\Phi_i(x,y)\,(i = 1, \ldots, p)$ are biregular functions. This shows that $\Phi(x,y)$ is a biregular function vector.

Definition 3.3 The integral

$$\Phi(t_1,t_2) = \lambda \int_{\partial\Omega_1 \times \partial\Omega_2} E_m(t_1,u) d\sigma_u \varphi(u,v) d\sigma_v E_k(t_2,v), (t_1,t_2) \in \partial\Omega_1 \times \partial\Omega_2, \tag{3.5}$$

which is called the singular integral in characteristic manifold $\partial\Omega_1 \times \partial\Omega_2$.

Now we consider

$$\Phi_\delta(t_1,t_2) = \lambda \int_{\partial\Omega_1 \times \partial\Omega_2 - \lambda_\delta} E_m(t_1,u) d\sigma_u \varphi(u,v) d\sigma_v E_k(t_2,v),$$

where λ_δ is the same as that in Definition 1.2.

Definition 3.4 If there exists $\lim_{\delta \to 0} \Phi_\delta(t_1,t_2) = I$, then I is called the Cauchy principle value of a singular integral, which is also called the Cauchy principle value of a Cauchy type integral denoted by $I = \Phi(t_1,t_2)$, where

$$\lim_{\delta\to 0} \Phi_\delta(t_1,t_2) = [\lim_{\delta\to 0} \Phi_{1\delta}(t_1,t_2), \ldots, \lim_{\delta\to 0} \Phi_{p\delta}(t_1,t_2)],\ I = (I_1, \ldots, I_p).$$

Next we introduce the singular integral operators

$$P_1\varphi = \frac{2}{\omega_m}\int_{\partial\Omega_1} E_m(t_1,u)d\sigma_u\varphi(u,t_2)$$

$$= [\frac{2}{\omega_m}\int_{\partial\Omega_1} E_m(t_1,u)d\sigma_u\varphi_1(u,t_2),\ldots,\frac{2}{\omega_m}E_m(t_1,u)d\sigma_u\varphi_p(u,t_2)],$$

$$P_2\varphi = \frac{2}{\omega_k}\int_{\partial\Omega_2}\varphi(t_1,v)d\sigma_v E_k(t_2,v)$$

$$= [\frac{2}{\omega_k}\int_{\partial\Omega_2}\varphi_1(t_1,v)d\sigma_v E_k(t_2,v),\ldots,\frac{2}{\omega_k}\int_{\partial\Omega_2}\varphi_p(t_1,v)d\sigma_v E_k(t_2,v)],$$

$$P_3\varphi = 4\Phi(t_1,t_2) = [4\Phi_1(t_1,t_2),\ldots,4\Phi_p(t_1,t_2)].$$

Theorem 3.3 (The Plemelj formula of function vectors with Cauchy type integrals) *Given the function vector $\varphi(u,v) = (\varphi_1(u,v),\ldots,\varphi_p(u,v)) \in H_p(\partial\Omega_1\times\partial\Omega_2,\beta)$ in (3.4), then for any $(t_1,t_2)\in\partial\Omega_1\times\partial\Omega_2$, we have*

$$\begin{cases} \Phi^{++}(t_1,t_2) = \frac{1}{4}[\varphi + P_1\varphi + P_2\varphi + P_3\varphi](t_1,t_2), \\ \Phi^{+-}(t_1,t_2) = \frac{1}{4}[-\varphi - P_1\varphi + P_2\varphi + P_3\varphi](t_1,t_2), \\ \Phi^{-+}(t_1,t_2) = \frac{1}{4}[-\varphi + P_1\varphi - P_2\varphi + P_3\varphi](t_1,t_2), \\ \Phi^{--}(t_1,t_2) = \frac{1}{4}[\varphi - P_1\varphi - P_2\varphi + P_3\varphi](t_1,t_2). \end{cases} \tag{3.6}$$

3.3 The existence of solutions of nonlinear boundary value problems for biregular function vectors

Suppose that the solution of Problem SR is as stated in (3.4), and substitute (3.6) into (3.3), then we have

$$\begin{aligned} & A(t_1,t_2)\{\frac{1}{4}[\varphi(t_1,t_2) + P_1\varphi(t_1,t_2) + P_2\varphi(t_1,t_2) + P_3\varphi(t_1,t_2)]\} \\ + & B(t_1,t_2)\{\frac{1}{4}[-\varphi(t_1,t_2) - P_1\varphi(t_1,t_2) + P_2\varphi(t_1,t_2) + P_3\varphi(t_1,t_2)]\} \\ + & C(t_1,t_2)\{\frac{1}{4}[-\varphi(t_1,t_2) + P_1\varphi(t_1,t_2) - P_2\varphi(t_1,t_2) + P_3\varphi(t_1,t_2)]\} \\ + & D(t_1,t_2)\{\frac{1}{4}[\varphi(t_1,t_2) - P_1\varphi(t_1,t_2) - P_2\varphi(t_1,t_2) + P_3\varphi(t_1,t_2)]\} \\ = & G(t_1,t_2)F_*[t_1,t_2,\Phi^{++}(t_1,t_2),\Phi^{+-}(t_1,t_2),\Phi^{-+}(t_1,t_2),\Phi^{--}(t_1,t_2)], \end{aligned}$$

and then

$$(A+B)(\varphi+P_1\varphi+P_2\varphi+P_3\varphi)+(C+D)(-\varphi+P_1\varphi$$
$$-P_2\varphi+P_3\varphi)+(B+D)(2\varphi-2P_1\varphi)-4B\varphi=4GF_*.$$

Let

$$F\varphi=(A+B)(\varphi+P_1\varphi+P_2\varphi+P_3\varphi)+(C+D)(-\varphi+P_1\varphi$$
$$-P_2\varphi+P_3\varphi)+(B+D)(2\varphi-2P_1\varphi)+(1-4B)\varphi-4GF_*.$$

Then Problem SR is reduced to finding the solution of singular integral equation $F\varphi=\varphi$.

Lemma 3.4 ([29]2)) *Let $\varphi(t_1,t_2)\in H(\partial\Omega_1\times\partial\Omega_2,\beta)$. Then there exists a positive constant J_1 independent of φ, such that*

$$||P_i\varphi||_\beta\le J_1||\varphi||_\beta\ (i=1,2),\ ||P_2\varphi+P_3\varphi||_\beta\le J_1||\varphi||_\beta. \tag{3.7}$$

Using Lemma 3.4 and the definition of the norm of a function vector, by calculation we can get the following theorem.

Theorem 3.5 *Let the function vector $\varphi(t_1,t_2)=(\varphi_1(t_1,t_2),\ldots,\varphi_p(t_1,t_2))\in H_p(\partial\Omega_1\times\partial\Omega_2,\beta)$. Then there exists a positive constant J_2 independent of φ, such that*

$$||P_i\varphi||_\beta\le J_2||\varphi||_\beta,\ i=1,2,3,\ ||2\varphi\pm 2P_i\varphi||_\beta\le J_2||\varphi||_\beta,\ i=1,2,$$
$$||\Phi^{\pm\pm}(t_1,t_2)||_\beta\le J_2||\varphi||_\beta,\ ||\pm\varphi\pm P_1\varphi+P_2\varphi+P_3\varphi||_\beta\le J_2||\varphi||_\beta,$$
$$||\Phi^{\pm\mp}(t_1,t_2)||_\beta\le J_2||\varphi||_\beta,\ ||\mp\varphi\pm P_1\varphi-P_2\varphi+P_3\varphi||_\beta\le J_2||\varphi||_\beta. \tag{3.8}$$

Theorem 3.6 *Let $A(t_1,t_2)$, $B(t_1,t_2)$, $C(t_1,t_2)$, $D(t_1,t_2)$, $G(t_1,t_2)\in H_p(\partial\Omega_1\times\partial\Omega_2,\beta)$, $F_*(t_1,t_2,\Phi^{(1)},\Phi^{(2)},\Phi^{(3)},\Phi^{(4)})$ be Hölder continuous about $t=(t_1,t_2)\in\partial\Omega_1\times\partial\Omega_2$ for any Clifford number $\Phi^{(1)},\Phi^{(2)},\Phi^{(3)},\Phi^{(4)}$, and for any $(t_1,t_2)\in\partial\Omega_1\times\partial\Omega_2$ about $\Phi^{(1)},\Phi^{(2)},\Phi^{(3)},\Phi^{(4)}$ satisfies the Lipschitz condition, i.e.*

$$|F_*(t_{11},t_{21},\Phi_1^{(1)},\Phi_1^{(2)},\Phi_1^{(3)},\Phi_1^{(4)})-F_*(t_{12},t_{22},\Phi_2^{(1)},\Phi_2^{(2)},\Phi_2^{(3)},\Phi_2^{(4)})|$$
$$\le\ J_3|(t_{11},t_{21})-(t_{12},t_{22})|^\beta+J_4|\Phi_1^{(1)}-\Phi_2^{(1)}|+J_5|\Phi_1^{(2)}-\Phi_2^{(2)}|$$
$$+J_6|\Phi_1^{(3)}-\Phi_2^{(3)}|+J_7|\Phi_1^{(4)}-\Phi_2^{(4)}|,$$

where $J_3,\ldots,J_7$ are positive constants independent of $t_{1j},t_{2j},\Phi_j^{(i)}$ $(i=1,2,3,4,\ j=1,2)$, and $F_*(0,0,0,0,0,0)=0$. If A, B, C, D satisfy

$\gamma = J_0[J_2(||A+B||_\beta + ||C+D||_\beta + ||B+D||_\beta) + ||1-4B||_\beta] < 1$, *and* $||G||_\beta < \delta$, *then when* $0 < \delta < \dfrac{M(1-\gamma)}{4J_0(J_8+J_9M)}$, *Problem SR is solvable, where M is a given positive number and* J_8, J_9 *are positive constants, such that* $||\varphi||_\beta \le M$, *and* $\|F_*\|_\beta \le J_8 + J_9||\varphi||_\beta$.

Proof We introduce a subset of the continuous function vector space $C_p(\partial\Omega_1 \times \partial\Omega_2)$ as $T = \{\varphi | \varphi \in H_p(\partial\Omega_1 \times \partial\Omega_2, \beta), ||\varphi||_\beta \le M\}$, and first prove that F maps the set T into itself.

By (3.6), (3.1), we have

$$\begin{aligned}
||F\varphi||_\beta &= ||(A+B)(\varphi + P_1\varphi + P_2\varphi + P_3\varphi) + (C+D)(-\varphi \\
&\quad +P_1\varphi - P_2\varphi + P_3\varphi) + (B+D)(2\varphi - 2P_1\varphi) + (1-4B)\varphi \\
&\quad -4GF_*||_\beta \le J_0||A+B||_\beta||\varphi + P_1\varphi + P_2\varphi \\
&\quad +P_3\varphi||_\beta + J_0||C+D||_\beta|| - \varphi + P_1\varphi \\
&\quad -P_2\varphi + P_3\varphi||_\beta + J_0||B+D||_\beta||2\varphi - 2P_1\varphi||_\beta \\
&\quad +J_0||1-4B||_\beta||\varphi||_\beta - 4J_0||G||_\beta||F_*||_\beta \\
&\le J_0||A+B||_\beta J_2||\varphi||_\beta + J_0||C+D||_\beta J_2||\varphi||_\beta \\
&\quad +J_0||B+D||_\beta J_2||\varphi||_\beta + J_0||1-4B||_\beta||\varphi||_\beta \\
&\quad -4J_0||G||_\beta||F_*||_\beta = J_0[J_2(||A+B||_\beta + ||C+D||_\beta \\
&\quad +||B+D||_\beta) + ||1-4B||_\beta]||\varphi||_\beta - 4J_0||G||_\beta||F_*||_\beta \\
&\le \gamma||\varphi||_\beta - 4J_0\delta||F_*||_\beta.
\end{aligned}$$

Moreover we consider $||F_*||_\beta$, and first discuss $C_p(F_*, \partial\Omega_1 \times \partial\Omega_2)$. By (3.8), we get

$$\begin{aligned}
|F_*| &= |F_*[t_1, t_2, \Phi^{++}(t_1,t_2), \Phi^{+-}(t_1,t_2), \Phi^{-+}(t_1,t_2), \Phi^{--}(t_1,t_2) \\
&\quad -F_*(0,0,0,0,0,0)| \\
&\le J_3|(t_1,t_2)|^\beta + J_4|\Phi^{++}(t_1,t_2)| + J_5|\Phi^{+-}(t_1,t_2)| \\
&\quad +J_6|\Phi^{-+}(t_1,t_2)| + J_7|\Phi^{--}(t_1,t_2)| \le J_3|(t_1,t_2)|^\beta \\
&\quad +J_4||\Phi^{++}(t_1,t_2)||_\beta + J_5||\Phi^{+-}(t_1,t_2)||_\beta \\
&\quad +J_6||\Phi^{-+}(t_1,t_2)||_\beta + J_7||\Phi^{--}(t_1,t_2)||_\beta \le J_3|(t_1,t_2)|^\beta
\end{aligned}$$

$$+J_4J_2||\varphi||_\beta + J_5J_2||\varphi||_\beta + J_6J_2||\varphi||_\beta$$

$$+J_7J_2||\varphi||_\beta \le J_3|(t_1,t_2)|^\beta + J_{10}||\varphi||_\beta,$$

so

$$C_p(F_*, \partial\Omega_1 \times \partial\Omega_2) = \max_{(t_1,t_2)\in\partial\Omega_1\times\partial\Omega_2} |F_*|$$

$$\le \max_{(t_1,t_2)\in\partial\Omega_1\times\partial\Omega_2} (J_3|(t_1,t_2)|^\beta + J_{10}||\varphi||_\beta) = J_{11} + J_{10}||\varphi||_\beta.$$

Next we discuss $H_p(F_*, \partial\Omega_1 \times \partial\Omega_2, \beta)$. Noting that

$$|F_*(t_{11}, t_{21}, \Phi^{++}(t_{11}, t_{21}), \Phi^{+-}(t_{11}, t_{21}), \Phi^{-+}(t_{11}, t_{21}), \Phi^{--}(t_{11}, t_{21}))$$
$$-F_*(t_{12}, t_{22}, \Phi^{++}(t_{12}, t_{22}), \Phi^{+-}(t_{12}, t_{22}), \Phi^{-+}(t_{12}, t_{22}), \Phi^{--}(t_{12}, t_{22}))|$$
$$\le J_3|(t_{11}, t_{21}) - (t_{12}, t_{22})|^\beta + J_4|\Phi^{++}(t_{11}, t_{21}) - \Phi^{++}(t_{12}, t_{22})|$$
$$+J_5|\Phi^{+-}(t_{11}, t_{21}) - \Phi^{+-}(t_{12}, t_{22})| + J_6|\Phi^{-+}(t_{11}, t_{21}) - \Phi^{-+}(t_{12}, t_{22})|$$
$$+J_7|\Phi^{--}(t_{11}, t_{21}) - \Phi^{--}(t_{12}, t_{22})| \le J_3|(t_{11}, t_{21}) - (t_{12}, t_{22})|^\beta$$
$$+J_4H_p(\Phi^{++}, \partial\Omega_1 \times \partial\Omega_2, \beta)|(t_{11}, t_{21}) - (t_{12}, t_{22})|^\beta$$
$$+J_5H_p(\Phi^{+-}, \partial\Omega_1 \times \partial\Omega_2, \beta)|(t_{11}, t_{21}) - (t_{12}, t_{22})|^\beta$$
$$+J_6H_p(\Phi^{-+}, \partial\Omega_1 \times \partial\Omega_2, \beta)|(t_{11}, t_{21}) - (t_{12}, t_{22})|^\beta$$
$$+J_7H_p(\Phi^{--}, \partial\Omega_1 \times \partial\Omega_2, \beta)|(t_{11}, t_{21}) - (t_{12}, t_{22})|^\beta$$
$$\le (J_3 + J_4||\Phi^{++}||_\beta + J_5||\Phi^{+-}||_\beta + J_6||\Phi^{-+}||_\beta + J_7||\Phi^{--}||_\beta)$$
$$|(t_{11}, t_{21}) - (t_{12}, t_{22})|^\beta$$
$$\le (J_3 + J_4J_2||\varphi||_\beta + J_5J_2||\varphi||_\beta + J_6J_2||\varphi||_\beta + J_7J_2||\varphi||_\beta)$$
$$|(t_{11}, t_{21}) - (t_{12}, t_{22})|^\beta \le (J_3 + J_{12}||\varphi||_\beta)|(t_{11}, t_{21}) - (t_{12}, t_{22})|^\beta,$$

and taking into account

$$F_*^1 = F_*(t_{11}, t_{21}, \Phi^{++}(t_{11}, t_{21}), \ldots, \Phi^{--}(t_{11}, t_{21})),$$

$$F_*^2 = F_*(t_{12}, t_{22}, \Phi^{++}(t_{12}, t_{22}), \ldots, \Phi^{--}(t_{12}, t_{22})),$$

then

$$H_p(F_*, \partial\Omega_1 \times \partial\Omega_2, \beta) = \sup_{\substack{(t_{1i},t_{2i})\in\partial\Omega_1\times\partial\Omega_2 \\ (t_{11},t_{21})\ne(t_{12},t_{22})}} \frac{|F_*^1 - F_*^2|}{|(t_{11}, t_{21}) - (t_{12}, t_{22})|^\beta}$$

$$\le J_3 + J_{12}||\varphi||_\beta,$$

hence we have

$$||F_*||_\beta = C_p(F_*, \partial\Omega_1 \times \partial\Omega_2) + H_p(F_*, \partial\Omega_1 \times \partial\Omega_2, \beta)$$
$$\leq J_{11} + J_{10}||\varphi||_\beta + J_3 + J_{12}||\varphi||_\beta \leq J_8 + J_9||\varphi||_\beta.$$

Furthermore from $\gamma = J_0[J_2(||A+B||_\beta + ||C+D||_\beta + ||B+D||_\beta) + ||1-4B||_\beta] < 1$, it follows that

$$||F\varphi||_\beta \leq \gamma||\varphi||_\beta - 4J_0\delta(J_8 + J_9||\varphi||_\beta)$$
$$\leq \gamma M - 4J_0\frac{M(1-\gamma)}{4J_0(J_8+J_9)M}(J_8 + J_9 M)$$
$$= \gamma M - M(1-\gamma) \leq M,$$

which shows that F maps the set T into itself.

By a similar method and through complicated calculations, we can prove that F is a continuous mapping, namely F continuously maps T into itself. In accordance with the Ascoli-Arzela theorem, we see that T is a compact set of continuous function vector space $C_p(\partial\Omega_1 \times \partial\Omega_2)$. This shows that F continuously maps a closed convex set T of $C_p(\partial\Omega_1 \times \partial\Omega_2)$ into itself, and $F(T)$ is also a compact set of $C_p(\partial\Omega_1 \times \partial\Omega_2)$. Hence by the Schauder fixed-point principle, we know that there exists at least one $\varphi_0 \in H_p(\partial\Omega_1 \times \partial\Omega_2, \beta)$ satisfying the singular integral equation (3.7). This proves that Problem SR has one solution.

Theorem 3.7 *Under the conditions as in Theorem 3.6, and if $F_* \equiv 1$, then problem SR has a unique solution.*

CHAPTER IV

BOUNDARY VALUE PROBLEMS OF SECOND ORDER PARTIAL DIFFERENTIAL EQUATIONS FOR CLASSICAL DOMAINS IN REAL CLIFFORD ANALYSIS

In this chapter, we first introduce the harmonic analysis in classical domains for several complex variables obtained by Luogeng Hua in 1958, and discuss the Cauchy formula, Poisson formula and boundary value problems of harmonic functions for classical domains in Luogeng Hua's sense. By using above results, I. N. Vekua's results about generalized analytic functions, and the tool of quasi-permutations (see Section 3, Chapter I), we discuss two boundary value problems for four kinds of complex partial differential equations of second order in four kinds of classical domains, and prove the existence and uniqueness of regular solutions for the problems and give their expressions in complex Clifford analysis. Finally we discuss a pseudo-modified boundary value problem in a ball for a kind of real partial differential equations of second order in real Clifford analysis and prove that the problem has a unique solution.

1 Harmonic Analysis in Classical Domains for Several Complex Variables

In function theory of several complex variables [89], a domain in $\mathbf{C}^n$ is called a symmetric domain, i.e. for every point of this domain, its symmetric point also belongs to the domain. In 1936, E.Cartan proved that a bounded symmetric domain can be divided into four classes, except 16 and 27 dimensional complex spaces. It is exactly said, except for the two special cases, every bounded symmetric domain in $\mathbf{C}^n$ must belong to one of these four classes of domains, or be equivalent to topology multiplication of some domains of these four classes of domains (see[8]). Luogeng Hua denoted by R_k(k=1,2,3,4) these four classical domains according to the matrix form. Until 1957, Qikeng Lu pointed out that $\mathrm{R}_3(\subset \mathbf{C}^4)$ and R_4 $(\subset \mathbf{C}^6)$ were equivalent to each other [47]2). In

1961, E. E. Berjiske-Sabilof [60] gave the definition of classical domains, namely those that can be analytically equivalent to a bounded domain and permit a transitive classical group. Consequently, the four classical domains firstly defined by Luogeng Hua were symmetric classical domains. Obviously there exists a symmetric classical domain.

The four classes of symmetric classical domains $R_k(1 \leq k \leq 4)$ defined by Luogeng Hua are the complex mn, $p(p+1)/2(\geq 3)$, $q(q-1)/2(\geq 6)$ and $N(\geq 5)$ dimensional respectively, where $R_1(m;n)$ is the matrix hyperbolic space satisfying $I^{(m)} - Z\overline{Z}' > 0$, $Z = (z_{ij})_{m\times n}(m \leq n)$, $I^{(m)}$ is the $m \times m$ unitary square matrix, where a square matrix > 0 means that this square matrix is positive definite; $\mathrm{R}_2(p)$ is the symmetric square matrix hyperbolic space satisfying $I^{(p)} - Z\overline{Z} > 0$, $Z = (z_{ij})_{p\times p}$, $p(p+1)/2 \geq 3$, $Z = Z'$; $R_3(q)$ is the oblique symmetric square matrix hyperbolic space satisfying $I^{(q)} + Z\overline{Z} > 0$, $Z = (z_{ij})_{q\times q}$, $q(q-1)/2 \geq 6$, $Z' = -Z$; $R_4(N)$ is the Lie ball hyperbolic space satisfying $|zz'|^2+1-2z\overline{z}' > 0$, $|zz'| < 1$, $z = (z_1, ..., z_N)$, $N \geq 5$. All $R_k(1 \leq k \leq 4)$ are bounded star circular domains with center at the origin [47]2). In function theory of one complex variable, we often study function theory in the unit disk $|z| < 1$, because in general all symmetric domains are equivalent to the unit disk, hence the above four classes of symmetric classical domains play an important role in function theory of several complex variables.

The geometries usually discussed by researchers in function theory of one complex variable include parabolic geometry, i.e. the complex plane $\mathbf{C}^1$ with measure $|dz|$; elliptic geometry, i.e. the Riemann ball with metrics $|dz|/(1+|z|^2)$; hyperbolic geometry, i.e. the unit disk with metrics $|dz|/(1-|z|^2)$. Their unitary curvatures are equal to 0, positive and negative respectively. In $\mathbf{C}^n$, the generalization of parabolic geometry is the geometry with metrics $\{\sum_{i=1}^n |dz_i|^2\}^{1/2}$. In order to discuss the geometric structure of boundary of symmetric classical domains, Qikeng Lu introduced hyperbolic metrics in symmetric classical domains, and proved that the symmetric classical domains and their boundaries possess a common geometric character([47]2),3)).

The boundary B_k of classical domain R_k, generally speaking, does not become a differential manifold, but B_k can be denoted as a sum of some subsets, in which every subset is transitive and invariable under the action of the motion group Γ^k. According to the theory of Lie groups, one can give a real analytic structure for every subset, such that it becomes a real analytic manifold. The interesting thing is the subspace of the closed classical domain $\overline{R}_k$, which is invariable under the action of the motion group Γ^k and its dimension is the lowest, and is a real homogenous compact analytic manifold. We call them the characteristic manifolds of R_k and denote them by $L_k\,(1 \le k \le 4)$. They play an important role in function theory in classical domains. For instance, the Cauchy integral formula for holomorphic functions in the classical domains defines the integral on a characteristic manifold, and does not need to be integrable on the whole boundary except for a few examples. This character essentially distinguishes it from with the Cauchy formula in function theory of one complex variable. The Poisson formula in classical domains has a similar property.

In order to find the Poisson kernel, from which the Cauchy kernel can be found, Qikeng Lu utilized the geometric character of symmetric classical domains and their boundaries. He first found the transposition relation to the volume element of characteristic manifold L_k under the action of Γ^k, and then he represented the volume element with the outer differential form, consequently the computation becomes simple. On the basis of this, we can prove the following mean value theorem.

Theorem 1.1 *Suppose that $R_{A_1}, R_{A_2}, ..., R_{A_k}$ are all domains in four classes of symmetric classical domains, N is the sum of dimensions of the classical domains, $R = R_{A_1} \times R_{A_2} \times \cdots \times R_{A_k}$, L_{A_i} is a characteristic manifold of R_{A_i}, $L = L_{A_1} \times L_{A_2} \times \cdots \times L_{A_k}$, and $f(z) = f(z_1, ..., z_N)$ is an analytic function in R and is continuous on $R \cup L$. Then we have*

$$f(0) = \frac{1}{V(L)} \int_L f(\xi)\dot{\xi},$$

where $\dot{\xi}$ is the volume element of L, and V(L) is the volume of L.

As an application of the mean value theorem, we shall derive the Poisson formula in classical domains R. Because every R_{A_i} is transitive under the action of Γ^{A_i}, we can conclude that R is transitive under the action of the direct product of these kinds of groups. Assume $z_0 \in R$, then there must exist an analytic transformation of this direct product as follows:

$$w = \Phi(z; z_0, \overline{z}_0), \tag{1.1}$$

which maps z_0 onto the origin. Let the inverse of this transformation be

$$z = \Phi^{-1}(w; z_0, \overline{z}_0). \tag{1.2}$$

Because every transformation of Γ^{A_i} is still analytic on the boundaries of R_{A_i}, it is evident that the transformations (1.1) and (1.2) are analytic on the boundary of R. We arbitrarily give an analytic function $f(z)$ in R, such that it is continuous on $R \cup L$, then

$$F(w) = f(\Phi^{-1}(w; z_0, \overline{z}_0))$$

is also analytic in R and continuous on $R \cup L$, and $F(0) = f(z_0)$. According to Theorem 1.1, we get

$$F(0) = \frac{1}{V(L)} \int_L F(\zeta)\dot{\zeta}, \tag{1.3}$$

where $\zeta = \Phi(\xi; z_0, \overline{z}_0)$, $\xi \in L$. Assume that through transformation (1.1), the integral possesses the relation

$$\int_L F(\zeta)\dot{\zeta} = \int_L f(\xi)P(z_0, \xi)\dot{\xi}, \tag{1.4}$$

then (1.3) can be rewritten as

$$f(z_0) = \frac{1}{V(L)} \int_L f(\xi)P(z_0, \xi)\dot{\xi},$$

where $z_0 \in R$. We call (1.4) the Poisson formula in the classical domains, and call $P(z_0, \xi)$ the Poisson kernel of R. By the computation, we can get that the Poisson kernel of every symmetric classical domain R_k $(1 \leq k \leq 4)$ can be written in the form

$$P_k(z, \xi) = \frac{|H_k(z, \overline{\xi})|^2}{H_k(z, \overline{z})}, \quad 1 \leq k \leq 4,$$

in which for the fixed $w \in R_k$, $H_k(z, \overline{w})$ in R_k is analytic concerning z, and $\overline{H_k(z, \overline{w})} = H_k(w, \overline{z})$. By using $H_k(z, \xi)$, we can derive the Cauchy formula of functions in R_k.

Let R_k $(1 \leq k \leq 4)$ be four classes of symmetric classical domains, and L_k $(1 \leq k \leq 4)$ be the characteristic manifolds of R_k. Given an arbitrary function $f(z)$ which is analytic in R_k and continuous on $R_k \cup L_k$, then for every $z \in R_k$, we have the Cauchy integral formula

$$f(z) = \frac{1}{V(L_k)} \int_{L_k} f(\xi)H_k(z, \overline{\xi})\dot{\xi}, \tag{1.5}$$

where $H_k(z, \overline{\xi})$ is called the Cauchy kernel of R_k. In fact, we construct the function

$$g(z) = f(z)[H_k(z, \overline{w})]^{-1},$$

when $w \in R_A$ is fixed, $g(z)$ is analytic in R_k and continuous on $R_k \cup L_k$; by using the Poisson formula (1.4), we get

$$g(z) = \frac{1}{V(L_k)} \int_{L_k} g(\xi) P_k(z, \xi) \dot{\xi} = \frac{1}{V(L_k)} \int_{L_k} g(\xi) \frac{|H_k(z, \overline{\xi})|^2}{H_k(z, \overline{z})} \dot{\xi},$$

that is

$$f(z)[H_k(z, \overline{w})]^{-1} = \frac{1}{V(L_k)} \int_{L_k} f(\xi)[H_k(\xi, \overline{w})]^{-1} \frac{|H_k(z, \overline{\xi})|^2}{H_k(z, \overline{z})} \dot{\xi}.$$

The above formula is valid for every point $w \in R_k$, especially for $w = z$, notice $H_k(z, \overline{z}) > 0$ and $\overline{H_k(z, \overline{\xi})} = H_k(\xi, \overline{z})$, then after eliminating $H_k(z, \overline{z})$ from two sides of the above formula, the formula (1.5) is concluded.

The above method about the Cauchy integral formula (1.5) was obtained by Qikeng Lu in 1963 [47]3); it is different from the method given by Luogeng Hua in 1958. In Hua's monograph [26]1), he first used the representation theory of groups to find the complete orthogonal and normal system of functions on the characteristic manifolds L_k $(1 \le k \le 4)$ of four classes of symmetric classical domains R_k $(1 \le k \le 4)$, and then he got the Cauchy integral formula.

In [26]1),2), Luogeng Hua and Qikeng Lu introduced the harmonic operators $\triangle$ $(1 \le k \le 4)$:

$$\triangle_1 = \sum_{j,k=1}^{m} \sum_{\alpha,\beta=1}^{n} (\delta_{jk} - \sum_{\lambda=1}^{n} z_{j\lambda}\overline{z}_{k\lambda})(\delta_{\alpha\beta} - \sum_{l=1}^{n} z_{l\alpha}\overline{z}_{l\beta}) \frac{\partial^2}{\partial z_{j\alpha} \partial \overline{z}_{k\beta}},$$

$$\delta_{\alpha\beta} = \begin{cases} 1, \ \alpha = \beta; \\ 0, \ \alpha \ne \beta; \end{cases}$$

$$\triangle_2 = \sum_{\alpha,\beta,\lambda,\mu=1}^{p} (\delta_{\lambda\mu} - \frac{1}{2} \sum_{\sigma=1}^{p} \frac{z_{\lambda\sigma}}{p_{\lambda\sigma}} \frac{\overline{z}_{\mu\sigma}}{p_{\mu\sigma}})(\delta_{\alpha\beta} - \frac{1}{2} \sum_{r=1}^{p} \frac{z_{\alpha r}}{p_{\alpha r}} \frac{\overline{z}_{\beta r}}{p_{\beta r}})$$
$$\times \frac{1}{2 p_{\lambda\alpha} p_{\mu\beta}} \cdot \frac{\partial^2}{\partial z_{\lambda\alpha} \partial \overline{z}_{\mu\beta}},$$

$$p_{\alpha\beta} = \begin{cases} \frac{1}{\sqrt{2}}, \ \alpha = \beta, \\ 1, \ \alpha \ne \beta; \end{cases}$$

$$\triangle_3 = \frac{1}{2} \sum_{\alpha,\beta,\lambda,\mu=1}^{q} (\delta_{\lambda\mu} - \sum_{\sigma=1}^{q} z_{\lambda\sigma}\overline{z}_{\mu\sigma})(\delta_{\alpha\beta} - \sum_{r=1}^{q} z_{\alpha r}\overline{z}_{\beta r}) q_{\lambda\alpha} q_{\mu\beta} \frac{\partial^2}{\partial z_{\lambda\alpha} \partial \overline{z}_{\mu\beta}},$$

$$q_{\alpha\beta} = \begin{cases} 0, \ \alpha = \beta, \\ 1, \ \alpha \neq \beta; \end{cases}$$

$$\Delta_4 = (1 + |z\, z'|^2 - 2z\,\overline{z}') \sum_{\alpha,\beta=1}^{N} (\delta_{\alpha\beta} - 2z_\alpha \overline{z}_\beta) \frac{\partial^2}{\partial z_\alpha \partial \overline{z}_\beta}$$

$$+2 \sum_{\alpha,\beta=1}^{N} (\overline{z}_\alpha - \overline{z\, z'} z_\alpha)(z_\beta - z\, z' \overline{z}_\beta) \frac{\partial^2}{\partial z_\alpha \partial \overline{z}_\beta}.$$

The function $f(z)$ in four classes of symmetric classical domains $R_k\,(1 \leq k \leq 4)$ satisfying $\Delta_k f(z) = 0$ and possessing twice continuous partial derivatives is called a harmonic function. After we prove that the Poisson kernels $P_k(z, \xi)$ of $R_k\,(1 \leq k \leq 4)$ are harmonic functions of $z \in R_k$, herein $\xi \in L_k$, it is easily seen that the Poisson integral

$$\frac{1}{V(L_k)} \int_{L_k} \varphi(\xi) P_k(z, \xi) \dot{\xi}$$

is also a harmonic function in $R_k\,(1 \leq k \leq 4)$, where $\varphi(\xi)$ is continuous on the characteristic manifolds $L_k\,(1 \leq k \leq 4))$ [32].

Applying the boundary properties and extremum principle of the Poisson integral in $R_k\,(1 \leq k \leq 4)$, we can prove the existence and uniqueness of solutions of the Dirichlet boundary value problem in classical domains $R_k\,(1 \leq k \leq 4)$ for harmonic functions of several complex variables [27].

Theorem 1.2 *Let $\varphi_k(\xi)$ be continuous on $L_k\,(1 \leq k \leq 4)$, $z \in \overline{R}_k \backslash L_k$. Then there exists a unique harmonic function $f_k(z)$ in R_k, such that $\lim_{z \to \zeta} f_k(z) = \varphi_k(\zeta)\,(1 \leq k \leq 4)$, if $\zeta \in L_k$, and $f_k(z)$ can be expressed as*

$$f_k(z) = \frac{1}{V(L_k)} \int_{L_k} \varphi_k(\xi) P_k(z, \xi) \dot{\xi},$$

where $P_k(z, \xi)$ is the Poisson kernel of classical domains $R_k\,(1 \leq k \leq 4)$.

2 The Dirichlet Problem of Second Order Complex Partial Differential Equations for Classical Domains in Complex Clifford Analysis

On the basis of the results of harmonic analysis in four classical domains by Luogeng Hua and Qikeng Lu as stated in the above section, and the results about generalized analytic functions by I. N. Vekua [77], in this section we shall discuss four kinds of complex partial differential equations of second order in complex Clifford analysis, and then we shall prove the existence and uniqueness of regular solutions for two boundary value problems of Dirichlet type in four classical domains and give their integral expressions (see [32]2)).

Denote by $p_k\,(1 \le k \le 4)$ the dimensions of four classical domains $R_k\,(1 \le k \le 4)$, that is $p_1 = mn$, $p_2 = p(p+1)/2$, $p_3 = q(q-1)/2$, $p_4 = N$, and by C^{p_k} the space of complex variables $z_1, ..., z_{p_k}$. When $k = 1$, set $(z_{11}, ..., z_{1n}, ..., z_{m1}, ..., z_{mn}) = (z_1, ..., z_{p_1})$, the rest can be given a similar notation. In this section, we consider the function $f(z)$ from C^{p_k} to complex Clifford algebra $A_{p_k}(C)\,(1 \le k \le 4)$.

In Section 3, Chapter 1, we divide the function $f(z) = f(z_1, ..., z_{p_k})$ into two parts:

$$f(z) = f^{(1)} + f^{(2)} = \sum_A{}' f'_A e'_A + \sum_A{}'' f''_A e''_A,$$

where A in the sum $\sum\limits_A{}'$ is chosen the first suffix, and A in $\sum\limits_A{}''$ is chosen the second suffix, and denote $f^{(i)} = J_i f\ (i = 1, 2)$.

Let $\sum\limits_A{}' \varphi'_A(\xi)e'_A$ on $L_k\,(1 \le k \le 4)$ be continuous; we shall seek a regular function $f(z) = \sum_A f_A(z)e_A$ in R_k satisfying the equation

$$\triangle_k f(z) = 0, \tag{2.1}$$

such that it is continuous on $\overline{R}_k\,(1 \le k \le 4)$ satisfying the boundary condition of Dirichlet type

$$f^{(1)}\Big|_{L_k} = J_1 f|_{L_k} = \sum_A{}' \varphi'_A(\xi)e'_A,\ \xi \in L_k. \tag{2.2}$$

The problem for complex partial differential equations of second order (2.1) in classical domains R_k will be called Problem $D_k\,(1 \le k \le 4)$, or Problem D_k for short.

Firstly, we find the integral expression of solutions for Problem D_k. Let $f(z)$ be a solution of Problem D_k; it is clear that $\sum\limits_A \triangle_k f_A(z)e_A = \triangle_k f(z) = 0$, and then $\triangle_k f_A(z) = 0$, $z \in R_k$. From (2.2) it follows that $f'_A(\xi) = \varphi'_A(\xi)\,(\xi \in L_k)$. By using Theorem 1.2, we see that $f'_A(z)$ may be uniquely expressed as

$$f'_A(z) = \frac{1}{V(L_k)} \int_{L_k} \varphi'_A(\xi)P_k(z,\xi)\dot{\xi}. \tag{2.3}$$

Because $f(z)$ is a regular solution of equation (2.1), similarly to Theorem 1.3 in Chapter II, the function in the complex Clifford analysis can be written as

$$\begin{cases} f''_{Az_1} = \sum\limits_{m=2}^{P_k} \delta'_{\overline{mA}} f'_{\overline{mA}z_m}, \\ f'_{Az_1} = \sum\limits_{m=2}^{P_k} \delta''_{\overline{mA}} f''_{\overline{mA}z_m}, \end{cases} \tag{2.4}$$

where denote by $f'_{\overline{mA}z_m}$ $(f''_{\overline{mA}z_m})$ the function $(f_{\overline{mA}})_{z_m}$, when the quasi-permutation $\overline{mA}$ is taken the first (second) suffix; $\delta'_{\overline{mA}}$ $(\delta''_{\overline{mA}})$ is the sign of corresponding quasi-permutations of first (second) suffix.

Next, we introduce the operators $T_1, \overline{T}_1$ (see [77]):

$$T_1 f_A(z) = \frac{1}{\pi} \int\limits_{\overline{G}_k} \frac{f_A(\xi_1, z_2, ..., z_{p_k})}{z_1 - \xi_1} d\sigma_{\xi_1},$$

$$\overline{T}_1 f_A(z) = \frac{1}{\pi} \int\limits_{\overline{G}_k} \frac{f_A(\xi_1, z_2, ..., z_{p_k})}{\overline{z}_1 - \overline{\xi}_1} d\sigma_{\xi_1},$$

where $1 \leq k \leq 4$, and integral variable ξ_1 taken over the section of $\overline{R}_k : \overline{G}_k = \overline{R}_k \cap \{\xi_2 = z_2, ..., \xi_{p_k} = z_{p_k}\}$, $z = (z_1, ..., z_{p_k}) \in R_k$, $d\sigma_{\xi_1}$ is the area element of $\overline{G}_k$. By [77], we see that from the first expression of (2.4), the following equalities hold:

$$f''_A = \overline{T}_1 \sum_{m=2}^{P_k} \delta'_{\overline{mA}} f'_{\overline{mA}z_m} + Q''_A(z), \ \partial_{z_1} Q''_A(z) = 0.$$

Let $R''_A = \sum\limits_{m=2}^{P_k} \delta'_{\overline{mA}} f'_{\overline{mA}z_m}$, then we have

$$f''_A(z) = \overline{T}_1 R''_A + Q''_A(z). \tag{2.5}$$

Moreover applying (2.3), we get

$$R''_A = \sum_{m=2}^{P_k} \delta'_{\overline{mA}} f'_{\overline{mA}z_m} = \frac{1}{V(L_k)} \int\limits_{L_k} \sum_{m=2}^{p_k} P_{k\,z_m}(z, \xi) \delta'_{\overline{mA}} \varphi'_{\overline{mA}}(\xi) \dot{\xi}. \tag{2.6}$$

Thus R''_A can be found from (2.6).

Substituting (2.5) into the second expression of (2.4), we have

$$\begin{aligned} f'_{Az_1} &= \sum_{m=2}^{P_k} \delta''_{\overline{mA}} [\overline{T}_1 R''_{\overline{mA}z_m}(z) + Q''_{\overline{mA}z_m}(z)] \\ &= \sum_{m=2}^{P_k} \delta''_{\overline{mA}} \overline{T}_1 R''_{\overline{mA}z_m}(z) + \sum_{m=2}^{P_k} \delta''_{\overline{mA}} Q''_{\overline{mA}z_m}(z); \end{aligned}$$

the above formula can be expressed as

$$f'_{Az_1} = H''_A(z) + \sum_{m=2}^{P_k} \delta''_{\overline{mA}} Q''_{\overline{mA}z_m}(z). \tag{2.7}$$

Noting that $f(z)$ is a solution of equation (2.1), we see that $f''_A(z)$ satisfies

$$\triangle_k f''_A(z) = 0, \;\; z \in R_k, 1 \le k \le 4. \tag{2.8}$$

Substituting (2.5) into (2.8), the equality

$$\triangle_k(\overline{T}_1 R''_A(z) + Q''_A(z)) = 0 \tag{2.9}$$

is derived. Applying Theorem 1.2, from (2.9) and (2.6), we can find $\overline{T}_1 R''_A(z) + Q''_A(z)$ and $R''_A(z)$, thus $\overline{T}_1 R''_A(z)$ can also be obtained. In addition, from (2.9) we can find the general solution of $Q''_A(z)$ satisfying $\partial_{z_1} Q''_A = 0$. In brief, the solution of Problem D_k for equation (2.1) in four classical domains includes the expression

$$f(z) = \sum_A{}' f'_A(z) e'_A + \sum_A{}'' f''_A(z) e''_A, \tag{2.10}$$

in which $f'_A(z), f''_A(z)$ are given by (2.3), (2.5), (2.6), and $Q''_A(z)$ in (2.5), given by (2.7) and (2.9), is an analytic function with respect to $\overline{z}_1$ [77].

Inversely, we verify that the function given by (2.3), (2.5), (2.6), (2.7), (2.9), (2.10) and

$$\partial_{z_1} Q''_A(z) = 0, \; z \in R_k, \; 1 \le k \le 4 \tag{2.11}$$

is just a solution of Problem D_k. In fact, by (2.5),(2.6),(2.11), we have

$$\begin{aligned} f''_{Az_1} &= (\overline{T}_1 R''_A(z) + Q''_A(z))_{z_1} = (\overline{T}_1 R''_A(z))_{z_1} \\ &= R''_A(z) = \sum_{m=2}^{p_k} \delta'_{\overline{mA}} f'_{\overline{mA}z_m}, \end{aligned}$$

and by (2.5), (2.7), we can derive

$$\begin{aligned} f'_{Az_1} &= H''_A(z) + \sum_{m=2}^{P_k} \delta''_{\overline{mA}} Q''_{\overline{mA}z_m}(z) \\ &= \sum_{m=2}^{P_k} \delta''_{\overline{mA}} \left(\overline{T}_1 R''_{\overline{mA}z_m}(z) + Q''_{\overline{mA}z_m}(z)\right) \\ &= \sum_{m=2}^{P_k} \delta''_{\overline{mA}} \left(\overline{T}_1 R''_{\overline{mA}}(z) + Q''_{\overline{mA}}(z)\right)_{z_m} \\ &= \sum_{m=2}^{P_k} \delta''_{\overline{mA}} f''_{\overline{mA}z_m}. \end{aligned}$$

In addition, from (2.4) we know that $f(z) = \sum_A{}' f'_A(z)e'_A + \sum_A{}'' f''_A(z)e''_A$ is regular in R_k. Because $f^{(1)}\big|_{L_k} = J_1 f(\xi) = \sum_A{}' f'_A(\xi)e'_A$, and by using Theorem 1.2, we conclude that (2.3) is the expression of solutions of boundary value problems for $\triangle_k f'_A(z) = 0\,(z \in R_k,\ 1 \le k \le 4)$. Hence $f'_A(\xi) = \varphi'_A(\xi)\,(\xi \in L_k)$, $\triangle_k f'_A(z) = 0$, and $f^{(1)}\big|_{L_k} = \sum_A{}' \varphi'_A(\xi)e'_A$. Finally by (2.9) and (2.5), we have $\triangle_k f''_A(z) = \triangle_k(\overline{T}_1 R''_A(z) + Q''_A(z)) = 0$. Consequently, for every arbitrary index $A, \triangle_k f_A(z) = 0\,(z \in R_k)$, we can derive $\triangle_k \sum_A f_A(z)e_A = \sum_A(\triangle_k f_A(z))e_A = 0$. Thus the function $f(z) = \sum_A f_A(z)e_A$ satisfies equation (2.1). This shows that $f(z)$ is a solution of Problem D_k in four classical domains $R_k\,(1 \le k \le 4)$ for equation (2.1). Therefore, we have the following theorem.

Theorem 2.1 *If $\varphi'_A(\xi)$ on characteristic manifolds L_k $(1 \le k \le 4)$ is continuous, then Problem D_k of Dirichlet type in four classical domains for equation (2.1) is solvable, and its solution can be expressed in the form*

$$f(z) = \sum_A{}' f'_A(z)e'_A + \sum_A{}'' f''_A(z)e''_A,$$

$$f'_A(z) = \frac{1}{V(L_k)}\int_{L_k} \varphi'_A(\xi)P_k(z,\xi)\dot{\xi},\ \ f''_A(z) = \overline{T}_1 R''_A(z) + Q''_A(z),$$

where

$$P_k(z,\xi) = \frac{|H_k(z,\overline{\xi})|^2}{H_k(z,\overline{z})},\ R''_A(z) = \frac{1}{V(L_k)}\int_{L_k}\sum_{m=2}^{p_k} P_{k\,z_m}(z,\xi)\delta'_{\overline{mA}}\varphi'_{\overline{mA}}(\xi)\dot{\xi},$$

when $w \in R_k$ is fixed, $H_k(z,\overline{w})$ in R_k is an analytic function with respect to z satisfying $\overline{H_k(z,\overline{w})} = H_k(w,\overline{z})\,(1 \le k \le 4)$, $Q''_A(z)$ is an analytic function about $\overline{z}_1$ satisfying $f'_{Az_1} = H''_A(z) + \sum_{m=2}^{P_k} \delta''_{\overline{mA}} Q''_{\overline{mA}z_m}(z)$, $\triangle_k(\overline{T}_1 R''_A(z) + Q''_A(z)) = 0\,(z \in R_k)$, and $\delta'_{\overline{mA}}$, $(\delta''_{\overline{mA}})$ is the sign of quasi-permutations, when $\overline{mA}$ is taken as the first (second) suffix; the operator $\overline{T}_1$ is stated as before, and can be found in [77].

In order to further investigate the uniqueness of the solution, we first consider the section domain of R_k. Denote by R_k^a the domain on the z_1-plane in R_k which is cut by the plane $S : (z_1, z_2 = a_2, z_3 = a_3, ..., z_{p_k} = a_{p_k})$, and L_k^a by the boundary of R_k^a. Set $b_a = (b_1, a_2, a_3, ..., a_{p_k})$, and let $b_a \in R_k^a$. From Section 1, we know that the four classical domains $R_k (1 \le k \le 4)$ are all circular domains, so R_k^a is also the circular domain on the z_1-plane. For a given complex constant $d''_{Aa} \in \mathbf{C}$, let $\sum\limits_A' \varphi'_A(\xi)e'_A$ on L_k and the real value function $\psi''_A(\xi_a)$ on L^a_k be continuous respectively, where $\xi_a = (\xi_1, a_2, a_3, ..., a_{p_k}) \in L^a_k\,(1 \le k \le 4)$. In the following, we find a regular function $f(z) = \sum\limits_A' f'_A(z)e'_A + \sum\limits_A'' f''_A(z)e''_A\,(z \in R_k)$, such that it is continuous in $\overline{R}_k$, and satisfies equation (2.1) and the pseudo-modified boundary conditions:

$$\begin{cases} J_1 f|_{L_k} = J_1 f(\xi) = \sum\limits_A' \varphi'_A(\xi)e'_A,\ \xi \in L_k, \\ \mathrm{Re} f''_A|_{L^a_k} = \psi''_A(\xi_a) + h''_A(\xi_a),\ \xi_a \in L^a_k, \\ f''_A(b_a) = d''_{Aa}, \end{cases}$$

in which $h''_A(\xi_a) \equiv h''_{Aa}\,(\xi_a \in L^a_k)$ is a real constant to be determined appropriately, and $\mathrm{Re} f''_A$ is the real part of f''_A. The above pseudo-modified problem in four classical domains will be called Problem D^*_k about equation (2.1), and denoted by Problem D^*_k for short.

Theorem 2.2 *Let a complex constant $d''_{Aa} \in \mathbf{C}$, and $\sum\limits_A' \varphi'_A(\xi)e'_A$ on L_k and the real-valued function $\psi''_A(\xi_a)$ on $L^a_k\,(1 \le k \le 4)$ be continuous respectively. Then Problem D^*_k has a unique solution, which can be given by (2.10), (2.3), (2.5), (2.6), (2.11), (2.7), (2.9) and*

$$\mathrm{Re} Q''_A(\xi_a) = -\mathrm{Re}[\overline{T}_1 R''_A(\xi_a)] + \psi''_A(\xi_a) + h''_A(\xi_a),\ \xi_a \in L^a_k, \tag{2.12}$$

$$Q''_A(b_a) = -\overline{T}_1 R''_A(b_a) + d''_{Aa}. \tag{2.13}$$

Proof On the basis of the proof of Theorem 2.1, it is sufficient to give the following supplement to the proof of Theorem 2.1. Firstly, we find the integral expression of the solution. Let $f(z)$ be the solution of Problem D^*_k. By (2.5) and the boundary condition, we have

$$\mathrm{Re}[\overline{T}_1 R''_A(\xi_a) + Q''_A(\xi_a)] = \mathrm{Re}[f''_A(\xi_a)] = \psi''_A(\xi_a) + h''_A(\xi_a),\ \xi_a \in L^a_k,$$

$$\overline{T}_1 R''_A(b_a) + Q''_A(b_a) = f''_A(b_a) = d''_{Aa}.$$

Using (2.11), we get $\partial_{\overline{z}} \overline{Q''_A(z)} = 0$, so $\overline{Q''_A(z_a)}$ on $\overline{R^a_k}$ satisfies

$$\begin{cases} \partial_{\overline{z}_1} \overline{Q''_A}(z_a) = 0,\ z_a \in R^a_k,\ 1 \le k \le 4, \\ \mathrm{Re}\overline{Q''_A(\xi_a)} = \mathrm{Re} Q''_A(\xi_a) \\ = -\mathrm{Re}[\overline{T}_1 R''_A(\xi_a)] + \psi''_A(\xi_a) + h''_A(\xi_a),\ \xi_a \in L^a_k, \end{cases} \tag{2.14}$$

$$\overline{Q''_A(b_a)} = -\overline{\overline{T}_1 R''_A(b_a) + d''_{Aa}}. \tag{2.15}$$

By the existence and uniqueness of the solution of the modified Dirichlet problem for analytic functions with one complex variable [80]7), and from (2.14),(2.15), we can find $\overline{Q''_A}(z_a)$ $(z_a \in R^a_k)$; moreover by the arbitrariness of a, we immediately obtain $\overline{Q''_A}(z)$ $(z \in R_k)$, and then $Q''_A(z)$ $(z \in R_k,\ 1 \le k \le 4)$ is found. In addition by (2.14) and (2.15), it is easy to get (2.12), (2.13). In brief, if $f(z)$ is a solution of Problem D^*_k, then we have the expressions (2.10), (2.3), (2.5), (2.6), (2.7), (2.9), (2.11), (2.12), and (2.13).

Secondly, we verify that the function $f(z)$ determined by (2.10),(2.3), (2.5),(2.6),(2.7),(2.9),(2.11),(2.12) and (2.13) is a solution of Problem D^*_k. In fact, by using (2.5) and (2.12), we have

$$\begin{gathered}\operatorname{Re} f''_A|_{L^a_k} = \{\operatorname{Re}[\overline{T}_1 R''_A + Q''_A]\}\Big|_{L^a_k} = \operatorname{Re}[\overline{T}_1 R''_A(\xi_a)] \\ +\operatorname{Re}[Q''_A(\xi_a)] = \psi''_A(\xi_a) + h''_A(\xi_a),\ \xi_a \in L^a_k,\end{gathered}$$

Moreover by (2.5) and (2.13), we get

$$f''_A(b_a) = \overline{T}_1 R''_A(b_a) + Q''_A(b_a) = d''_{Aa}.$$

Hence the above function $f(z)$ is just a solution of Problem D^*_k.

Finally we prove the uniqueness of solutions of Problem D^*_k $(1 \le k \le 4)$. Let $f_i(z)$ $(i = 1, 2)$ be two solutions of Problem D^*_k. Denoting $F(z) = f_1(z) - F_2(z)$, it is easy to see that $F(z)$ is a solution of the corresponding homogeneous problem (Problem D^*_{k0}), and the solution is regular in R_k and satisfies

$$J_1 F|_{L_k} = \sum_A{}' \varphi'_A(\xi)e'_A - \sum_A{}' \varphi'_A(\xi)e'_A = 0,\ \xi \in L_k.$$

For convenience, in the following we still denote $f(z)$ by $F(z)$, i.e. $F(z) = \sum_A{}' f'_A(z)e'_A + \sum_A{}'' f''_A(z)e''_A$. Noting that $F(z)$ is a solution of Problem D^*_{k0}, it is clear that for every arbitrary index A, every $f'_A|_{L_k}$ satisfies $\triangle_k f'_A(z) = 0\,(z \in R_k)$. From $J_1 F|_{L_k} = \sum_A{}' f'_A(z)e'_A|_{L_k} = 0$, it follows that $f'_A|_{L_k} = 0$. By using Theorem 1.2, we get $f'_A(z) \equiv 0\,(z \in R_k)$. Similarly we have $J_1 F(z) \equiv 0\,(z \in R_k)$, and then $f'_{\overline{m}A} \equiv 0\,(z \in R_k)$. By the definition of R''_A, we immediately obtain $R''_A \equiv 0\,(z \in R_k)$. Consequently,

$$f''_A(z) = \overline{T}_1 R''_A + Q''_A = Q''_A(z),\ z \in R_k. \tag{2.16}$$

Moreover, since $F(z)$ is a solution of Problem D^*_{k0}, by (2.11), (2.14), (2.15), we have

$$\begin{cases} \partial_{\bar{z}_1}\overline{Q''_A(z)} = \overline{\partial_{z_1}Q''_A(z)} = 0, \ z \in R^a_k, \\ \mathrm{Re}\overline{Q''_A(\xi_a)} = h''_A(\xi_a), \ \xi \in L^a_k, \\ \overline{Q''_A(b_a)} = 0, \ b_a \in R^a_k. \end{cases}$$

In accordance with the existence and uniqueness of solutions for the modified Dirichlet problem for analytic functions [80]7), we see that $\overline{Q''_A(z)} = 0$, $(z \in R^a_k)$. Noting the arbitrariness of a, we have $\overline{Q''_A(z)} \equiv 0\,(z \in R_k)$, and then $Q''_A(z) \equiv 0\,(z \in R_k)$. Using (2.16), we get $f''_A(z) \equiv 0\,(z \in R_k)$, hence $J_2F(z) \equiv 0\,(z \in R_k)$, and $F(z) \equiv 0\,(z \in R_k)$, this shows that $f_1(z) \equiv f_2(z)\,(z \in R_k)$. Therefore Problem D^*_k has at most one solution. The proof of Theorem 2.2 is finished.

3 A Pseudo-Modified Boundary Value Problem of Second Order Real Partial Differential Equations for a Hyperball in Real Clifford Analysis

In this section, we discuss the first kind of function $f(x) : \mathbf{R}^n \to \mathcal{A}_n(\mathbf{R})$ in Clifford analysis, which was introduced in Section 2, Chapter 1 (see [29]3)).

In Section 1, Chapter 2, we not only rewrite the members $\sum\limits_A x_A e_A$ in $\mathcal{A}$ as $\sum\limits_A x_A e_A = \sum\limits_B I_B e_B$, where $x_A \in \mathbf{R}$, $I_B \in \mathbf{C}$, $B = \{\alpha_1, \alpha_2, ..., \alpha_h\} \subseteq \{1, 3, 4, ..., n\}\,(1 \le \alpha_1 < \alpha_2 < \cdots < \alpha_h \le n)$, but also use the "quasi-permutation" to give the sufficient and necessary condition of the regular function $f(x)$:

$$\begin{cases} \overline{\partial}_{12} I''_B = \sum\limits_{m=3}^{n} \delta'_{\overline{mB}} \overline{I}'_{\overline{mB}x_m}, \\ \overline{\partial}_{12} I'_B = \sum\limits_{m=3}^{n} \delta''_{\overline{mB}} \overline{I}''_{\overline{mB}x_m}, \end{cases} \tag{3.1}$$

where the operator

$$\overline{\partial}_{i-1\,i} = e_{i-1}\frac{\partial}{\partial x_{i-1}} + e_i\frac{\partial}{\partial x_i};\ \partial_{i-1\,i} = e_{i-1}\frac{\partial}{\partial x_{i-1}} - e_i\frac{\partial}{\partial x_i},\ 2 \le i \le n.$$

Now we consider the partial differential equation of second order

$$(\overline{\partial}_{12}+\partial_{12})[(1-xx')^{2-n}(\overline{\partial}_{12}+\partial_{12})f(x)]$$

$$-\sum_{i=2}^{n}(\overline{\partial}_{i-1\,i}-\partial_{i-1\,i})[(1-xx')^{2-n}(\overline{\partial}_{i-1\,i}-\partial_{i-1\,i})f(x)]=0,\ x\in R^n, \tag{3.2}$$

where $x'=x^T$ is represented by the transposition of $x=(x_1,x_2,...,x_n)$.

Denote by $D:xx'<1$ the unit ball, and $L:xx'=1$ the unit sphere; its surface area is $\omega_n=2\pi^{\frac{n}{2}}/\Gamma(\frac{n}{2})$. Suppose that $\sum_B'' u_B''(\xi)e_B''$ on L is continuous. Then we shall find a regular function $f(z)=f^{(1)}+f^{(2)}=\sum_B' I_B'e_B'+\sum_B'' I_B''e_B''$, which is continuous in $\overline{D}$ and $f^{(2)}$ (in Section 3, Chapter 1, we call it the second part of f) in D satisfies (3.2) and

$$f^{(2)}\Big|_L=J_2f(\xi)=\sum_B{}'' u_B''(\xi)e_B''\ \text{on}\ L:\xi\xi'=1.$$

The above boundary value problem in the unit ball will be called Problem A for equation (3.2).

In order to give the integral expression of the solution of Problem A, let

$$f(z)=\sum_B{}' I_B'e_B'+\sum_B{}'' I_B''e_B'' \tag{3.3}$$

be the solution of the above problem. Then we can derive

$$(\overline{\partial}_{12}+\partial_{12})\left[(1-xx')^{2-n}(\overline{\partial}_{12}+\partial_{12})f^{(2)}(x)\right]$$

$$-\sum_{i=2}^{n}(\overline{\partial}_{i-1\,i}-\partial_{i-1\,i})\left[(1-xx')^{2-n}(\overline{\partial}_{i-1\,i}-\partial_{i-1\,i})f^{(2)}(x)\right]=0,$$

that is

$$\frac{\overline{\partial}_{12}+\partial_{12}}{2}\left[(1-xx')^{2-n}\frac{\overline{\partial}_{12}+\partial_{12}}{2}f^{(2)}(x)\right]$$

$$+\sum_{i=2}^{n}\frac{(-e_i)(\overline{\partial}_{i-1\,i}-\partial_{i-1\,i})}{2}$$

$$\times\left[(1-xx')^{2-n}\frac{(-e_i)(\overline{\partial}_{i-1\,i}-\partial_{i-1\,i})}{2}f^{(2)}(x)\right]=0,$$

hence

$$\frac{\partial}{\partial x_i}\left[(1-xx')^{2-n}\frac{\partial}{\partial x_i}f^{(2)}(x)\right]+\sum_{i=2}^{n}\frac{\partial}{\partial x_i}\left[(1-xx')^{2-n}\frac{\partial}{\partial x_i}f^{(2)}(x)\right]=0,$$

namely

$$\sum_{i=1}^{n} \frac{\partial}{\partial x_i}\left[(1-xx')^{2-n}\frac{\partial}{\partial x_i}\left(\sum_B{}'' I''_B e''_B\right)\right] = 0.$$

Thus we have

$$\sum_B{}'' \left\{\sum_{i=1}^{n} \frac{\partial}{\partial x_i}\left[(1-xx')^{2-n}\frac{\partial}{\partial x_i} I''_B\right] e''_B\right\} = 0,$$

and then

$$(1-xx')^n \sum_{i=1}^{n} \frac{\partial}{\partial x_i}\left[(1-xx')^{2-n}\frac{\partial I''_B}{\partial x_i}\right] = 0. \tag{3.4}$$

Using the above results, we see that I''_B is a solution of the Dirichlet problem for equation (3.4), and satisfies $I''_B(\xi) = u''_B(\xi)\,(\xi \in L)$. Applying the result of Luogeng Hua [26]2), we get

$$I''_B(x) = \frac{1}{\omega_n} \underbrace{\int\cdots\int}_{\xi\xi'=1} \left(\frac{1-xx'}{1-2x\xi'+xx'}\right)^{n-1} u''_B(\xi)\dot{\xi}, \tag{3.5}$$

where $P(x,\xi) = \left(\dfrac{1-xx'}{1-2x\xi'+xx'}\right)^{n-1}$ is the Poisson kernel.

Let

$$z_{12} = x_1 + x_2 e_2,\ \xi_{12} = \xi_1 + \xi_2 e_2,$$

$$\partial_{z_{12}} = \frac{1}{2}\left(\frac{\partial}{\partial x_1} - e_2\frac{\partial}{\partial x_2}\right),\ \partial_{\overline{z}_{12}} = \frac{1}{2}\left(\frac{\partial}{\partial x_1} + e_2\frac{\partial}{\partial x_2}\right).$$

From the above formula, we can conclude $\partial_{12} = 2\partial_{z_{12}}$, $\overline{\partial}_{12} = 2\partial_{\overline{z}_{12}}$, moreover we introduce the operators T_{12}, $\overline{T}_{12}$ as follows:

$$T_{12} f_B(x) = \frac{1}{\pi}\int\int_{\xi_1^2+\xi_2^2<1-x_3^2-\cdots-x_n^2} \frac{f_B(\xi_1,\xi_2,x_3,x_4,...,x_n)}{z_{12}-\zeta_{12}} d\xi_1 d\xi_2,$$

$$\overline{T}_{12} f_B(x) = \frac{1}{\pi}\int\int_{\xi_1^2+\xi_2^2<1-x_3^2-\cdots-x_n^2} \frac{f_B(\xi_1,\xi_2,x_3,x_4,...,x_n)}{\overline{z}_{12}-\overline{\zeta}_{12}} d\xi_1 d\xi_2,$$

in which $xx' < 1$. By the results in [77] and the second formula of (3.1), we get

$$I'_B = \frac{1}{2} T_{12} \sum_{m=3}^{n} \delta''_{mB} \overline{I}''_{mBx_m} + Q'_B(x), \tag{3.6}$$

herein $Q'_B(x)$ satisfies

$$\partial_{\bar{z}_{12}} Q'_B(x) = 0. \tag{3.7}$$

Since $(x\xi')_{x_m} = \xi_m$, $(xx')_{x_m} = 2x_m$, we see that

$$P_{x_m} = \frac{2(n-1)(1-xx')^{n-2}[2x_m(x\xi'-1)+\xi_m(1-xx')]}{(1-2x\xi'+xx')^n}.$$

Consequently, by (3.5) we obtain

$$R'_B(x) = \frac{(n-1)(1-xx')^{n-2}}{\omega_n} \underbrace{\int \cdots \int}_{\xi\xi'=1} \sum_{m=3}^{n} \frac{[2x_m(x\xi'-1)+\xi_m(1-xx')]}{(1-2x\xi'+xx')^n} \delta''_{\overline{mB}} \overline{u''_{\overline{mB}}}(\xi)\dot{\xi}, \tag{3.8}$$

and then

$$I'_B(x) = T_{12} R'_B(x) + Q'_B(x). \tag{3.9}$$

Putting (3.9) into the first formula of (3.1), we can conclude

$$\overline{\partial}_{12} I''_B(x) = \sum_{k=3}^{n} \delta'_{\overline{kB}} [\overline{T}_{12} \overline{R'_{\overline{kB}x_k}}(x) + \overline{Q'_{\overline{kB}x_k}}(x)]. \tag{3.10}$$

By (3.8), (3.5), we know that $R'_{\overline{kB}}$ and I''_B can be obtained by the known function u''_B, hence in (3.10), only $Q'_{\overline{kB}}(x)$ is unknown. So it suffices to find $Q'_B(x)$ from (3.10), (3.7).

If n is an even integer, then by a substitution analogous to $z_{12} = x_1 + x_2 e_2$, we can write $Q'_{\overline{KB}x_k}(x)$ in (3.10) as a complex overdetermined system of first order about Q'_B; for simplicity, we can assume that (3.10) is the complex form. Hence by the result of W. Tutschke [75], for the overdetermined systems (3.10) and (3.7) with appropriate conditions, is called Condition C, we can find Q'_B. Because of page limitations, we do not recount this in detail. Moreover we can obtain I'_B by (3.9), and then the solution of Problem A:

$$f(z) = \sum_B{}' I'_B e'_B + \sum_B{}'' I''_B e''_B$$

is found. Thus we have the following theorem.

Theorem 3.1. *Let* $\sum_B{}'' u''_B e''_B$ *on* L *be continuous and* n *be an even integer. Then if Condition* C *is satisfied, Problem* A *is solvable, and its* *solution can be expressed as*

$$f(z) = \sum_B{}' I'_B(x)e'_B + \sum_B{}'' I''_B(x)e''_B,$$

$$I''_B(x) = \frac{1}{\omega_n} \underbrace{\int \cdots \int}_{\xi\xi'=1} P(x,\xi)u''_B(\xi)\dot{\xi},$$

$$I'_B(x) = T_{12}R'_B(x) + Q'_B(x),$$

where $P(x,\xi) = \left(\dfrac{1-xx'}{1-2x\xi'+xx'}\right)^{n-1}$ *and*

$$R'_B(x) = \frac{1}{2\omega_n} \underbrace{\int \cdots \int}_{\xi\xi'=1} \sum_{m=3}^{n} P_{x_m}\delta''_{\overline{mB}}\overline{u''_{\overline{mB}}(\xi)}\dot{\xi},$$

in which Q'_B *is a generalized analytic function about* z_{12} *and satisfies*

$$\overline{\partial}_{12}I''_B(x) = \sum_{k=3}^{n} \delta'_{\overline{kB}}[\overline{T}_{12}\overline{R'_{\overline{kB}x_k}(x)} + \overline{Q'_{\overline{kB}x_k}(x)}],$$

where $\delta_{\overline{mB}}$ *is the sign of quasi-permutation* $\overline{mB}$.

In order to discuss the uniqueness of the solution of Problem A, we first consider the sectional domain of D. Denote by $G_a : x_a x'_a = x_1^2 + x_2^2 + \sum_{m=3}^{n} a_m^2 < 1$ the domain in D and on the x_1x_2-plane, which is cut by the plane $S : (x_1, x_2, x_3 = a_3, \ldots, x_n = a_n)\,(n \geq 3)$, where $x_a = (x_1, x_2, a_3, a_4, \ldots, a_n)$, and by $\Gamma_a : \xi_a\xi'_a = 1$ the boundary of G_a with center at $O_a = (0, 0, a_3, a_4, \ldots, a_n)$. Now we give a fixed $d'_{Ba} \in \mathbf{C}$, and assume that $\sum_B{}'' u''_B(\xi)e''_B$ on L and $\varphi'_B(\xi_a)$ on Γ_a are continuous respectively. In the following we seek a regular function

$$f(x) = \sum_B{}' I'_B e'_B + \sum_B{}'' I''_B e''_B \, (x \in D),$$

such that it is continuous in $\overline{D}$, and satisfies equation (3.2) and the pseudo-modified boundary conditions

$$\begin{cases} J_2 f|_L = J_2 f(\xi) = \sum_B{}'' u''_B(\xi)e''_B, \ \xi \in L, \\ \mathrm{Re} I'_B|_{\Gamma_a} = \varphi'_B(\xi_a) + h'_B(\xi_a), \ \xi_a \in \Gamma_a, \\ I'_B(O_a) = d'_{Ba}, \end{cases} \tag{3.11}$$

where $h'_B(\xi_a) = h'_{Ba}$ $(\xi_a \in \Gamma_a)$ is an undetermined real constant. The above pseudo-modified boundary value problem in the unit ball is denoted by Problem A^* about equation (3.2).

In 2003 Huang Min, Huang Sha have proved the existence and uniqueness of Problem A^* (see Advances in Natural Science, (2003) 13(4), (Chinese)).

Theorem 3.2 *Under the same conditions as in Theorem 3.1, if $\varphi'_B(\xi_a)$ on Γ_a is continuous, and $d'_{Ba} \in C$ is given, then Problem A^* has a unique solution, which can be obtained by (3.3),(3.5),(3.9),(3.8),(3.10) and*

$$\mathrm{Re}Q'_B(\xi_a) = -\mathrm{Re}[T_{12}R'_B(\xi_a)] + \varphi'_B(\xi_a) + h'_B(\xi_a),\ \xi_a \in \Gamma_a, \tag{3.12}$$

$$Q'_B(O_a) = -T_{12}R'_B(O_a) + d'_{Ba}, \partial_{\overline{z}_{12}}Q'_B(x) = 0,\ x \in D. \tag{3.13}$$

(see [29]3)).

Proof On the basis of the proof of Theorem 3.1, it is evident that we only need give the following supplement to the proof of Theorem 3.1.

Firstly, we need to find the integral expression of the solution. Let $f(z)$ be a solution of problem A^*. By (3.9) and the pseudo-modified boundary condition (3.11), we get

$$\mathrm{Re}[T_{12}R'_B(\xi_a) + Q'_B(\xi_a)] = \mathrm{Re}[I'_B(\xi_a)] = \varphi'_B(\xi_a) + h'_B(\xi_a),\ \xi_a \in \Gamma_a,$$
$$T_{12}R'_B(O_a) + Q'_B(O_a) = I'_B(O_a) = d'_{Ba},$$

and $\partial_{\overline{z}_{12}}Q'_B(x) = 0\,(x \in D)$, hence $Q'_B(x_a)$ on $\overline{G_a}$ satisfies (3.12),(3.13), and $Q'_B(x_a)$ is an analytic function about z_{12}. Using the existence and uniqueness of the solution of the pseudo-modified boundary value problem for analytic functions [80]7), we can obtain an analytic function $Q'_B(x_a)\,(x_a \in G_a)$ from (3.12) and (3.13). Due to the arbitrariness of a, we can get $Q'_B(x)\,(x \in D)$. In other words, if $f(x)$ is a solution of Problem A^*, then from Theorem 3.1 we can obtain the solutions of Problem A^*, which can be expressed by (3.3), (3.5), (3.9), (3.8), (3.10), (3.12) and (3.13).

Secondly, we prove that the function $f(x)$ determined by (3.3), (3.5), (3.9), (3.8), (3.10), (3.12) and (3.13) is a solution of Problem A^*. In fact, by (3.9) and (3.12) we have $\mathrm{Re}I'_B|_{\Gamma_a} = \{\mathrm{Re}[T_{12}(R'_B) + Q'_B]\}|_{\Gamma_a} =$

$\mathrm{Re}[T_{12}R'_B(\xi_a)] + \mathrm{Re}[Q'_B(\xi_a)] = \varphi'_B(\xi_a) + h'_B(\xi_a), \xi_a \in \Gamma_a$. Moreover by (3.9) and (3.13), we get $I'_B(O_a) = T_{12}R'_B(O_a) + Q'_B(O_a) = d'_{Ba}$. Hence, the above function $f(x)$ is just the solution of the pseudo-modified problem (Problem A^*).

Finally, we prove that the solution of Problem A^* is unique. Let $f_1(x), f_2(x)$ be two solutions of Problem A^*. It is clear that $F(x) = f_1(x) - f_2(x)$ is a solution of the corresponding homogenous problem (Problem A_0^*), which is regular in D and satisfies $F^{(2)}|_L = J_2F(\xi) = \sum_B'' u''_B(\xi)e''_B - \sum_B'' u''_B(\xi)e''_B = 0$.

For convenience, in the following, we still adopt the notation about $f(x)$ as in the proof of Theorem 3.1. For instance, we still denote

$$F(x) = \sum_B{}' I'_B e'_B + \sum_B{}'' I''_B e''_B$$

and so on. Because $F(x)$ is a solution of Problem A_0^* satisfying (3.2), consequently, for every B, I''_B satisfies equation (3.2). That is to say, every I''_B in D is harmonic. Then by $F^{(2)}|_L = \sum_B'' I''_B e''_B|_L = 0$, the equality $I''_B|_L = 0$ is derived. By the uniqueness of the solution of the Dirichlet problem for harmonic functions about (3.2) [26]2), we get $I''_B = 0$ and $I''_{\overline{mB}} \equiv 0$ in D. By using (3.5), (3.8), it is easy to see that $R'_B(x) = \frac{1}{2}\sum_{m=3}^{n} \delta''_{\overline{mB}} \overline{I}''_{\overline{mB}x_m}$, and $R'_B(x) \equiv 0$ in D. In addition by (3.9), we get

$$I'_B = T_{12}R'_B(x) + Q'_B(x) = Q'_B(x). \tag{3.14}$$

Because $F(x)$ is the solution of the corresponding homogenous problem (Problem A_0^*), from (3.12) and (3.13), it follows that

$$\begin{cases} \partial_{\bar{z}_{12}} Q'_B(x_a) = 0, \ x_a \in G_a, \\ \mathrm{Re}Q'_B(\xi_a) = h'_B(\xi_a), \ \xi_a \in \Gamma_a, \\ Q'_B(O_a) = 0, \ O_a \in \Gamma_a. \end{cases}$$

At last by the existence and the uniqueness of the solution to the modified Dirichlet problem for analytic functions [80]7), we know $Q'_B(x_a) \equiv 0\,(x_a \in G_a)$. Noting the arbitrariness of a, we have $Q'_B(x) \equiv 0\,(x \in D)$. Hence by (3.14), we get $I'_B(x) \equiv 0\,(x \in D)$. So $J_1F \equiv 0\,(x \in D)$, and $F(x) \equiv 0\,(x \in D)$, i.e. $f_1(x) \equiv f_2(x)\,(x \in D)$. This completes the proof.

In this chapter, we used quasi-permutation as a tool to construct the close connection between harmonic analysis in classical domains for several complex variables functions and Clifford analysis, in particular the boundary value problems, which symbolically establishes a bridge between complex analysis and Clifford analysis.

CHAPTER V

INTEGRALS DEPENDENT ON PARAMETERS AND SINGULAR INTEGRAL EQUATIONS IN REAL CLIFFORD ANALYSIS

In the above chapters, we have discussed two kinds of functions in real and complex analysis: $f : R^n \to \mathcal{A}_n(R)$ and $f : C^n \to \mathcal{A}_n(C)$. In the first section of this chapter, we shall introduce the third kind of function $f : R^p \to \mathcal{A}_q(C)$ and Cauchy's estimates of three integrals dependent on parameters [29]11). In the second section, firstly we give the definition of two kinds of singular integrals with Cauchy's kernel in real Clifford analysis. Moreover we find some singular integrals which can be exchanged the integral order. Finally we prove three kinds of Poincaré-Bertrand transformation formulas of singular integrals with Cauchy's kernel [29]12). On the basis of the transformation formulas as stated in the first section, in the third section we shall prove the composition formula and the inverse formula of singular integrals with Cauchy's kernel [29]13). In the fourth section we introduce the Fredholm theory of a kind of singular integral equations in real Clifford analysis by using transformation formulas, and discuss the regularization operator [29]9). In the last section according to the method of unite resolution, we define generalized integrals in the sense of M. Spivak [73] on an open manifold for unbounded functions in real Clifford analysis, and discuss the solvability and series expression of solutions for second kinds of generalized integral equations with exchangeable factors, and give the error estimate of the approximate calculation [67].

1 Cauchy's Estimates of Integrals with One Parameter in Real Clifford Analysis

Since 1965, many scholars have studied functions, integrals, and boundary value problems in real and complex Clifford analysis and other related problems. R. Delanghe in [12], F. Brackx et al in [6] and other scholars investigated the function $f : R^n \to \mathcal{A}_n(R)$ (the function which appeared early is called the first kind function). Zhengyuan Xu in [86] and Sha Huang in [29]3) discussed some boundary value problems in real Clifford analysis. F. Sommon in [72]2), J. Ryan in [68]2), and Sha Huang and Yuying Qiao in [32]2) studied theoretic properties for functions $f : C^n \to \mathcal{A}_n(C)$ (which we call the second kind functions) and corresponding boundary value problems for canonical domains in complex Clifford analysis. Kimiro Somo in [36] discussed functions $f : R^p \to \mathcal{A}_q(C)$ (which we call the third kind functions) and obtained Cauchy's estimates for monogenic and harmonic functions and their partial derivatives, which are the generalizations of Sommon's result and Aranissiann's result. In this section we discuss Cauchy's estimate for the partial derivatives of the above three integrals dependent on parameters, which are developments of the results in [36], [2], [72]3).

1.1 Cauchy's estimates of Cauchy's Kernel

Lemma 1.1 ([36]) *Let $m, n \in N$, $m^{(n)} = \Gamma(m-n+1)/\Gamma(m+1)$, $N_0 = N \cup \{0\}$, $\alpha, \beta^1, ..., \beta^n \in N_0^m$ and $\alpha = (\alpha_1, ..., \alpha_m)$, $\gamma = (\gamma_1, ..., \gamma_n) \in R^n$, $|\gamma| = \gamma_1 + \cdots + \gamma_n$, $\alpha! = \alpha_1!...\alpha_m!$, where N is the set of natural numbers. Then*

$$\sum_{\beta^1+\cdots+\beta^n=\alpha} \frac{\alpha!}{\beta^1!\cdots\beta^n!}\gamma_1^{(|\beta^1|)}\cdots\gamma_n^{(|\beta^n|)} = |\gamma|^{(|\alpha|)}. \tag{1.1}$$

Proof Differentiate the equality

$$\{(x_1 + \cdots + x_m)t\}^{\gamma_1}\cdots\{(x_1 + \cdots + x_m)t\}^{\gamma_n} = \{(x_1 + \cdots + x_m)t\}^{|\gamma|}$$

for $|\alpha|$ times with respect to t. Let $t = (x_1 + \cdots + x_m)^{-1}$, then expand it and form coefficients of x^α. Thus we obtain the identity (1.1).

Lemma 1.2 ([36]) *If $p - 1 \in N$, $n \in Z$, $\alpha \in N_0^p$, $x \in R^p$, where Z is the set of positive integrals, then*

$$\left|\left(\frac{\partial}{\partial x}\right)^\alpha x^n\right| \le \left|n^{(|\alpha|)}\right| |x|^{n-|\alpha|}. \tag{1.2}$$

Proof When $\alpha = 0$ or $n = 0$, it is clear that (1.2) holds; therefore, in what follows we use the inductive method on α and show (1.2) for any $n \ne 0$. Assume that (1.2) holds for any $n \in Z$, and $\beta \in N_0^p$ such that $0 \le \beta \le \alpha$. Note that x^k belongs to R^p for any $k \in Z$; then if $n > 0$, by

Corollary 1.4, Chapter 1, for any $j = 1, ..., p$ we get

$$\left|\left(\frac{\partial}{\partial x}\right)^{\alpha} \frac{\partial}{\partial x_j} x^n\right| = \left|\left(\frac{\partial}{\partial x}\right)^{\alpha} \sum_{k=0}^{n-1} x^k e_j x^{n-k-1}\right|$$
$$\leq \sum_{k=0}^{n-1} \sum_{\beta+\gamma=\alpha} \frac{\alpha!}{\beta!\gamma!} \left|\left(\frac{\partial}{\partial x}\right)^{\beta} x^k\right| \left|\left(\frac{\partial}{\partial x}\right)^{\gamma} x^{n-k-1}\right|,$$

where $\beta,\, \gamma \in N_0^p$. From the hypothesis it follows that

$$\left|\left(\frac{\partial}{\partial x}\right)^{\alpha} \frac{\partial}{\partial x_j} x^n\right| \leq \sum_{k=0}^{n-1} \sum_{\beta+\gamma=\alpha} \frac{\alpha!}{\beta!\gamma!} k^{(|\beta|)} (n-k-1)^{(|\gamma|)} |x|^{n-|\alpha|-1}.$$

By Lemma 1.1, we obtain

$$\left|\left(\frac{\partial}{\partial x}\right)^{\alpha} \frac{\partial}{\partial x_j} x^n\right| \leq \sum_{k=0}^{n-1} (n-1)^{(|\alpha|)} |x|^{n-|\alpha|-1} = n^{(|\alpha|+1)} |x|^{n-|\alpha|-1}. \quad (1.3)$$

If $n < 0$, we set $m = -n$, then by Corollary 1.4, Chapter 1, for any $j = 1, ..., p$ we have

$$\left|\left(\frac{\partial}{\partial x}\right)^{\alpha} \frac{\partial}{\partial x_j} \frac{1}{x^m}\right|$$
$$= \left|\left(\frac{\partial}{\partial x}\right)^{\alpha} \sum_{k=0}^{m-1} \frac{1}{x^k} \overline{e}_j \left(\frac{\partial}{\partial x_j} \frac{1}{x \overline{e}_j}\right) \frac{1}{x^{m-k-1}}\right|$$
$$\leq \sum_{k=0}^{m-1} \left|\left(\frac{\partial}{\partial x}\right)^{\alpha} \left\{\frac{1}{x^k} \overline{e}_j \frac{-1}{(x\overline{e}_j)^2} \frac{1}{x^{m-k-1}}\right\}\right|$$
$$\leq \sum_{k=0}^{m-1} \sum_{\beta+\gamma+\delta+\varepsilon=\alpha} \frac{\alpha!}{\beta!\,\gamma!\,\delta!\,\varepsilon!} \left|\left(\frac{\partial}{\partial x}\right)^{\beta} \frac{1}{x^k}\right|$$
$$\times \left|\left(\frac{\partial}{\partial x}\right)^{\gamma} \frac{1}{x}\right| \left|\left(\frac{\partial}{\partial x}\right)^{\delta} \frac{1}{x}\right| \left|\left(\frac{\partial}{\partial x}\right)^{\varepsilon} \frac{1}{x^{m-k-1}}\right|,$$

where $\beta,\, \gamma,\, \delta,\, \varepsilon \in N_0^p$. From the hypothesis it follows that

$$\left|\left(\frac{\partial}{\partial x}\right)^{\alpha} \frac{\partial}{\partial x_j} \frac{1}{x^m}\right|$$

$$\leq \sum_{k=0}^{m-1} \sum_{\beta+\gamma+\delta+\varepsilon=\alpha} \frac{\alpha!}{\beta!\gamma!\delta!\varepsilon!} \frac{(-k)^{(|\beta|)}(-1)^{(|\gamma|)}(-1)^{(|\delta|)}(-m+k+1)^{(|\varepsilon|)}}{(-1)^{|\alpha|}|x|^{m+|\alpha|+1}}.$$

By Lemma 1.1 we obtain

$$\begin{aligned} &\left|\left(\frac{\partial}{\partial x}\right)^{\alpha} \frac{\partial}{\partial x_j} \frac{1}{x^m}\right| \\ \leq\ & \sum_{k=0}^{m-1} \frac{|(-m-1)^{(|\alpha|)}|}{|x|^{m+|\alpha|+1}} = \left|n^{(|\alpha|+1)}\right| |x|^{n-|\alpha|-1}. \end{aligned} \tag{1.4}$$

Thus, by (1.3) and (1.4), the assertion is true for any $n \neq 0$ and any $\alpha \in N_0^p$.

Theorem 1.3 *Suppose that m, $n \in N_0$, $p-1 \in N$, $\alpha \in N_0^p$, and $x \in R^p\backslash\{0\}$. Then*

$$\left|\left(\frac{\partial}{\partial x}\right)\left(\frac{1}{x^m}\frac{1}{|x|^n}\right)\right| \leq \frac{|(-m-n)^{(|\alpha|)}|}{|x|^{m+|\alpha|+n}}. \tag{1.5}$$

Proof We prove this theorem by the inductive method on α. When $\alpha = 0$, (1.5) is clearly true. Assume that (1.5) is true for $n \in N_0$, $m \in N_0$, and $\beta \in N_0^p$ such that $0 \leq \beta \leq \alpha$; for any $j = 1, ..., p$ and any $\gamma \in N_0^p$ such that $0 \leq \gamma \leq \alpha$, we obtain

$$\begin{aligned} &\left|\left(\frac{\partial}{\partial x}\right)^{\gamma} \frac{\partial}{\partial x_j} \frac{1}{|x|^n}\right| = \left|\left(\frac{\partial}{\partial x}\right)^{\gamma} \frac{-nx_j}{|x|^{n+2}}\right| \\ &= \left|\left(\frac{\partial}{\partial x}\right)^{\gamma} \frac{n[e_j\overline{x}]_1}{|x|^n x\overline{x}}\right| = \left|\left(\frac{\partial}{\partial x}\right)^{\gamma} \frac{\frac{n}{2}(e_j\overline{x} + x\overline{e}_j)}{|x|^n x\overline{x}}\right| \\ &= \frac{n}{2}\left|e_j\left(\frac{\partial}{\partial x}\right)^{\gamma} \frac{1}{x|x|^n} + \left(\frac{\partial}{\partial x}\right)^{\gamma} \frac{1}{\overline{x}|x|^n}\overline{e}_j\right| \\ &\leq n\left|\left(\frac{\partial}{\partial x}\right)^{\gamma} \frac{1}{x|x|^n}\right|. \end{aligned}$$

By the hypothesis, we have

$$\left|\left(\frac{\partial}{\partial x}\right)^{\gamma} \frac{\partial}{\partial x_j} \frac{1}{|x|^n}\right| \leq \frac{n|(-n-1)^{(|\gamma|)}|}{|x|^{|\gamma|+n+1}}.$$

Furthermore, from the definition of sign $(m)^{(n)}$ we get

$$\left|(-n)^{(|\gamma|+1)}\right| = \frac{(|\gamma|+1+n-1)!}{(n-1)!}$$

$$= n\frac{(|\gamma|+n)!}{n!} = n\left|(-n-1)^{(|\gamma|)}\right|. \tag{1.6}$$

Similarly we have

$$\left|(-m)^{(|\gamma|+1)}\right| = (|\gamma|+m)\left|(-m)^{(|\gamma|)}\right|, \tag{1.7}$$

$$\left|(-m)^{(|\gamma|)}\right| = \frac{(-m)^{(|\gamma|)}}{(-1)^{|\gamma|}}, \tag{1.8}$$

and

$$\left|\left(\frac{\partial}{\partial x}\right)^{\gamma}\frac{\partial}{\partial x_j}\frac{1}{|x|^n}\right| \leq \frac{|(-n)^{(|\gamma|+1)}|}{|x|^{|\gamma|+n+1}},\ 0 \leq \gamma \leq \alpha. \tag{1.9}$$

Again from Corollary 1.4, Chapter I, it follows for any $j = 1, ..., p$, the following inequality holds:

$$\left|\left(\frac{\partial}{\partial x}\right)^{\beta}\frac{\partial}{\partial x_j}\frac{1}{x^m}\right|$$

$$= \left|\left(\frac{\partial}{\partial x}\right)^{\beta}\sum_{k=0}^{m-1}\frac{1}{x^k}\overline{e}_j\left(\frac{\partial}{\partial x_j}\frac{1}{x\overline{e}_j}\right)\frac{1}{x^{m-k-1}}\right|$$

$$\leq \sum_{k=0}^{m-1}\left|\left(\frac{\partial}{\partial x}\right)^{\beta}\left\{\frac{1}{x^k}\overline{e}_j\frac{-1}{(x\overline{e}_j)^2}\frac{1}{x^{m-k-1}}\right\}\right|$$

$$\leq \sum_{k=0}^{m-1}\sum_{\beta^1+\cdots+\beta^4=\beta}\frac{\beta!}{\beta^1!\cdots\beta^4!}\left|\left(\frac{\partial}{\partial x}\right)^{\beta^1}\frac{1}{x^k}\right|$$

$$\times\left|\left(\frac{\partial}{\partial x}\right)^{\beta^2}\frac{1}{x}\right|\left|\left(\frac{\partial}{\partial x}\right)^{\beta^3}\frac{1}{x}\right|\left|\left(\frac{\partial}{\partial x}\right)^{\beta^4}\frac{1}{x^{m-k-1}}\right|,$$

where $\beta^i \in N_0^p$, $i = 1, ..., 4$. By means of Lemmas 1.1, 1.2 and (1.6), we can obtain

$$\begin{aligned}
&\left|\left(\frac{\partial}{\partial x}\right)^{\beta}\frac{\partial}{\partial x_j}\frac{1}{x^m}\right| \\
&\le \sum_{k=0}^{m-1} B\frac{\beta!}{\beta^1!\ldots\beta^4!}\frac{(-k)^{(|\beta^1|)}(-1)^{(|\beta^2|)}(-1)^{(|\beta^3|)}(-m+k+1)^{(|\beta^4|)}}{(-1)^{|\beta|}|x|^{m+|\beta|+1}} \qquad (1.10)\\
&\le \sum_{k=0}^{m-1}\frac{|(-m-1)^{(|\beta|)}|}{|x|^{m+|\beta|+1}} = \frac{m|(-m-1)^{(|\beta|)}|}{|x|^{m+|\beta|+1}} = \frac{|(-m)^{(|\alpha|+1)}|}{|x|^{m+|\alpha|+1}}
\end{aligned}$$

where $B = \sum_{\beta^1+\cdots+\beta^4=\beta}$. From (1.9), (1.10), Lemma 1.2 and hypothesis of induction, we get

$$\begin{aligned}
&\left|\left(\frac{\partial}{\partial x}\right)^{\alpha}\frac{\partial}{\partial x_j}\frac{1}{x^m}\frac{1}{|x|^n}\right| \\
= &\left|\left(\frac{\partial}{\partial x}\right)^{\alpha}\left[\left(\frac{\partial}{\partial x_j}\frac{1}{x^m}\right)\frac{1}{|x|^n} + \frac{1}{x^m}\left(\frac{\partial}{\partial x_j}\frac{1}{|x|^n}\right)\right]\right| \\
\le &\sum_{\beta+\gamma=\alpha}\frac{\alpha!}{\beta!\gamma!}\left\{\left|\left(\frac{\partial}{\partial x}\right)^{\beta}\frac{\partial}{\partial x_j}\frac{1}{x^m}\right|\left|\left(\frac{\partial}{\partial x}\right)^{\alpha}\frac{1}{|x|^n}\right|\right. \\
&\left.+\left|\left(\frac{\partial}{\partial x}\right)^{\beta}\frac{1}{x^m}\right|\left|\left(\frac{\partial}{\partial x}\right)^{\alpha}\frac{\partial}{\partial x_j}\frac{1}{|x|^n}\right|\right\} \\
\le &\sum_{\beta+\gamma=\alpha}\frac{\alpha!}{\beta!\gamma!}\frac{|(-m)^{(|\beta|+1)}(-n)^{(|\gamma|)}| + |(-m)^{(|\beta|)}(-n)^{(|\gamma|+1)}|}{|x|^{m+|\alpha|+n+1}}.
\end{aligned}$$

Again by (1.7), (1.8) and Lemma 1.1, we can derive the inequality

$$\begin{aligned}
&\left|\left(\frac{\partial}{\partial x}\right)\frac{\partial}{\partial x_j}\frac{1}{x^m}\frac{1}{|x|^n}\right| \\
&\le \sum_{\beta+\gamma=\alpha}\frac{\alpha!}{\beta!\gamma!}\frac{(-m)^{(|\beta|+1)}(-n)^{(|\gamma|)} + (-m)^{(|\beta|)}(-n)^{(|\gamma|+1)}}{(-1)^{|\alpha|+1}|x|^{m+|\alpha|+n+1}}
\end{aligned}$$

$$\leq \sum_{\beta+\gamma=\alpha} \frac{\alpha!}{\beta!\gamma!} \frac{(-m)^{(|\beta|)}(-n)^{(|\gamma|)}(|\beta|+m+|\gamma|+n)}{(-1)^{|\alpha|}|x|^{m+n+|\beta|+|\gamma|+1}}$$

$$\leq \frac{(-m-n)^{(|\alpha|)}(|\alpha|+m+n)}{(-1)^{|\alpha|}|x|^{m+n+|\alpha|+1}}$$

$$= \frac{(-m-n)^{(|\alpha|+1)}}{(-1)^{|\alpha|+1}|x|^{m+n+|\alpha|+1}} = \frac{|(-m-n)^{(|\alpha|+1)}|}{|x|^{m+n+|\alpha|+1}}.$$

Theorem 1.3 is an improvement of the corresponding results in [36], [72]3).

1.2 Cauchy's estimates of three integrals dependent on a parameter in Clifford analysis

We discuss the third kind of function $f : R^p \to \mathcal{A}_q(C)$. Let $B_r = \{x \mid x \in R^p, |x| < r\}$, $\partial B_r = \{x \mid x \in R^p, |x| = r\}$. The set C^1-function in ∂B_r with values in $\mathcal{A}_q(C)$ is denoted by $C^1(\partial B_r, \mathcal{A}_q(C))$. If $f, u \in C^1(\partial B_r, \mathcal{A}_q(C))$, we discuss the following three integrals depending on one parameter

$$F_{mn}(x) = \frac{1}{\omega_p}\int_{\partial B_r} f(y) d\sigma_y \frac{1}{(y-x)^m |y-x|^n},$$

$$\Phi_n(x) = \frac{1}{\omega_p}\int_{\partial B_r} \frac{u(y)}{|y-x|^n} dS_y, \; U_n(x) = \frac{r^2-|x|^2}{r}\Phi_n(x),$$

where $B_{r_1} = \{x \in R^p, |x| < r_1 < r\}$, $x \in \overline{B}_{r_1} = B_{r_1} \bigcup \partial B_{r_1}$.

Theorem 1.4 *Let $p-1, q-1 \in N$ with $p \leq q$, $m, n \in N_0$, $\alpha \in N_0^p$, $\overline{B}_r \subset R^p$, and $f(y) \in C^1(\partial B_r, \mathcal{A}_q(C))$. Then*

$$\left|\left(\frac{\partial}{\partial x}\right)^{\alpha} F_{mn}(0)\right| \leq \frac{|(-m-n)^{(|\alpha|)}|}{|\gamma|^{m+|\alpha|+n-p+1}} M_r,$$

where $M_r = \sup_{y\in\partial B_r} |f(y)|$.

Proof When $x \in \overline{B}_{r_1}$, we obtain

$$\begin{aligned}
&\left(\frac{\partial}{\partial x}\right)^{\alpha} F_{mn}(x)\\
&= \frac{1}{\omega_p}\int_{\partial B_r} f(y)d\sigma_y \left(\frac{\partial}{\partial x}\right)^{\alpha} \frac{1}{(y-x)^m|y-x|^n}\\
&= \frac{1}{\omega_p}\int_{\partial B_r} f(y)d\sigma_y \left(\frac{\partial}{\partial y}\right)^{\alpha} \frac{(-1)^{|\alpha|}}{(y-x)^m|y-x|^n}.
\end{aligned}$$

From Corollary 1.4, Chapter 1 and Lemma 1.1, we can find

$$\begin{aligned}
&\left|\left(\frac{\partial}{\partial x}\right)^{\alpha} F_{mn}(0)\right|\\
&= \frac{1}{\omega_p}\int_{\partial B_r} |f(y)||d\sigma_y| \left|\left(\frac{\partial}{\partial y}\right)^{\alpha} \frac{(-1)^{|\alpha|}}{y^m|y|^n}\right|\\
&\leq M_r \frac{1}{\omega_p}\int_{\partial B_r} |d\sigma_y| \frac{|(-m-n)^{(|\alpha|)}|}{|\gamma|^{m+n+|\alpha|}}\\
&= M_r \frac{|(-m-n)^{(|\alpha|)}|}{|\gamma|^{m+n+|\alpha|-p+1}} \frac{1}{\omega_p}\int_{\partial B_r} |d\sigma_y| \frac{1}{\gamma^{p-1}}\\
&= \frac{|(-m-n)^{(|\alpha|)}|}{|\gamma|^{m+n+|\alpha|-p+1}} M_r.
\end{aligned}$$

Theorem 1.4 is the generalization of Theorem 2 in [36].

Corollary 1.5 *Let $p-1, q-1 \in N$ with $p \leq q$, $\alpha \in N_0^p$, and f be right regular in a neighborhood of $\overline{B}_r \subset R^p$ with values in $\mathcal{A}_q(C)$. Then*

$$\left|\left(\frac{\partial}{\partial x}\right)^{\alpha} f(0)\right| \leq \frac{|(-p+1)|^{(|\alpha|)}}{|\gamma|^{|\alpha|}} M_r,$$

where $M_r = \sup\limits_{|y|=r} |f(y)|$.

Proof By Theorem 1.4 and Cauchy's integral formula [36], we have

$$\begin{aligned}
f(x) &= \frac{1}{\omega_p}\int_{\partial B_r} f(y)d\sigma_y \frac{\overline{y}-\overline{x}}{|y-x|^p}\\
&= \frac{1}{\omega_p}\int_{\partial B_r} f(y)d\sigma_y \frac{1}{(y-x)|y-x|^{p-2}},
\end{aligned}$$

which is similar to the Cauchy's integral formula in Section 2, Chapter 1, hence the result in Corollary 1.5 is valid.

Theorem 1.6 *Suppose that $p-1,\ q-1 \in N$ with $p \le q$, $n \in N_0$, $\alpha \in N_0^p$, and $u \in C^1\ (\partial B_r,\ \mathcal{A}_q(C))$, $\overline{B}_r \subset R^p$. Then*

$$\left|\left(\frac{\partial}{\partial x}\right)^{\alpha}\Phi_n(0)\right| \le \frac{|(-n)^{(\alpha)}|N_r}{|\gamma|^{n+|\alpha|-p+1}},$$

where $N_r = \sup\limits_{y\in\partial B_r} |u(y)|$.

Proof By Theorem 1.4, we immediately get the result in Theorem 1.6.

Theorem 1.7 *Under the conditions as in Theorem 1.6, we have*

$$\left|\left(\frac{\partial}{\partial x}\right)^{\alpha} U_n(0)\right| \le \frac{|(-n+1)^{(|\alpha|)}+|\alpha||(-n+1)^{(|\alpha|-1)}|}{|\gamma|^{n+|\alpha|-p}} N_r.$$

Proof Since $|y-x|^2=(y-x)(\overline{y}-\overline{x})$, from the kernel function, when $|y|>|x|$, we can find

$$\frac{|y|^2-|x|^2}{|y-x|^n} = y\frac{1}{(y-x)|y-x|^{n-2}} + \frac{1}{(\overline{y}-\overline{x})|y-x|^{n-2}}\overline{x}.$$

Putting this into $U_n(x)$, we get

$$U_n(x) = \frac{1}{\omega_p\, r}\int_{\partial B_r} u(y)y\frac{1}{(y-x)|y-x|^{n-2}}dS_y$$
$$+\frac{1}{\omega_p\, r}\int_{\partial B_r} u(y)\frac{\overline{x}}{(\overline{y}-\overline{x})|y-x|^{n-2}}dS_y.$$

Moreover we have

$$\left(\frac{\partial}{\partial x}\right)^{\alpha} U_n(x) = \frac{1}{\omega_p\, r}\int_{\partial B_r} u(y)y\left(\frac{\partial}{\partial x}\right)^{\alpha}\frac{1}{(y-x)|y-x|^{n-2}}dS_y$$
$$+\frac{1}{\omega_p\, r}\int_{\partial B_r} u(y)\left(\frac{\partial}{\partial x}\right)^{\alpha}\frac{\overline{x}}{(\overline{y}-\overline{x})|y-x|^{n-2}}dS_y.$$

Therefore,

$$\left(\frac{\partial}{\partial x}\right)^{\alpha} U_n(o) = \frac{1}{\omega_p\, r}\int_{\partial B_r} u(y)y\left(\frac{\partial}{\partial x}\right)^{\alpha}\frac{(-1)^{(|\alpha|)}}{y|y|^{n-2}}dS_y$$
$$+\sum_{j=1,\alpha\ge\lambda_j}^{p}\frac{1}{\omega_p\, r}\int_{\partial B_r} u(y)\alpha_j\left(\frac{\partial}{\partial x}\right)^{\alpha-\lambda_j}\frac{(-1)^{|\alpha|-1}}{\overline{y}|y|^{n-2}}\overline{e}_j dS_y,$$

where $\lambda_j = (\delta_{j1}, ..., \delta_{jp}) \in N_0^p$, $j = 1, ..., p$. On the basis of Corollary 1.4, Chapter 1 and Theorem 1.3, we can find

$$\begin{aligned}
&\left|\left(\frac{\partial}{\partial x}\right)^{\alpha} U_n(o)\right| \\
&= \frac{1}{\omega_p r}\int_{\partial B_r} |u(y)| \left\{ \frac{|(-n+1)^{(|\alpha|)}|}{\gamma^{|\alpha|+n-1}} r \right. \\
&\quad \left. + \sum_{j=1,\lambda_j\geq 1}^{p} \alpha_j \frac{|(-n+1)^{(|\alpha|-1)}}{\gamma^{|\alpha|+n-2}} \right\} dS_y \\
&\leq \frac{|(-n+1)^{(|\alpha|)}|+|\alpha||(-n+1)^{(|\alpha|-1)}|}{\gamma^{|\alpha|+n-p}} N_r \frac{1}{\omega_p} \int_{\partial B_r} \frac{1}{r^{p-1}} dS_y \\
&\leq \frac{|(-n+1)^{(|\alpha|)}| + |\alpha||(-n+1)^{(|\alpha|-1)}|}{\gamma^{|\alpha|+n-p}} N_r.
\end{aligned}$$

Theorem 1.7 is an improvement of the corresponding results in [36], [72]3).

Corollary 1.8 *Let $p-1$, $q-1 \in N$ with $p \leq q$, $\alpha \in N_0^p$, and $u \in C^1$ $(\partial B_r, \mathcal{A}_q(C))$ be harmonic in a neighborhood of $\overline{B}_r \subset R^p$. Then*

$$\left|\left(\frac{\partial}{\partial x}\right)^{\alpha} U_n(0)\right| \leq \frac{|(-p+1)^{(|\alpha|)} + |\alpha|(-p+1)^{(|\alpha|-1)}}{|\gamma|^{|\alpha|}} N_r.$$

Proof By Poisson's integral formula of harmonic functions (see [36])

$$U(x) = \frac{r^2 - |x|^2}{\omega_p r} \int_{\partial B_r} \frac{u(y)}{|y-x|^p} dS_y$$

and Theorem 1.7, we immediately get the result in Corollary 1.8.

2 Three Kinds of Poincaré-Bertrand Transformation Formulas of Singular Integrals with a Cauchy's Kernel in Real Clifford Analysis

Tongde Zhong and Sheng Gong have studied the Poincaré-Bertrand (P-B) transformation formulas and singular integrals in several complex variables (see [89], [88]2), [88]3), [88]4), [20]). In 1992, M. R. Kandmamov gave a new proof of P-B transformation formulas in several complex variables functions (see [35]). Under the illumination of [88]2), [35], on the basis of the results of Section 1, Chapter 3, we shall prove three kinds of P-B transformation formulas of singular integrals with Cauchy's kernel in real Clifford analysis.

2.1 Preparation

Let $\Omega_1 \subset R^m$, $\Omega_2 \subset R^k$ be differentiable, oriented, bounded, compact manifolds, whose dimensions are m, k respectively, and $\partial\Omega_i$ be the boundary of Ω_i, $i = 1, 2$. $\partial\Omega_i (i = 1, 2)$ are oriented Liapunov surfaces, and let the orientation be coordinated with that of Ω_i. Let $\partial\Omega = \partial\Omega_1 \times \partial\Omega_2$. Denote by $\partial\Omega_\eta$ the variable $\eta = (\eta_1, \eta_2)$ on $\partial\Omega$, and set $f(\eta^1, \eta^2; \xi^1, \xi^2) \in H(\partial\Omega, \beta)$, $0 < \beta < 1$ (see Section 1, Chapter 3). For convenience, we give the signs

$$\overline{A} = \frac{\overline{\eta^1} - \overline{\zeta^1}}{\omega_m|\eta^1 - \zeta^1|^m}, \quad \overline{B} = \frac{\overline{\eta^2} - \overline{\zeta^2}}{\omega_k|\eta^2 - \zeta^2|^k},$$
$$\overline{C} = \frac{\overline{\xi^1} - \overline{\eta^1}}{\omega_m|\xi^1 - \eta^1|^m}, \quad \overline{D} = \frac{\overline{\xi^2} - \overline{\eta^2}}{\omega_k|\xi^2 - \eta^2|^k},$$
$$\overline{E} = f(\eta^1, \eta^2; \xi^1, \xi^2).$$

In Section 1, Chapter 3, we have defined the Cauchy principal value of singular integrals with the Cauchy's kernel on the character manifold, which is called the one time singular integral. Similarly to several dimensional singular integrals (see [88]3)), we must give a precise definition of the two times singular integral as follows.

Definition 2.1 The two times singular integral is defined as

$$\begin{aligned}
&\int_{\partial\Omega_\eta} \overline{A} d\sigma_{\eta^1} d\sigma_{\eta^2} \overline{B} \int_{\partial\Omega_\xi} d\sigma_{\xi^1} \overline{C}\,\overline{E}\bar{D} d\sigma_{\xi^2} \\
&= \int_{\partial\Omega_\eta} \overline{A} d\sigma_{\eta^1} \left[\int_{\partial\Omega_\xi} d\sigma_{\xi^1} \bar{C}\bar{E}\bar{D} d\sigma_{\xi^2}\right] d\sigma_{\eta^2} \overline{B} \\
&= \int_{\partial\Omega_{1\eta^1}} \overline{A} d\sigma_{\eta^1} \left[\int_{\partial\Omega_{2\eta^2}} d\sigma_{\eta^2} \overline{B} \int_{\partial\Omega_\xi} d\sigma_{\xi^1} \bar{C}\bar{E}\bar{D} d\sigma_{\xi^2}\right] \\
&= \int_{\partial\Omega_{1\eta^1}} \overline{A} d\sigma_{\eta^1} \left\{\int_{\partial\Omega_{2\eta^2}} \left[\int_{\partial\Omega_{1\xi^1}} d\sigma_{\xi^1} \overline{C} \left(\int_{\partial\Omega_{2\xi^2}} \bar{E}\bar{D} d\sigma_{\xi^2}\right)\right] d\sigma_{\eta^2} \overline{B}\right\}.
\end{aligned} \tag{2.1}$$

Similarly to several dimensional singular integrals (see [88]3)), we also give the definition of a two times singular integral which has to be changed in integral order as follows

$$\begin{aligned}
&\int_{\partial\Omega_\xi} d\sigma_{\xi^1} d\sigma_{\xi^2} \int_{\partial\Omega_\eta} \overline{A} d\sigma_{\eta^1} \bar{C}\bar{E}\bar{D} d\sigma_{\eta^2} \overline{B} \\
&= \int_{\partial\Omega_\xi} d\sigma_{\xi^1} [\int_{\partial\Omega_\eta} \overline{A} d\sigma_{\eta^1} \bar{C}\bar{E}\bar{D} d\sigma_{\eta^2} \overline{B}] d\sigma_{\xi^2} \\
&= \int_{\partial\Omega_{1\xi^1}} d\sigma_{\xi^1} [\int_{\partial\Omega_{2\xi^2}} d\sigma_{\xi^2} \int_{\partial\Omega_\eta} \overline{A} d\sigma_{\eta^1} \bar{C}\bar{E}\bar{D} d\sigma_{\eta^2} \overline{B}] \\
&= \int_{\partial\Omega_{1\xi^1}} d\sigma_{\xi^1} \{\int_{\partial\Omega_{2\xi^2}} [\int_{\partial\Omega_{1\eta^1}} \overline{A} d\sigma_{\eta^1} \overline{C} (\int_{\partial\Omega_{2\eta^2}} \bar{E}\bar{D} d\sigma_{\eta^2} \overline{B})] d\sigma_{\xi^2}\},
\end{aligned} \tag{2.2}$$

where $\int_{\partial\Omega_{i\xi^i}(\partial\Omega_{i\eta^i})}$ expresses the integral over $\partial\Omega_i (i = 1, 2)$ for the variable $\xi^i(\eta^i)$, and the integrals are all in the sense of Cauchy principal values (see Chapter I). In general, (2.1), (2.2) are not the same. For discussing their relation, we give the Poincaré-Bertrand transformation formulas. From Chapters I and III, it is clear that the following results are valid:

$$\begin{cases}
\|f\, g\|_\beta \le J_1 \|f\|_\beta \|g\|_\beta, \\
(J_1 \textit{ a positive constant}, \ f, g \in H(\partial\Omega, \beta)) \\
\int_{\partial\Omega_{1\eta^1}} \overline{A} d\sigma_{\eta^1} = \int_{\partial\Omega_{2\eta^2}} d\sigma_{\eta^2} \overline{B} = \frac{1}{2}, \ \zeta \in \partial\Omega; \\
\overline{\partial}_{\eta^1} \overline{A} = \overline{A}\, \overline{\partial}_{\eta^1} = \overline{\partial}_{\eta^2} \overline{B} = \overline{B}\, \overline{\partial}_{\eta^2} = 0, \ \zeta^i \ne \eta^i, \ i = 1.2; \\
(\overline{\eta^1} - \overline{\zeta^1}) \overline{\partial}_{\eta^1} = m, \ (\overline{\xi^1} - \overline{\eta^1}) \overline{\partial}_{\eta^1} = -m, \\
(\overline{\eta^2} - \overline{\zeta^2}) \overline{\partial}_{\eta^2} = k, \ (\overline{\xi^2} - \overline{\eta^2}) \overline{\partial}_{\eta^2} = -k; \\
\int_{B(\zeta^1, \varepsilon)} d\eta^1 = \frac{\omega_n \varepsilon^m}{m}, \ \int_{B(\xi^2, \varepsilon)} d\eta^2 = \frac{\omega_k \varepsilon^k}{k}, \\
\textit{here } B(\zeta^1, \varepsilon) : |\eta^1 - \zeta^1| \le \varepsilon, \ B(\xi^2, \varepsilon) : |\eta^2 - \xi^2| \le \varepsilon.
\end{cases} \tag{2.3}$$

Theorem 2.1 *The following equalities hold:*

$$h_{1\eta^1}(\zeta^1,\, \xi^1) = \int_{\partial\Omega_{1\eta^1}} \overline{A} d\sigma_{\eta^1} \overline{C} = 0,$$

$$h_{2\eta^2}(\xi^2,\, \zeta^2) = \int_{\partial\Omega_{2\eta^2}} \overline{D} d\sigma_{\eta^2} \overline{B} = 0,$$

$$\int_{\partial\Omega_{1\eta^1}} \overline{A}\,\overline{C} d\sigma_{\eta^1} = 0, \int_{\partial\Omega_{2\eta^2}} d\sigma_{\eta^2} \overline{D}\,\overline{B} = 0,$$

where $\zeta^1,\, \xi^1 \in \partial\Omega_1,\, \zeta^1 \neq \xi^1,\, \zeta^2,\, \xi^2 \in \partial\Omega_2,\, \zeta^2 \neq \xi^2$.

Proof By means of the Cauchy principal value, we have

$$h_{1\eta^1}(\zeta^1,\, \xi^1) = \lim_{\varepsilon\to 0} \int_{\partial\Omega_{1\eta^1}\backslash(B(\zeta^1,\varepsilon)\bigcup B(\xi^1,\varepsilon))} \overline{A} d\sigma_{\eta^1} \overline{C} = B_0.$$

Here ∂ is the boundary, and the orientation of Ω_1, $B(\zeta^1,\varepsilon)$, $B(\xi^1,\varepsilon)$ $\Omega_1\backslash(B(\zeta^1,\varepsilon)\bigcup B(\xi^1,\varepsilon))$ on the communal part is harmonious, and the orientations of $\partial\Omega_1$, $\partial[\Omega_1\backslash(B(\zeta^1,\varepsilon)\bigcup B(\xi^1,\varepsilon))]$, $\partial B(\zeta^1,\varepsilon)$, $\partial B(\xi^1,\varepsilon)$ are the induced orientations of Ω_1, $\Omega_1\backslash(B(\zeta^1,\varepsilon)\bigcup B(\xi^1,\varepsilon))$, $B(\zeta^1,\varepsilon)$, $B(\xi^1,\varepsilon)$ respectively. Similarly to [36], we rewrite B_0 as a sum of the following three terms.

$$B_0 = \lim_{\varepsilon\to 0} \int_{\partial[\Omega_1\backslash(B(\zeta^1,\varepsilon)\bigcup B(\xi^1,\varepsilon))]} \overline{A} d\sigma_{\eta^1} \overline{C}$$

$$+\lim_{\varepsilon\to 0} \int_{\Omega_1\bigcap\partial B(\zeta^1,\varepsilon)} \overline{A} d\sigma_{\eta^1} \overline{C} + \lim_{\varepsilon\to 0} \int_{\Omega_1\bigcap\partial B(\xi^1,\varepsilon)} \overline{A} d\sigma_{\eta^1} \overline{C} = B_1 + B_2 + B_3.$$

By using the Stokes theorem (see Section 2, Chapter 1 or Theorem 9.2 in [6]) and (2.3), the limit

$$B_1 = \lim_{\varepsilon\to 0} \int_{\Omega_1\backslash(B(\zeta^1,\varepsilon)\bigcup B(\xi^1,\varepsilon))} [\overline{A}(\overline{\partial}_{\eta^1}\overline{C}) + (\overline{A}\,\overline{\partial}_{\eta^1})\overline{C}] d\sigma_{\eta^1} = 0 \qquad (2.4)$$

is derived. Noting that $|\eta^1 - \zeta^1| = \varepsilon$ on $\partial B(\zeta^1,\varepsilon)$ and using the Stokes theorem and formula (2.3), we get

$$B_2 = \frac{1}{2} \lim_{\varepsilon\to 0} \int_{\partial B(\zeta^1,\varepsilon)} \overline{A} d\sigma_{\eta^1} \overline{C} = \frac{1}{2} \lim_{\varepsilon\to 0} \frac{1}{\varepsilon^m \omega_m} \int_{\partial B(\zeta^1,\varepsilon)} (\overline{\eta^1} - \overline{\zeta^1}) d\sigma_{\eta^1} \overline{C}$$

$$= \frac{1}{2} \lim_{\varepsilon\to 0} \frac{1}{\varepsilon^m \omega_m} \int_{B(\zeta^1,\varepsilon)} [(\overline{\eta^1} - \overline{\zeta^1})(\overline{\partial}_{\eta^1}\overline{C}) + ((\overline{\eta^1} - \overline{\zeta^1})\overline{\partial}_{\eta^1})\overline{C}] d\sigma_{\eta^1}$$

$$= \frac{1}{2} \lim_{\varepsilon\to 0} \frac{1}{\varepsilon^m \omega_m} \int_{B(\zeta^1,\varepsilon)} m\overline{C} d\sigma_{\eta^1}.$$

When $\varepsilon \to 0$, $\eta^1 \to \zeta^1$, we can write $\overline{C} = \overline{H} + O_A(\varepsilon)$, here

$$\lim_{\varepsilon\to 0} O_A(\varepsilon) = 0,\ \overline{H} = \frac{\overline{\xi_1} - \overline{\zeta^1}}{\omega_m|\xi^1 - \zeta^1|^m}.$$

Thus

$$B_2 = \frac{1}{2}\lim_{\varepsilon\to 0}\frac{1}{\varepsilon^m\omega_m}(m)\overline{H}\int_{B(\zeta^1,\varepsilon)} d\eta^1 = \frac{1}{2}\lim_{\varepsilon\to 0}\frac{m\overline{H}\varepsilon^m\omega_m}{\varepsilon^m\omega_m m} = \frac{\overline{H}}{2}. \tag{2.5}$$

Similarly when $\varepsilon \to 0$, $(\overline{\zeta^1} - \overline{\eta^1})\overline{\partial}_{\eta^2} = -m$, we know

$$B_3 = \frac{-\overline{H}}{2}. \tag{2.6}$$

Combining (2.4), (2.5), (2.6), we get $h_{1\eta^1}(\zeta^1, \xi^1) = 0$. Moreover we can prove

$$h_{2\eta^2}(\xi^2,\, \zeta^2) = \int_{\partial\Omega_{2\eta^2}} \overline{D} d\sigma_{\eta^2}\overline{B} = 0,$$

and

$$\int_{\partial\Omega_{1\eta^1}} \overline{A}\,\overline{C} d\sigma_{\eta^1} = 0,\ \int_{\partial\Omega_{2\eta^2}} d\sigma_{\eta^2}\overline{D}\,\overline{B} = 0.$$

Theorem 2.2 *Let $\varphi(\xi^1,\, \xi^2) \in H(\partial\Omega,\, \beta)$, $0 < \beta < 1$. Then*

$$\int_{\partial\Omega_{1\eta^1}} \varphi(\xi^1,\, \xi^2)\overline{A} d\sigma_{\eta^1}\overline{C} = \int_{\partial\Omega_{2\eta^2}} \overline{D} d\sigma_{\eta^2}\overline{B}\varphi(\xi^1,\, \xi^2) = 0,$$

where $\zeta^1,\, \xi^1 \in \partial\Omega_1$, $\zeta^1 \neq \xi^1$, $\zeta^2,\, \xi^2 \in \partial\Omega_2$, $\zeta^2 \neq \xi^2$.

Proof Similarly to the proof of Theorem 2.1, it suffices to notice that

$$\overline{\partial}_{\eta^i}\varphi(\xi^1,\, \xi^2) = \varphi(\xi^1,\, \xi^2)\overline{\partial}_{\eta^i} = 0,\ i = 1, 2.$$

Theorem 2.3 *The limit*

$$\lim_{\delta\to 0}\int_{\sigma_\delta(\zeta^1,\eta^1)} \overline{A} d\sigma_{\eta^1} = \lim_{\delta\to 0}\int_{\sigma_\delta(\zeta^2,\eta^2)} d\sigma_{\eta^2}\overline{B} = 0$$

is valid, in which the orientation of $\sigma_\delta(\zeta^k, \eta^k) = \partial\Omega_{k\eta^k} \bigcap B(\zeta^k, \delta)$, $\zeta^k \in \partial\Omega_k$, $k = 1, 2$, Ω_k, $B(\zeta^k, \delta)$ is coordinated, and the orientation of $\partial\Omega_k$ is the inductive orientation of Ω_k.

Proof We only prove the first formula. Letting $0 < \overline{\delta} < \delta$, $\sigma_{\overline{\delta}}(\zeta^1, \eta^1) = \Omega_{1\eta^1} \bigcap B(\zeta^1, \overline{\delta})$ and using (2.3), we have

$$\lim_{\delta \to 0} \int_{\sigma_\delta(\zeta^1,\eta^1)} \overline{A} d\sigma_{\eta^1} = \lim_{\delta \to 0} \left[\lim_{\overline{\delta} \to 0} \int_{(\sigma_\delta(\zeta^1,\eta^1)) \backslash (\sigma_{\overline{\delta}}(\zeta^1,\eta^1))} \overline{A} d\sigma_{\eta^1} \right]$$

$$= \lim_{\delta \to 0} \left[\lim_{\overline{\delta} \to 0} \int_{(\partial\Omega_{1\eta^1} \backslash \sigma_{\overline{\delta}}(\zeta^1,\eta^1)) \backslash (\partial\Omega_{1\eta^1} \backslash \sigma_\delta(\zeta^1,\eta^1))} \overline{A} d\sigma_{\eta^1} \right]$$

$$= \lim_{\overline{\delta} \to 0} \int_{(\partial\Omega_{1\eta^1} \backslash \sigma_{\overline{\delta}}(\zeta^1,\eta^1))} \overline{A} d\sigma_{\eta^1} - \lim_{\delta \to 0} \int_{\partial\Omega_{1\eta^1} \backslash \sigma_\delta(\zeta^1,\eta^1)} \overline{A} d\sigma_{\eta^1}$$

$$= \frac{1}{2} - \frac{1}{2} = 0.$$

Similarly we can prove the second formula.

2.2 Some singular integrals whose integral order can be exchanged

In the following, $\varphi(\eta^i, \xi^i) \in H(\partial\Omega_i, \beta)$ $(i = 1, 2,\ 0 < \beta < 1)$ mean that the function φ about η^i and ξ^i all belong to $H(\partial\Omega_i, \beta)$, $i = 1, 2$ (see Section 3, Chapter II, [88]3)).

Theorem 2.4 *Suppose that* $\varphi(\eta^1, \xi^1) \in H(\partial\Omega_1, \beta)$, $0 < \beta < 1$, $\zeta^1 \in \partial\Omega_1$. *Then*

$$\int_{\partial\Omega_{1\eta^1}} \overline{A} d\sigma_{\eta^1} \int_{\partial\Omega_{1\xi^1}} d\sigma_{\xi^1} \overline{C}[\varphi(\eta^1, \xi^1) - \varphi(\xi^1, \xi^1)]$$

$$= \int_{\partial\Omega_{1\xi^1}} d\sigma_{\xi^1} \int_{\partial\Omega_{1\eta^1}} \overline{A} d\sigma_{\eta^1} \overline{C}[\varphi(\eta^1, \xi^1) - \varphi(\xi^1, \xi^1)];$$

$$\int_{\partial\Omega_{1\eta^1}} \overline{A} d\sigma_{\eta^1} \int_{\partial\Omega_{1\xi^1}} d\sigma_{\xi^1} \overline{C}[\varphi(\xi^1, \xi^1) - \varphi(\eta^1, \eta^1)]$$

$$= \int_{\partial\Omega_{1\xi^1}} d\sigma_{\xi^1} \int_{\partial\Omega_{1\eta^1}} \overline{A} d\sigma_{\eta^1} \overline{C}[\varphi(\xi^1, \xi^1) - \varphi(\eta^1, \eta^1)];$$

$$\Theta = \int_{\partial\Omega_{1\eta^1}} \overline{A} d\sigma_{\eta^1} \int_{\partial\Omega_{1\xi^1}} d\sigma_{\xi^1} \overline{C}[\varphi(\eta^1, \eta^1) - \varphi(\zeta^1, \zeta^1)]$$

$$= \int_{\partial\Omega_{1\xi^1}} d\sigma_{\xi^1} \int_{\partial\Omega_{1\eta^1}} \overline{A} d\sigma_{\eta^1} \overline{C}[\varphi(\eta^1, \eta^1) - \varphi(\zeta^1, \zeta^1)] = \Theta'.$$

Proof We only prove the third formula, because the other formulas can be similarly proved. Let $\Theta = \Theta_0 + \Theta_\delta$, $\Theta' = \Theta'_0 + \Theta'_\delta$, where

$$\Theta_0 = \int_{\partial\Omega_{1\eta^1}} \overline{A} d\sigma_{\eta^1} \int_{\partial\Omega_{1\xi^1}\backslash\sigma_\delta(\eta^1,\xi^1)} d\sigma_{\xi^1} \overline{C}[\varphi(\eta^1,\eta^1) - \varphi(\zeta^1,\zeta^1)],$$

$$\Theta_\delta = \int_{\partial\Omega_{1\eta^1}} \overline{A} d\sigma_{\eta^1} \int_{\sigma_\delta(\eta^1,\xi^1)} d\sigma_{\xi^1} \overline{C}[\varphi(\eta^1,\eta^1) - \varphi(\zeta^1,\zeta^1)],$$

$$\Theta'_0 = \int_{\partial\Omega_{1\xi^1}\backslash\sigma_\delta(\eta^1,\xi^1)} d\sigma_{\xi^1} \int_{\partial\Omega_{1\eta^1}} \overline{A} d\sigma_{\eta^1} \overline{C}[\varphi(\eta^1,\eta^1) - \varphi(\zeta^1,\zeta^1)],$$

$$\Theta'_\delta = \int_{\sigma_\delta(\eta^1,\xi^1)} d\sigma_{\xi^1} \int_{\partial\Omega_{1\eta^1}} \overline{A} d\sigma_{\eta^1} \overline{C}[\varphi(\eta^1,\eta^1) - \varphi(\zeta^1,\zeta^1)],$$

in which the orientations of $\sigma_\delta(\eta^1, \xi^1) = \partial\Omega_{1\xi^1} \cap B(\eta^1,\delta)$, Ω_1 and $B(\eta^1,\delta)$ are all coordinate, and the orientation of $\partial\Omega_1$ is the inductive orientation of Ω_1. From Section 1, Chapter III and [29]2), we know that Θ_0, Θ'_0 are integrals in the normal sense, by using the Fubini Theorem [7], [6], [19], whose integral order can be exchanged, i.e. $\Theta_0 = \Theta'_0$, hence $|\Theta - \Theta'| \le |\Theta_\delta| + |\Theta'_\delta|$. In addition, from Section 1, Chapter III (or see [29]2)) we can get

$$\int_{\partial\Omega_{1\eta^1}} \left|\overline{A} d\sigma_{\eta^1}[\varphi(\eta^1,\eta^1) - \varphi(\zeta^1,\zeta^1)]\right| \le N,$$

where N is a positive constant. From Theorem 2.3, we know when δ is small enough, there exists $\varepsilon > 0$ independent of η^1 such that $\left|\int_{\sigma_\delta(\eta^1,\xi^1)} d\sigma_{\xi^1}\overline{C}\right| \le 2\varepsilon/N$, so

$$|\Theta_\delta| \le \frac{\varepsilon}{2}. \tag{2.7}$$

Next we consider Θ'_δ. By using Theorem 2.1, we can substitute $\varphi(\zeta^1, \zeta^1)$ by $\varphi(\xi^1, \xi^1)$ in Θ'_δ, thus Θ'_δ can be written as

$$\Theta'_\delta = \int_{\sigma_\delta(\eta^1,\xi^1)} d\sigma_{\xi^1} \int_{\partial\Omega_{1\eta^1}} \overline{A} d\sigma_{\eta^1} \overline{C}[\Psi_1 + \Psi_2 + \Psi_3],$$

in which

$$\Psi_1 = \varphi(\eta^1,\eta^1) - \varphi(\eta^1,\xi^1) - \varphi(\zeta^1,\eta^1) + \varphi(\zeta^1,\xi^1),$$

$$\Psi_2 = \varphi(\eta^1,\xi^1) - \varphi(\xi^1,\xi^1),\ \Psi_3 = \varphi(\zeta^1,\eta^1) - \varphi(\zeta^1,\xi^1).$$

Since

$$\int_{\partial\Omega_{1\eta^1}} \overline{A} d\sigma_{\eta^1} \overline{C}[\Psi_1 + \Psi_2 + \Psi_3]$$

$$= \lim_{\delta \to 0} \int_{\partial\Omega_{1\eta^1} \setminus \sigma_\delta(\zeta^1, \eta^1)} \overline{A} d\sigma_{\eta^1} \overline{C}[\Psi_1 + \Psi_2 + \Psi_3], \tag{2.8}$$

where $\sigma_\delta(\zeta^1, \eta^1) = \partial\Omega_{1\eta^1} \bigcap B(\zeta^1, \delta)$, we consider

$$\Theta''_\delta = \int_{\sigma_\delta(\eta^1, \xi^1)} d\sigma_{\xi^1} \int_{\partial\Omega_{1\eta^1} \setminus \sigma_\delta(\zeta^1, \eta^1)} \overline{A} d\sigma_{\eta^1} \overline{C}[\Psi_1 + \Psi_2 + \Psi_3]$$
$$= P_1 + P_2 + P_3.$$

From Section 1, Chapter III or [29]2), we have $|\Psi_1| \leq J_2 \rho_1^{\beta/2} \rho_2^{\beta/2}$, herein J_2 is a positive constant, $\rho_1 = |\eta^1 - \xi^1|$, $\rho_2 = |\eta^1 - \zeta^1|$, $|d\sigma_{\xi^1}| \leq L_1 \rho_1^{m-2} d\rho_1$, $|d\sigma_{\eta^1}| \leq L_2 \rho_2^{m-2} d\rho_2$, L_1, L_2 are positive constants. Thus

$$|P_1| \leq L_1 L_2 J_2 \int_0^\delta \rho_1^{-m+1} \rho_1^{\frac{\beta}{2}} \rho_1^{m-2} d\rho_1 \int_{\partial\Omega_{1\eta^1} \setminus \sigma_\delta(\zeta^1, \eta^1)} \rho_2^{-m+1} \rho_2^{\frac{\beta}{2}} \rho_2^{m-2} d\rho_2$$
$$= L_1 L_2 J_2 \int_0^\delta \rho_1^{\frac{\beta}{2}-1} d\rho_1 \int_{\partial\Omega_{1\eta^1} \setminus \sigma_\delta(\zeta^1, \eta^1)} \rho_2^{\frac{\beta}{2}-1} d\rho_2 \leq J_3 \delta^{\frac{\beta}{2}}, \tag{2.9}$$

$$|P_2| \leq J_4 \int_0^\delta \rho_1^{-m+1} \rho_1^{\beta} \rho_1^{m-2} d\rho_1 \int_{\partial\Omega_{1\eta^1} \setminus \sigma_\delta(\zeta^1, \eta^1)} \rho_2^{-m+1} \rho_2^{m-2} d\rho_2$$
$$= J_4 \int_0^\delta \rho_1^{\beta-1} d\rho_1 \int_{\partial\Omega_{1\eta^1} \setminus \sigma_\delta(\zeta^1, \eta^1)} \rho_2^{-1} d\rho_2 \leq J_5 \delta^{\beta}. \tag{2.10}$$

Similarly we can get

$$|P_3| \leq J_6 \delta^{\beta}, \tag{2.11}$$

in which $J_i \, (i = 3, 4, 5, 6)$ are positive constants. From (2.8)–(2.11), we see that if δ is small enough, the inequality

$$|\Theta'_\delta| \leq \frac{\varepsilon}{2} \tag{2.12}$$

is derived. From (2.7), (2.12), it follows that $|\Theta - \Theta'| < \varepsilon$. Due to the arbitrariness of ε, the equality $\Theta = \Theta'$ is obvious.

Similarly we can prove

Theorem 2.5 *Let $\varphi(\eta^2, \xi^2) \in H(\partial\Omega_2, \beta), 0 < \beta < 1, \zeta^2 \in \partial\Omega_2$. Then*

$$\int_{\partial\Omega_{2\eta^2}} d\sigma_{\eta^2}\overline{B}\int_{\partial\Omega_{2\xi^2}}[\varphi(\eta^2,\xi^2)-\varphi(\xi^2,\xi^2)]\overline{D}d\sigma_{\xi^2}$$
$$=\int_{\partial\Omega_{2\xi^2}} d\sigma_{\xi^2}\int_{\partial\Omega_{2\eta^2}}[\varphi(\eta^2,\xi^2)-\varphi(\xi^2,\xi^2)]\overline{D}d\sigma_{\eta^2}\overline{B};$$
$$\int_{\partial\Omega_{2\eta^2}} d\sigma_{\eta^2}\overline{B}\int_{\partial\Omega_{2\xi^2}}[\varphi(\xi^2,\xi^2)-\varphi(\eta^2,\eta^2)]\overline{D}d\sigma_{\xi^2}$$
$$=\int_{\partial\Omega_{2\xi^2}} d\sigma_{\xi^2}\int_{\partial\Omega_{2\eta^2}}[\varphi(\xi^2,\xi^2)-\varphi(\eta^2,\eta^2)]\overline{D}d\sigma_{\eta^2}\overline{B};$$
$$\int_{\partial\Omega_{2\eta^2}} d\sigma_{\eta^2}\overline{B}\int_{\partial\Omega_{2\xi^2}}[\varphi(\eta^2,\eta^2)-\varphi(\zeta^2,\zeta^2)]\overline{D}d\sigma_{\xi^2}$$
$$=\int_{\partial\Omega_{2\xi^2}} d\sigma_{\xi^2}\int_{\partial\Omega_{2\eta^2}}[\varphi(\eta^2,\eta^2)-\varphi(\zeta^2,\zeta^2)]\overline{D}d\sigma_{\eta^2}\overline{B}.$$

Theorem 2.6 *If* $f(\eta^1,\eta^2;\,\xi^1,\,\xi^2)\in H(\partial\Omega,\,\beta)$, $0<\beta<1$, $\zeta^1\in\partial\Omega_1$, $\zeta^2\in\partial\Omega_2$, *then the following integrals order can be exchanged for the integral order.*

$$\int_{\partial\Omega_\eta}\overline{A}d\sigma_{\eta^1}d\sigma_{\eta^2}\overline{B}\int_{\partial\Omega_\xi}d\sigma_{\xi^1}\overline{C}[f(\eta^1,\eta^2;\xi^1,\xi^2)-f(\eta^1,\eta^2;\eta^1,\xi^2)$$
$$-f(\eta^1,\eta^2;\xi^1,\eta^2)+f(\eta^1,\eta^2;\eta^1,\eta^2)]\overline{D}d\sigma_{\xi^2},$$
$$\int_{\partial\Omega_\eta}\overline{A}d\sigma_{\eta^1}d\sigma_{\eta^2}\overline{B}\int_{\partial\Omega_\xi}d\sigma_{\xi^1}\overline{C}[f(\eta^1,\eta^2;\xi^1,\eta^2)-f(\eta^1,\eta^2;\eta^1,\eta^2)$$
$$-f(\eta^1,\zeta^2;\xi^1,\zeta^2)+]f(\eta^1,\zeta^2;\eta^1,\zeta^2)]\overline{D}d\sigma_{\xi^2},$$
$$\int_{\partial\Omega_\eta}\overline{A}d\sigma_{\eta^1}d\sigma_{\eta^2}\overline{B}\int_{\partial\Omega_\xi}d\sigma_{\xi^1}\overline{C}[f(\eta^1,\eta^2;\eta^1,\xi^2)-f(\eta^1,\eta^2;\eta^1,\eta^2)$$
$$-f(\zeta^1,\eta^2;\zeta^1,\xi^2)+f(\zeta^1,\eta^2;\zeta^1,\eta^2)]\overline{D}d\sigma_{\xi^2},$$
$$\int_{\partial\Omega_\eta}\overline{A}d\sigma_{\eta^1}d\sigma_{\eta^2}\overline{B}\int_{\partial\Omega_\xi}d\sigma_{\xi^1}\overline{C}[f(\eta^1,\eta^2;\eta^1,\eta^2)-f(\zeta^1,\eta^2;\zeta^1,\eta^2)$$
$$-f(\eta^1,\zeta^2;\eta^1,\zeta^2)+f(\zeta^1,\zeta^2;\zeta^1,\zeta^2)]\overline{D}d\sigma_{\xi^2}.$$

2.3 Three kinds of Poincaré-Bertrand transformation formulas of two times singular integrals

Theorem 2.7 *Let* $\overline{E}=f(\eta^1,\eta^2;\,\xi^1,\,\xi^2)\in H(\partial\Omega,\,\beta)$, $0<\beta<1$, $\zeta^1\in$

$\partial\Omega_1$, $\zeta^2 \in \partial\Omega_2$. *Then*

$$\begin{aligned}\triangle_1 &= \int_{\partial\Omega_\eta} \overline{A} d\sigma_{\eta^1} d\sigma_{\eta^2} \overline{B} \int_{\partial\Omega_\xi} d\sigma_{\xi^1} \overline{C}\bar{E}\bar{D} d\sigma_{\xi^2} \\ &= \int_{\partial\Omega_\xi} d\sigma_{\xi^1} d\sigma_{\xi^2} \int_{\partial\Omega_\eta} \bar{A} d\sigma_{\eta^1} \bar{C}\bar{E}\bar{D} d\sigma_{\eta^2} \overline{B} \\ &+ \frac{1}{4}\left[\int_{\partial\Omega_{1\xi^1}} d\sigma_{\xi^1} \int_{\partial\Omega_{1\eta^1}} \overline{A} d\sigma_{\eta^1} \overline{C} f(\eta^1, \zeta^2; \xi^1, \zeta^2)\right. \\ &+ \left.\int_{\partial\Omega_{2\xi^2}} d\sigma_{\xi^2} \int_{\partial\Omega_{2\eta^2}} f(\zeta^1, \eta^2; \zeta^1, \xi^2) \overline{D} d\sigma_{\eta^2} \overline{B}\right] \\ &+ \frac{1}{16} f(\zeta^1, \zeta^2; \zeta^1, \zeta^2) = \triangle_2.\end{aligned}$$

Proof Transform $\triangle_1$ into the form

$$\triangle_1 = \int_{\partial\Omega_\eta} \overline{A} d\sigma_{\eta^1} d\sigma_{\eta^2} \overline{B} \int_{\partial\Omega_\xi} d\sigma_{\xi^1} \overline{C}(\Phi_1 + \cdots + \Phi_{11}) \bar{D} d\sigma_{\xi^2} = D_1 + \cdots + D_{11},$$

where

$$\Phi_1 = f(\eta^1, \eta^2; \xi^1, \xi^2) - f(\eta^1, \eta^2; \eta^1, \xi^2) - f(\eta^1, \eta^2; \xi^1, \eta^2) + f(\eta^1, \eta^2; \eta^1, \eta^2),$$

$$\Phi_2 = f(\eta^1, \eta^2; \xi^1, \eta^2) - f(\eta^1, \eta^2; \eta^1, \eta^2) - f(\eta^1, \zeta^2; \xi^1, \zeta^2) + f(\eta^1, \zeta^2; \eta^1, \zeta^2),$$

$$\Phi_3 = f(\eta^1, \eta^2; \eta^1, \xi^2) - f(\eta^1, \eta^2; \eta^1, \eta^2) - f(\zeta^1, \eta^2; \zeta^1, \xi^2) + f(\zeta^1, \eta^2; \zeta^1, \eta^2),$$

$$\Phi_4 = f(\eta^1, \eta^2; \eta^1, \eta^2) - f(\zeta^1, \eta^2; \zeta^1, \eta^2) - f(\eta^1, \zeta^2; \eta^1, \zeta^2) - f(\zeta^1, \zeta^2; \zeta^1, \zeta^2),$$

$$\Phi_5 = f(\eta^1, \zeta^2; \xi^1, \zeta^2) - f(\xi^1, \zeta^2; \xi^1, \zeta^2),$$

$$\Phi_6 = f(\zeta^1, \eta^2; \zeta^1, \xi^2) - f(\zeta^1, \xi^2; \zeta^1, \xi^2),$$

$$\Phi_7 = f(\xi^1, \zeta^2; \xi^1, \zeta^2) - f(\eta^1, \zeta^2; \eta^1, \zeta^2),$$

$$\Phi_8 = f(\zeta^1, \xi^2; \zeta^1, \xi^2) - f(\zeta^1, \eta^2; \zeta^1, \eta^2),$$

$$\Phi_9 = f(\eta^1, \zeta^2; \eta^1, \zeta^2) - f(\zeta^1, \zeta^2; \zeta^1, \zeta^2),$$

$$\Phi_{10} = f(\zeta^1, \eta^2; \zeta^1, \eta^2) - f(\zeta^1, \zeta^2; \zeta^1, \zeta^2),$$

$$\Phi_{11} = f(\zeta^1, \zeta^2; \zeta^1, \zeta^2).$$

For every D_i, $i = 1, ..., 4$, we exchange their integral orders by using Theorem 2.6 and then merge them, and for every D_i, $i = 5, ..., 10$, we exchange their integral orders by using (2.3), Theorem 2.4, 2.5 and then merge them. Moreover for Φ_{11}, by using (2.3) we obtain

$$\triangle_1 = \int_{\partial\Omega_\xi} d\sigma_{\xi^1} d\sigma_{\xi^2} \int_{\partial\Omega_\eta} \overline{A} d\sigma_{\eta^1} \overline{C}(\Phi_1 + \cdots + \Phi_4)\bar{D} d\sigma_{\eta^2} \overline{B}$$

$$+\frac{1}{4}\int_{\partial\Omega_{1\xi_1}} d\sigma_{\xi^1} \int_{\partial\Omega_{1\eta^1}} \overline{A} d\sigma_{\eta^1} \bar{C}(\Phi_5 + \Phi_7 + \Phi_9)$$

$$+\frac{1}{4}\int_{\partial\Omega_{2\xi_2}} d\sigma_{\xi^2} \int_{\partial\Omega_{2\eta^2}} (\Phi_6 + \Phi_8 + \Phi_{10})\overline{D} d\sigma_{\eta^2} \bar{B}$$

$$+\frac{1}{16} f(\zeta^1, \zeta^2; \zeta^1, \zeta^2)$$

$$= \int_{\partial\Omega_\xi} d\sigma_{\xi^1} d\sigma_{\xi^2} \int_{\partial\Omega_\eta} \overline{A} d\sigma_{\eta^1} \overline{C}\left[f(\eta^1, \eta^2; \xi^1, \xi^2) - f(\eta^1, \zeta^2; \xi^1, \zeta^2)\right.$$

$$\left. - f(\zeta^1, \eta^2; \zeta^1, \xi^2) + f(\zeta^1, \zeta^2; \zeta^1, \zeta^2)\right] \overline{D} d\sigma_{\eta^2} \overline{B}$$

$$+\frac{1}{4}\int_{\partial\Omega_{1\xi_1}} d\sigma_{\xi^1} \int_{\partial\Omega_{1\eta^1}} \overline{A} d\sigma_{\eta^1} \bar{C}\left[f(\eta^1, \zeta^2; \xi^1, \zeta^2) - f(\zeta^1, \zeta^2; \zeta^1, \zeta^2)\right]$$

$$+\frac{1}{4}\int_{\partial\Omega_{2\xi_2}} d\sigma_{\xi^2} \int_{\partial\Omega_{2\eta^2}} \left[f(\zeta^1, \eta^2; \zeta^1, \xi^2) - f(\zeta^1, \zeta^2; \zeta^1, \zeta^2)\right] \overline{D} d\sigma_{\eta^2} \bar{B}$$

$$+\frac{1}{16} f(\zeta^1, \zeta^2; \zeta^1, \zeta^2).$$

Finally, applying Theorem 2.1 we can get

$$\begin{aligned} \triangle_1 = & \int_{\partial\Omega_\xi} d\sigma_{\xi^1} d\sigma_{\xi^2} \int_{\partial\Omega_\eta} \overline{A} d\sigma_{\eta^1} \overline{C} f(\eta^1, \eta^2; \xi^1, \xi^2)\overline{D} d\sigma_{\eta^2} \bar{B} \\ & +\frac{1}{4}\int_{\partial\Omega_{1\xi_1}} d\sigma_{\xi^1} \int_{\partial\Omega_{1\eta^1}} \overline{A} d\sigma_{\eta^1} \bar{C} f(\eta^1, \zeta^2; \xi^1, \zeta^2) \\ & +\frac{1}{4}\int_{\partial\Omega_{2\xi_2}} d\sigma_{\xi^2} \int_{\partial\Omega_{2\eta^2}} f(\zeta^1, \eta^2; \zeta^1, \xi^2)\overline{D} d\sigma_{\eta^2} \bar{B} \\ & +\frac{1}{16} f(\zeta^1, \zeta^2; \zeta^1, \zeta^2) = \triangle_2. \end{aligned}$$

Theorem 2.8 *Let* $\overline{G} = b(\eta^1, \eta^2)$, $\overline{F} = f(\xi^1, \xi^2) \in H(\partial\Omega, \beta)$, $0 < \beta < 1$, $\zeta^1 \in \partial\Omega_1$, $\zeta^2 \in \partial\Omega_2$. *Then*

$$\int_{\partial\Omega_\eta} \overline{A} d\sigma_{\eta^1} d\sigma_{\eta^2} \overline{B} \int_{\partial\Omega_\xi} \overline{G} d\sigma_{\xi^1} \overline{C} \bar{F} \bar{D} d\sigma_{\xi^2}$$

$$
\begin{aligned}
= & \int_{\partial\Omega_\xi} d\sigma_{\xi^1} d\sigma_{\xi^2} \int_{\partial\Omega_\eta} \bar{A}\overline{G} d\sigma_{\eta^1} \bar{C}\bar{F}\bar{D} d\sigma_{\eta^2} \overline{B} \\
& + \frac{1}{4}\left[\int_{\partial\Omega_{1\xi^1}} d\sigma_{\xi^1} \int_{\partial\Omega_{1\eta^1}} \overline{A} b(\eta^1, \zeta^2) d\sigma_{\eta^1} \overline{C} f(\xi^1, \zeta^2)\right. \\
& \left. + \int_{\partial\Omega_{2\xi^2}} d\sigma_{\xi^2} \int_{\partial\Omega_{2\eta^2}} b(\zeta^1, \eta^2) f(\zeta^1, \xi^2) \overline{D} d\sigma_{\eta^2} \overline{B}\right] \\
& + \frac{1}{16} b(\zeta^1, \zeta^2) f(\zeta^1, \zeta^2).
\end{aligned}
$$

Similar to the proof of Theorem 2.7, we can prove Theorems 2.8 and 2.9.

Theorem 2.9 *Let* $\overline{E} = f(\eta^1, \eta^2; \xi^1, \xi^2) \in H(\partial\Omega, \beta)$, $0 < \beta < 1$, $\zeta^1 \in \partial\Omega_1$, $\zeta^2 \in \partial\Omega_2$. *Then*

$$
\begin{aligned}
& \int_{\partial\Omega_\eta} \overline{A} d\sigma_{\eta^1} d\sigma_{\eta^2} \overline{B} \int_{\partial\Omega_\xi} \overline{C} d\sigma_{\xi^1} \overline{E} d\sigma_{\xi^2} \overline{D} \\
= & \int_{\partial\Omega_\xi} d\sigma_{\xi^1} d\sigma_{\xi^2} \int_{\partial\Omega_\eta} \bar{A}\overline{C} d\sigma_{\eta^1} \bar{E} d\sigma_{\eta^2} \overline{D}\,\overline{B} \\
& + \frac{1}{4}\left[\int_{\partial\Omega_{1\xi^1}} d\sigma_{\xi^1} \int_{\partial\Omega_{1\eta^1}} \overline{A}\,\overline{C} d\sigma_{\eta^1} f(\eta^1, \zeta^2; \xi^1, \zeta^2)\right. \\
& \left. + \int_{\partial\Omega_{2\xi^2}} d\sigma_{\xi^2} \int_{\partial\Omega_{2\eta^2}} f(\zeta^1, \eta^2; \zeta^1, \xi^2) d\sigma_{\eta^2} \overline{D}\,\overline{B}\right] \\
& + \frac{1}{16} f(\zeta^1, \zeta^2; \zeta^1, \zeta^2).
\end{aligned}
$$

3 The Composition Formula and Inverse Formula of Singular Integrals with a Cauchy's Kernel in Real Clifford Analysis

In this section, we first prove the composition formula and the inverse formula of singular integrals with Cauchy's kernel by using the Poincaré-Bertrand transformation formula, and then we give the second proof method for the composition formula and the inverse formula by using the Plemelj formula for Cauchy's integrals.

We suppose $\overline{E} = F(\xi^1, \xi^2; \zeta^1, \zeta^2) = f(\xi^1, \xi^2)$ in Theorem 2.9, and use Theorem 2.1, then the composition formula of singular integrals with Cauchy's kernel can be obtained.

Theorem 3.1 *Let $f(\xi^1, \xi^2) \in H(\partial\Omega, \beta)$, $0 < \beta < 1$, $\zeta^i \in \partial\Omega_i$, $i = 1, 2$. Then*

$$\int_{\partial\Omega_\eta} \overline{A} d\sigma_{\eta^1} d\sigma_{\eta^2} \overline{B} \int_{\partial\Omega_\xi} \overline{C} d\sigma_{\xi^1} f(\xi^1, \xi^2) d\sigma_{\xi^2} \overline{D} = \frac{1}{16} f(\zeta^1, \zeta^2). \tag{3.1}$$

Denote the operator

$$Wf = 4 \int_{\partial\Omega_\xi} \overline{C} d\sigma_{\xi^1} f(\xi^1, \xi^2) d\sigma_{\xi^2} \overline{D},$$

then the composition formula can written as

$$W^2 f = W(Wf) = f. \tag{3.2}$$

By using the composition formula, the inverse formula for singular integrals with Cauchy's kernel can also be obtained.

Theorem 3.2 *If $f(\xi^1, \xi^2) \in H(\partial\Omega, \beta)$, $0 < \beta < 1$, and we write*

$$Wf = g(\eta^1, \eta^2), \tag{3.3}$$

then

$$Wg = f(\zeta^1, \zeta^2). \tag{3.4}$$

Proof From (3.3), (3.2), we get (3.4) and

$$Wg = W(Wf) = f(\zeta^1, \zeta^2).$$

Inversely, we have (3.3) from (3.4), and (3.3), (3.4) are called the inverse formulas for singular integrals with Cauchy's kernel on characteristic manifolds. Obviously it is equivalent to the composition formula. If (3.3) is seen as a singular integral equation with the unknown function f, then from inverse formula (3.4), we can derive that the equation has a unique solution.

As a corollary, we can get the inverse formula on the smooth closed manifold as follows:

$$\begin{cases} g(\eta^1) = 2 \displaystyle\int_{\partial\Omega_{1\xi^1}} \overline{C} d\sigma_{\xi^1} f(\xi^1), \\ f(\zeta^1) = \displaystyle\int_{\partial\Omega_{1\eta^1}} \overline{A} d\sigma_{\eta^1} g(\eta^1). \end{cases} \tag{3.5}$$

From Theorem 1.1 in Section 1, Chapter III, we know that Cauchy type integral

$$F(x,y)=\int_{\partial\Omega_\xi}\frac{\overline{\xi^1}-\overline{x}}{\omega_m|\xi^1-x|^m}d\sigma_{\xi^1}f(\xi^1,\,\xi^2)d\sigma_{\xi^2}\frac{\overline{\xi^2}-\overline{y}}{\omega_k|\xi^2-y|^k},$$
$$x\notin\partial\Omega_1,\ y\notin\partial\Omega_2$$

is a biregular function in $\Omega_1^\pm\times\Omega_2^\pm$, and $F(x,\infty)=F(\infty,y)=F(\infty,\infty)=0$. Using Theorem 1.1 (the Plemelj formula) in Section 1, Chapter III, it is easy to see that

$$\begin{aligned}F^{++}(\eta^1,\eta^2)+F^{+-}(\eta^1,\eta^2)+F^{-+}(\eta^1,\eta^2)+F^{--}(\eta^1,\eta^2)\\ =g(\eta^1,\eta^2),\ (\eta^1,\eta^2)\in\partial\Omega.\end{aligned}\tag{3.6}$$

From (3.3), we can get $g(\eta^1,\eta^2)=Wf$. If we consider another biregular function $Q(x,y)$ in $\Omega_1^\pm\times\Omega_2^\pm$,

$$Q(x,y)=\begin{cases}F(x,y),\ (x,y)\in\Omega_1^+\times\Omega_2^+\text{ or }\Omega_1^-\times\Omega_2^-\\ -F(x,y),\ (x,y)\in\Omega_1^+\times\Omega_2^-\text{ or }\Omega_1^-\times\Omega_2^+,\end{cases}\tag{3.7}$$

then (3.6) can be written as

$$\begin{aligned}Q^{++}(\eta^1,\eta^2)-Q^{+-}(\eta^1,\eta^2)-Q^{-+}(\eta^1,\eta^2)+Q^{--}(\eta^1,\eta^2)\\ =g(\eta^1,\eta^2),\end{aligned}\tag{3.8}$$

where $(\eta^1,\eta^2)\in\partial\Omega$, i.e. $Q(x,y)$ is a solution of (3.8). From Theorem 1.1 (Plemelj formula) of Section 1, Chapter III again, we see that the Cauchy type integral

$$Q_1(x,y)=\int_{\partial\Omega_\eta}\frac{\overline{\eta^1}-\overline{x}}{\omega_m|\eta^1-x|^m}d\sigma_{\eta^1}g(\eta^1,\,\eta^2)d\sigma_{\eta^2}\frac{\overline{\eta^2}-\overline{y}}{\omega_k|\eta^2-y|^k}$$

is a solution of (3.8). Let $Q_2(x,y)=Q(x,y)-Q_1(x,y)$. From condition (3.8), we know that Q_2 satisfies the homogeneous boundary condition of (3.8):

$$\begin{aligned}Q_2^{++}(\eta^1,\eta^2)-Q_2^{+-}(\eta^1,\eta^2)-Q_2^{-+}(\eta^1,\eta^2)+Q_2^{--}(\eta^1,\eta^2)\\ =0,\ (\eta^1,\eta^2)\in\partial\Omega.\end{aligned}$$

This shows that $Q_2(x,y)$ possesses the jump degree zero, hence $Q_2^{\pm\pm}(x,y)$ can be extended to a biregular function through each other's $\partial\Omega$, namely $Q_2(x,y)$ is a biregular function in $R^m\times R^k$ and $Q_2(x,\infty)=$

$Q_2(\infty, y) = Q_2(\infty, \infty) = 0$. From the Liouville theorem (see [19]), we get $Q_2 \equiv 0$, thus $Q(x, y) \equiv Q_1(x, y)$. From the above Plemelj formula again, we have

$$\begin{aligned} &F^{++}(\eta^1, \eta^2) - F^{+-}(\eta^1, \eta^2) - F^{-+}(\eta^1, \eta^2) + F^{--}(\eta^1, \eta^2) \\ &\quad = f(\eta^1, \eta^2),\ (\eta^1, \eta^2) \in \partial\Omega, \\ &Q^{++}(\eta^1, \eta^2) + Q^{+-}(\eta^1, \eta^2) + Q^{-+}(\eta^1, \eta^2) + Q^{--}(\eta^1, \eta^2) \\ &\quad = Wg(\eta^1, \eta^2),\ (\eta^1, \eta^2) \in \partial\Omega. \end{aligned}$$

From (3.7), the definition of $g(\eta^1, \eta^2)$ and two formulas as stated above, we obtain

$$f(\eta^1, \eta^2) = Wg(\eta^1, \eta^2) = W(Wf) = W^2 f,$$

which shows that the inverse formula is true.

4 The Fredholm Theory of a Kind of Singular Integral Equations in Real Clifford Analysis

In this section we deal with a kind of integral equations with quasi-Cauchy kernel in real Clifford analysis. Firstly we write the condition for an integral equation, which can be reduced to the Fredholm type equation, and find the regularization operator and prove the regularization theorem. Let $b(\eta^1, \eta^2)$, $c(\eta^1, \eta^2)$, $\varphi(\eta^1, \eta^2) \in H(\partial\Omega, \beta)$, $0 < \beta < 1$, and introduce the singular integral operator with Cauchy kernel and quasi-Cauchy kernel K, L:

$$(Kf)(\eta^1, \eta^2) = 4 \int_{\partial\Omega_\xi} d\sigma_{\xi^1} \overline{C} f(\xi^1, \xi^2) \overline{D} d\sigma_{\xi^2},$$

$$(Lf)(\eta^1, \eta^2) = \int_{\partial\Omega_\xi} L^1(\xi^1, \eta^1) d\sigma_{\xi^1} f(\xi^1, \xi^2) d\sigma_{\xi^2} L^2(\xi^2, \eta^2),$$

where the Cauchy kernels $\overline{C}$, $\overline{D}$ have been given in Section 2, and the quasi-Cauchy kernel is defined as

$$L^1(\xi^1, \eta^1) = \frac{l^1(\xi^1, \eta^1)}{|\xi^1 - \eta^1|^{m-1-r_1}},\ L^2(\xi^2, \eta^2) = \frac{l^2(\xi^2, \eta^2)}{|\xi^2 - \eta^2|^{k-1-r_1}},$$

where $l^i(\xi^i, \eta^i) \in H(\partial\Omega, \alpha_i)$, $0 < \alpha_i < 1$, $r_i > 0$, $i = 1, 2$. We consider the singular integral equation with quasi-Cauchy kernel in $H(\partial\Omega, \beta)$:

$$Sf = f + bKf + cLf = \varphi, \tag{4.1}$$

in which $\partial\Omega$ is the regularizing manifold. The following regularization theorem about (4.1) can be obtained.

Theorem 4.1 *Let $(1-b^2)$ in (4.1) be invertible on $\partial\Omega$, and $\|1/(1-b^2)\| < Y$, Y a positive constant. Then when $\|b\|_\beta$ is small enough, (4.1) can be transformed into a Fredholm type equation.*

Proof Firstly we give the operator M : $M\psi \equiv 1(\psi - bK\psi)/(1-b^2), \psi \in H(\partial\Omega, \beta)$, and prove that M is a regularization operator of (4.1). In fact, when M acts on the two sides of (4.1), the right-hand side of (4.1) becomes a function g such that $M\varphi = g$ and the left-hand side becomes

$$M(Sf) = \frac{1}{1-b^2}[f + cLf - bK(bKf) - bK(cLf)].$$

By applying Theorem 2.8 to $K(bKf)$ in the above formula and exchanging the integral order of $K(cLf)$ (from Theorems 2.4, 2.5, 2.6, we see that the order can be exchanged), hence (4.1) possesses the form

$$\begin{aligned} &[M(SF)](\zeta^1, \zeta^2) \\ = \ & f + \frac{c}{1-b^2}Lf - \frac{16b}{1-b^2}Uf - \frac{4b}{1-b^2}Jf - \frac{4b}{1-b^2}Qf - \frac{4b}{1-b^2}Vf \qquad (4.2) \\ = \ & g(\zeta^1, \zeta^2), \end{aligned}$$

in which

$$(Uf)(\zeta^1,\zeta^2) = \int_{\partial\Omega_\xi} d\sigma_{\xi^1} d\sigma_{\xi^2} \int_{\partial\Omega_\eta} \bar{A} b(\eta^1,\eta^2) d\sigma_{\eta^1} \bar{C} f(\xi^1,\xi^2) \overline{D} d\sigma_{\eta^2} \overline{B},$$

$$(Jf)(\zeta^1,\zeta^2) = \int_{\partial\Omega_{1\xi^1}} d\sigma_{\xi^1} \int_{\partial\Omega_{1\eta^1}} \overline{A} b(\eta^1, \zeta^2) d\sigma_{\eta^1} \overline{C} f(\xi^1,\xi^2),$$

$$(Qf)(\zeta^1,\zeta^2) = \int_{\partial\Omega_{2\xi^2}} d\sigma_{\xi^2} \int_{\partial\Omega_{2\eta^2}} b(\zeta^1,\zeta^2) f(\zeta^1,\xi^2) \overline{D} d\sigma_{\eta^2} \overline{B},$$

$$\begin{aligned} &(Vf)(\zeta^1,\zeta^2) = \int_{\partial\Omega_\xi} d\sigma_{\xi^1} d\sigma_{\xi^2} \\ &\times \int_{\partial\Omega_\eta} \bar{A} c(\eta^1,\eta^2) L^1(\xi^1,\eta^1) d\sigma_{\eta^1} f(\xi^1,\xi^2) d\sigma_{\eta^2} L^2(\xi^2,\eta^2) \overline{B}. \end{aligned}$$

We write the kernel of U, J, Q, V, L as U_1, J_1, Q_1, V_1, L_1, herein

$$U_1(\zeta^1, \zeta^2; \xi^1, \xi^2) = \int_{\partial\Omega_\eta} \bar{A} b(\eta^1, \eta^2) d\sigma_{\eta^1} \bar{C} \overline{D} d\sigma_{\eta^2} \overline{B},$$

$$J_1(\zeta^1, \zeta^2, \xi^1) = \int_{\partial\Omega_{1\eta^1}} \overline{A}b(\eta^1, \zeta^2)d\sigma_{\eta^1}\overline{C},$$

$$Q_1(\zeta^1, \zeta^2, \xi^2) = \int_{\partial\Omega_{2\eta^2}} b(\zeta^1, \eta^2)\overline{D}d\sigma_{\eta^2}\overline{B},$$

$$V_1(\zeta^1, \zeta^2; \xi^1, \xi^2)$$

$$= \int_{\partial\Omega_\eta} \bar{A}c(\eta^1, \eta^2)L^1(\xi^1, \eta^1)d\sigma_{\eta^1}d\sigma_{\eta^2}L^2(\xi^2, \eta^2)\overline{B},$$

$$L_1(\zeta^1, \zeta^2; \xi^1, \xi^2) = L^1(\xi^1, \zeta^1)L^2(\xi^2, \zeta^2).$$

Moreover setting $v(\eta^1, \eta^2) = b(\eta^1, \eta^2) - b(\eta^1, \xi^2) - b(\xi^1, \eta^2) + b(\xi^1, \xi^2)$, and using Theorem 2.2, we know that U_1 can be transformed into

$$U_1 = \int_{\partial\Omega_\eta} \bar{A}v(\eta^1, \eta^2)d\sigma_{\eta^1}\bar{C}\overline{D}d\sigma_{\eta^2}\overline{B}.$$

From Section 1, Chapter III or [29]2), we get

$$|v| \le J_2|\eta^1 - \xi^1|^{\frac{\beta}{2}}|\eta^2 - \xi^2|^{\frac{\beta}{2}} \tag{4.3}$$

where J_2 is a positive constant. Hence

$$|U_1|$$
$$\le J_3 \int_{\partial\Omega_\eta} \frac{|d\sigma_{\eta^1}||d\sigma_{\eta^2}|}{|\eta^1 - \zeta^1|^{m-1}|\eta^1 - \xi^1|^{m-1-\frac{\beta}{2}}|\eta^2 - \zeta^2|^{k-1-\frac{\beta}{2}}|\eta^2 - \zeta^2|^{k-1}}, \tag{4.4}$$

in which J_3 is a positive constant. Using the Hadamard Theorem (see [19]), we see that when $m > \frac{\beta}{2} + 1$, $k > \frac{\beta}{2} + 1$, the estimate

$$|U_1| \le \frac{J_4}{|\xi^1 - \zeta^1|^{m-1-\frac{\beta}{2}}|\xi^2 - \zeta^2|^{k-1-\frac{\beta}{2}}}, \tag{4.5}$$

is derived, where J_4 is a positive constant. Similarly, we have

$$|V_1|$$
$$\le J_5 \int_{\partial\Omega_\eta} \frac{|d\sigma_{\eta^1}||d\sigma_{\eta^2}|}{|\eta^1 - \zeta^1|^{m-1}|\eta^1 - \xi^1|^{m-1-r_1}|\eta^2 - \zeta^2|^{k-1}|\eta^2 - \xi^2|^{k-1-r_2}}$$
$$\le \frac{J_6}{|\xi^1 - \zeta^1|^{m-1-r_1}|\xi^2 - \zeta^2|^{k-1-r_2}}. \tag{4.6}$$

where J_5, J_6 are positive constants. Using Theorem 2.2, we know that J_1 can be changed as

$$J_1 = \int_{\partial\Omega_{1\eta^1}} \overline{A} b_1(\eta^1, \zeta^2) d\sigma_{\eta^1} \overline{C},$$

in which $b_1(\eta^1, \zeta^2) = b(\eta^1, \zeta^2) - b(\zeta^1, \zeta^2)$, and

$$|b_1| \leq J_7 |\eta^1 - \zeta^1|^\beta, \tag{4.7}$$

herein J_7 is a positive constant. By the Hadamard Theorem and (4.7), when $m > \beta + 1$, we have

$$\begin{aligned} |J_1| &\leq J_8 \int_{\partial\Omega_{1\eta_1}} \frac{|d\sigma_{\eta^1}|}{|\eta^1 - \zeta^1|^{m-1-\beta} |\eta^1 - \xi^1|^{m-1}} \\ &\leq \frac{J_9}{|\zeta^1 - \xi^1|^{m-1-\beta}}. \end{aligned} \tag{4.8}$$

Next we discuss Q_1. We can get, when $k > \beta + 1$,

$$|Q_1| \leq \frac{J_{10}}{|\xi^2 - \zeta^2|^{k-1-\beta}}. \tag{4.9}$$

For L_1 obviously we have

$$|L_1| \leq \frac{J_{11}}{|\xi^1 - \zeta^1|^{m-1-r_1} |\xi^2 - \zeta^2|^{k-1-r_2}}, \tag{4.10}$$

in which J_8, J_9, J_{10}, J_{11} are all positive constants. Combining (4.5), (4.6), (4.8), (4.9) and (4.10), it is easy to see that (4.2) is a weak singular equation. By the Hadamard Theorem, when ζ^1, $\xi^1 \in \partial\Omega_1$ and μ_1, $\mu_2 > 0$, we get

$$\int_{\partial\Omega_{1\eta^1}} \frac{|d\sigma_{\eta^1}|}{|\eta^1 - \zeta^1|^{m-1-\mu_1} |\eta^1 - \xi^1|^{m-1-\mu_2}} \leq \frac{J_{12}}{|\zeta^1 - \xi^1|^{m-1-(\mu_1+\mu_2)}}, \tag{4.11}$$

where $m > \mu_1 + \mu_2 + 1$, J_{12} is a positive constant. In addition for ζ^2, $\xi^2 \in \partial\Omega_2$, the same estimate can be concluded. Similarly by the Hadamard theorem, when $\zeta, \xi \in \partial\Omega$ and $\mu_i > 0$, $1 \leq i \leq 4$, we can obtain

$$\begin{aligned} &\int_{\partial\Omega_\eta} \frac{|d\sigma_{\eta^1}||d\sigma_{\eta^2}|}{|\eta^1 - \zeta^1|^{m-1-\mu_1} |\eta^1 - \xi^1|^{m-1-\mu_2} |\eta^2 - \zeta^2|^{k-1-\mu_3} |\eta^2 - \xi^2|^{m-1-\mu_4}} \\ &\quad \leq \frac{J_{13}}{|\zeta^1 - \xi^1|^{m-1-(\mu_1+\mu_2)} |\zeta^2 - \xi^2|^{k-1-(\mu_3+\mu_4)}}, \end{aligned} \tag{4.12}$$

where $m > \mu_1 + \mu_2 + 1$, $k > \mu_3 + \mu_4 + 1$, J_{13} is a positive constant. Thus from (4.5), (4.11), (4.12), we see that the reiterative kernel $U_1^{(p)}$ of U_1 for p times satisfies the inequality (see [46])

$$U_1^{(p)} \leq \frac{J_{14}}{|\zeta^1 - \xi^1|^{m-1-\frac{p\beta}{2}} |\zeta^2 - \xi^2|^{k-1-\frac{p\beta}{2}}}; \tag{4.13}$$

here J_{14} is a positive constant. Moreover the reiterative kernels of V_1, J_1, Q_1, L_1 for p times satisfy the similar inequality. Because of $\zeta^i \in \partial\Omega$ $(i = 1, 2)$, the integrals

$$\int_{\partial\Omega_{1\eta^1}} \frac{|d\sigma_{\eta^1}|}{|\eta^1 - \zeta^1|^{m-1-\mu_1}}, \ 0 < \mu_1 < 1,$$

$$\int_{\partial\Omega_{2\eta^2}} \frac{|d\sigma_{\eta^2}|}{|\eta^2 - \zeta^2|^{k-1-\mu_3}}, \ 0 < \mu_3 < 1$$

are uniformly bounded, hence the mixed reiterative kernels of U_1, V_1, J_1, Q_1, L_1 for p times satisfy the similar inequality (see [46]).

From (4.13) we see that for the positive integer p satisfying $k - 1 - \frac{p\beta}{2} \leq 0$, $m - 1 - \frac{p\beta}{2} \leq 0$, the reiterative kernel $U_1^{(p)}$ of U_1 for p times are all bounded functions, it is sufficient to assume $p \geq \max\{[2(m-1)/\beta] + 1, [2(k-1)/\beta] + 1\}$. Similarly we can consider the reiterative kernels and mixed reiterative kernels of U_1, V_1, J_1, Q_1, L_1 for p times. We know if p is large enough the reiterative kernels and mixed reiterative kernels for p times are all bounded functions. This shows that (4.2) is a Fredholm equation, and the Fredholm theorem holds for it.

Finally we prove that there exists an inverse of M. In fact, from Section 1, Chapter III or [29]2), we know that $\|K\Psi\|_\beta \leq J_{15}\|\Psi\|_\beta$, where J_{15} is a positive constant, hence when $\|b\|_\beta$ is small enough, the operator $bK\Psi$ is a compact operator. Moreover from the hypothesis $\|1/(1 - b^2)\|_\beta < Y$, we see that when $\|b\|_\beta$ is small enough, there exists an inverse operator M^{-1} (see [46]). This shows that (4.1) is equivalent to the Fredholm equation (4.2).

Remark 1 When $\|b\|_\beta$, $\|c\|_\beta$, $\|\varphi\|_\beta$ are appropriately small, from Section 1, Chapter III (or see [29]2)), we know that (4.1) is solvable

Remark 2 When a is invertible, the singular integral equation (see [19])

$$af + bKf + cLf = \varphi, \tag{4.14}$$

can be rewritten as

$$f + a^{-1}bKf + a^{-1}cLf = a^{-1}\varphi.$$

Hence from Theorem 4.1, we know that under some condition (4.14) is equivalent to a Fredholm equation.

5 Generalized Integrals and Integral Equations in Real Clifford Analysis

By using the method of resolution of the identity, in this section we define the generalized integrals in the sense of M. Spivak (see [73]) on open manifold for unbounded functions in real Clifford analysis, and discuss the solvability and the series expression of solutions for the second kind of generalized integral equations. Finally we give the error estimate for the approximate calculation.

Let $\Omega \subset R^n$ be an n-dimensional bounded manifold. We consider a class of functions belonging to $C_\Omega(\mathcal{A})$, where the functions are defined in Ω and with values in the real Clifford space $\mathcal{A}_n(R)$.

Now we give the definition of resolution of the identity.

Let Ω be as stated before, θ be an open covering on Ω, and for every $U \subset \theta$, we have $U \subset \Omega$. Then we call θ a permissible open covering on Ω. Thus there must exist a group of sets Φ of function φ belonging to C^∞; φ is defined on an open set including Ω, and satisfies

1. For every $x \in \Omega$, we have $0 \le \varphi(x) \le 1$.
2. For every $x \in \Omega$, there exists an open set V including x such that there exists a finite $\varphi \in \Phi$ which isn't equal to 0, V.
3. For every $x \in \Omega$, we have $\sum\limits_{\varphi\in\Phi} \varphi(x) = 1$.
4. For every $\varphi \in \Phi$, there exists an open set $\overline{U}$ belonging to θ such that φ is equal to 0 on a closed subset in U.

If $\Phi \subset C^\infty$ satisfies $1-3$, then Φ is called a resolution of the identity on Ω. If Φ also satisfies 4, then Φ is said to be a resolution of the identity belonging to θ. If Φ is a resolution of the identity belonging to θ on Ω, f is a function from Ω to R and f is bounded in an open set of each point in Ω, the measure of the set $\{x : f \text{ is discontinuous on } x\}$ is 0, then for any $\varphi \in \theta$, $\varphi|f|$ on Ω is integrable. If $\sum_{\varphi\in\Phi} \int_\Omega \varphi|f|$ converges, then f is said to be a generalized integrable function on Ω.

We consider the function class $C_\Omega(\mathcal{A})$, where the functions are defined in Ω and with values in the Clifford space.

Definition 5.1 Let $f(x) \in C_\Omega(\mathcal{A})$, θ be a permissible open covering on Ω, $\Phi(\subset C^\infty)$ be a resolution of the identity belonging to θ on Ω so that for any $\varphi \in \theta$, $\varphi|f|$ be integrable on Ω. If $\sum\limits_{\varphi\in\Phi} \int_\Omega \varphi|f|$ as series is convergent (see [73]), we say that f is integrable as a generalized integral in the sense of M. Spivak. The sum of the series is called the integral of f on Ω. All the generalized integrable functions in $C_\Omega(\mathcal{A})$ can be written as $I_\Omega(\mathcal{A})$.

In the following, the generalized integrable functions are considered such functions in the sense of M. Spivak.

Lemma 5.1 *If $\Omega \in R^n$ is as stated above, then $f(x) = \Sigma f_A(x)e_A \in I_\Omega(\mathcal{A})$, if and only if each $f_A(x)$ is generalized integrable on Ω.*

Lemma 5.2 *If $\Omega \in R^n$, $I_\Omega(\mathcal{A})$ is as stated above, then for any $f,\ g \in I_\Omega(\mathcal{A})$ and Clifford number $\lambda \in \mathcal{A}_n(R)$ we have $f + g \in I_\Omega(\mathcal{A})$ and $\lambda f \in I_\Omega(\mathcal{A})$.*

Proof Let $f, g \in I_\Omega(\mathcal{A})$, i.e. there exist a permissible open covering θ_1 on Ω and a resolution of the identity Φ_1 belonging to θ_1, such that f is bounded in an open set of each point in Ω. Hence, for any $\varphi \in \theta_1$, $\varphi|f|$ is integrable, and $\sum_{\varphi\in\Phi}\int_\Omega \varphi|f|$ as a series is convergent. At the same time, there exists another permissible open covering θ_2 on Ω and a resolution of the identity $\Phi_2 \in \theta_2$, such that g is bounded in one open set of each point in Ω, and then for any $\varphi \in \theta_1$, $\varphi|f|$ is integrable, and $\sum_{\varphi\in\Phi}\int_\Omega \varphi|g|$ as a series is convergent. Make a permissible open covering $\theta = \theta_1 \bigcup \theta_2 = \{U|U = u_1 \bigcup u_2, u_i \in \theta_i,\ i = 1,2\}$ on Ω and a resolution of the identity $\Phi = \{\varphi|\varphi = \frac{\varphi_1+\varphi_2}{2},\ \varphi_i \in \Phi_i,\ i = 1,2\}$ on Ω, then

$$\sum_{\varphi\in\Phi}\int_\Omega \varphi|f+g| \le \sum_{\varphi\in\Phi}\left[\int_\Omega \varphi|f| + \int_\Omega \varphi|g|\right]$$

$$\le \sum_{\varphi_1\in\Phi_1}\left[\int_\Omega \varphi_1|f| + \int_\Omega \varphi_1|g|\right] + \sum_{\varphi_2\in\Phi_2}\left[\int_\Omega \varphi_2|f| + \int_\Omega \varphi_2|g|\right],$$

and each integral on the right-hand side of the inequality is convergent. Hence the integral on the left-hand side is also convergent. That is to say that $f+g$ is generalized integrable. In the same way, if λ is a real Clifford number, then λf is generalized integrable too. The proof is completed.

Definition 5.2 Let $f(x,y) = \sum_A f_A(x,y)e_A$, $(x,y) \in \Omega \times \Omega$,where each $f_A(x,y) : \Omega \times \Omega \to R^1$ is a real function. If $f_A(x,y)$ is square integrable for each variable (the other variable is looked on as a constant), i.e. $|f|^2$ looked on as a function of x or y is generalized integrable and $\int_\Omega \int_\Omega |f(x,y)|^2 dxdy < \infty$. Then we define $f(x,y)$ as a square generalized integrable function. All the square generalized integrable functions can be written as $I_{\Omega\times\Omega}(\mathcal{A})$.

Lemma 5.3 *If $f(x,y), g(x,y) \in I_{\Omega\times\Omega}(\mathcal{A})$ and $\Phi^2 \in I_\Omega(\mathcal{A})$, then*

1. *About u (here x,y are seen as constants) $f(x,u)g(u,y) \in I_\Omega(\mathcal{A})$.*
2. *$\int_\Omega f(x,u)g(u,y)du \in I_{\Omega\times\Omega}(\mathcal{A})$.*
3. *$\int_\Omega f(x,u)\Phi^2(u)du \in I_\Omega(\mathcal{A})$.*

Definition 5.3 Let $K(x,y) = \sum_A K_A(x,y)e_A \in I_{\Omega\times\Omega}(\mathcal{A})$ be a function. We define the corresponding kernel as

$$K^0(x,y) = \sum_A K_A(x,y)e_A h_A,$$

where $h_A = h_{e_A} = h_{e_{r_1}\ldots e_{r_h}} = h_{e_{r_1}}\ldots h_{e_{r_h}}$, $A = \{r_1, \ldots, r_h\}$, and each h_{e_i} is a transformation

$$h_{e_i}(e_j) = \begin{cases} e_j, & i = j, \\ -e_j, & i \neq j, \end{cases}$$

and define h_A as a left exchange factor.

Definition 5.4 Define the equation

$$\varphi(x) - \lambda \int_\Omega K^0(x,u)\varphi(u)du = f(x) \tag{5.1}$$

as a second kind integral equation, where $\varphi(x)$ is an unknown function, $K^0(x,y)$ is a function defined in Definition 5.3, f is a known function satisfying $f^2 \in I_\Omega$, and $\lambda(\in \mathcal{A}_n(R))$ is a real Clifford constant. In the following, we find a solution of equation (5.1).

We find a solution of equation (5.1) by using a successive iteration. Let

$$\begin{cases} \varphi_0(x) = f(x), \\ \varphi_m(x) = f(x) + \lambda \int_\Omega K^0(x,u)\varphi_{m-1}(u)du, \; m = 1, 2.... \end{cases} \tag{5.2}$$

It is easy to prove that if the sequence of functions $\{\varphi_m(x)\}$ on Ω uniformly converges to a function, then the function is a solution of equation (5.1).

In order to study the property of $\{\varphi_m(x)\}$, we give the following definition.

Definition 5.5 If $K(x, y) \in I_{\Omega\times\Omega}(\mathcal{A})$ is as stated before, we define

$$K_2(x, y) = \int_\Omega K(x, t)K^0(t, y)dt,$$

$$K_m(x, y) = \int_\Omega K(x, t)K_{m-1}(t, y)dt,\ m \geq 3,$$

as the reiterative kernel of $K(x, y)$ $(m \geq 2)$ for m times.

Theorem 5.4 *If $K^0(x, y)$, $K(x, y)$, λ, $f(x)$ is as stated above, then*

$$\begin{cases} K^0(x, u)f(x) = f(x)K(x, u), \\ K^0(x, u)\lambda = \lambda K(x, u). \end{cases} \tag{5.3}$$

By the definition of the exchange factor, the theorem is easy to prove. Consider the sequence of functions $\{\varphi_m(x)\}$ as follows:

$$\varphi_1(x) = f(x) + \lambda \int_\Omega K^0(x, t)\varphi_0(t)dt,$$

$$\begin{aligned} \varphi_2(x) &= f(x) + \lambda \int_\Omega K^0(x, t)\varphi_1(t)dt \\ &= f(x) + \lambda \int_\Omega K^0(x, t) \left[f(t) + \lambda \int_\Omega K^0(t, u)f(u)du\right] dt \\ &= f(x) + \lambda \int_\Omega K^0(x, u)f(u) + (\lambda)^2 \int_\Omega \left[\int_\Omega K(x, t)K^0(t, u)f(u)du\right] dt \\ &= f(x) + \lambda \int_\Omega K^0(x, u)f(u) + (\lambda)^2 \int_\Omega K_2(x, u)f(u)du, \end{aligned}$$

$$\varphi_3(x) = \ldots.$$

In general, we have

$$\varphi_n(x) = f(x) + \lambda \int_\Omega K^0(x, u)f(u)du + \sum_{m=2}^{n} (\lambda)^m \int_\Omega K_m(x, u)f(u)du,$$

$$n = 1, 2, \ldots. \tag{5.4}$$

Theorem 5.5 *If $f(x) \in I_\Omega(\mathcal{A})$ and*

$$\int_\Omega |f(u)|^2 dv_u = H^2, \ |K(x,u)| \le M(u),$$

which is valid for each $x \in \Omega$, $u \in \Omega$ and $M(u)$ is a generalized square integrable function $\int_\Omega |M(u)|^2 dv_u = L^2$, where dv_u is the volume element of Ω, then

$$|\int_\Omega K_m(x,u)f(u)du| \le HL^m J_1^m, \ m = 1, 2, \ldots. \tag{5.5}$$

Proof If $m = 1$, we denote $K(x,u) = K^0(x,u)$ and have

$$|\int_\Omega K_1(x,u)f(u)du| \le \int_\Omega |K_1(x,u)f(u)|dv_u \le J_1 \int_\Omega |M(u)|\,|f(u)|dv_u$$

$$\le J_1 \left[\int_\Omega |M(u)|^2 dv_u\right]^{\frac{1}{2}} \left[\int_\Omega |f(u)|^2 dv_u\right]^{\frac{1}{2}} = J_1 HL.$$

Suppose that the estimate (5.5) is true for $m-1$; we shall prove that the estimate (5.5) is also true for m, i.e.

$$|\int_\Omega K_m(x,u)f(u)du| = |\int_\Omega \int_\Omega K(x,t)K_{(m-1)}(t,u)dt f(u)du|$$

$$= |\int_\Omega K(x,t)[\int_\Omega K_{(m-1)}(t,u)f(u)du]dt|$$

$$\le J_1 \left[\int_\Omega |K(x,t)|^2 dv_t\right]^{\frac{1}{2}} \left[\int_\Omega |\int_\Omega K_{(m-1)}(t,u)f(u)du|^2 dv_t\right]^{\frac{1}{2}}$$

$$\le J_1 \left[\int_\Omega |M(t)|^2 dv_t\right]^{\frac{1}{2}} \left[\int_\Omega J_1^{2(m-1)} H^2 dv_t\right]^{\frac{1}{2}} \le HL^m J_1^m.$$

Theorem 5.6 *Under the same result as in Theorem 5.2, then for the real Clifford number λ satisfying $|\lambda| < 1/J_1^2 L$, equation (5.1) has a unique solution, and the solution is the limit of the sequence of functions: (5.4).*

Proof From Theorem 5.4, it is easy to see that the above sequence of functions uniformly converges on Ω. Obviously its limit function is a solution of equation (5.1). In the following, we will prove the uniqueness of the solution. If there exists $\lambda \in \mathcal{A}_n(R)$, and $|\lambda| < 1/J_1^2 L$, then there exist two solutions $\varphi_1(x), \varphi_2(x)$ of equation (5.1), i.e.

$$\varphi_1(x) - \lambda \int_\Omega K^0(x,u)\varphi_1(u)du = f(x),$$

$$\varphi_2(x) - \lambda \int_\Omega K^0(x,y)\varphi_2(u)du = f(x).$$

Let the first equality be subtracted from the second one, then we get

$$\varphi_1(x) - \varphi_2(x) - \lambda \int_\Omega K^0(x,u)[\varphi_1(u) - \varphi_2(u)]du = 0.$$

Denote $\omega(x) = \varphi_1(x) - \varphi_2(x)$, then we have

$$\omega(x) = \lambda \int_\Omega K^0(x,u)\omega(u)du,$$

and then

$$|\omega(x)|^2 \le J_1^2|\lambda|^2 \int_\Omega |K^0(x,u)|^2 dv_u \int_\Omega |\omega(u)|^2 dv_u.$$

Making the volume integration go to x on Ω in both sides of the inequality, i.e.

$$\int_\Omega |\omega(x)|^2 dv_x J_1^2 \le L^2|\lambda|^2 \int_\Omega |\omega(u)|^2 dv_u,$$

we get

$$(1 - J_1^2 L^2 |\lambda|^2) \int_\Omega |\omega(u)|^2 dv_x \le 0,$$

and

$$\int_\Omega |\omega(u)|^2 dv_x = 0,\ \varphi_1 = \varphi_2.$$

The proof of Theorem 5.3 is finished.

Theorem 5.7 *Under the same conditions described in Theorem 5.4, we substitute $\{\varphi_n\}$ in (5.4) into the accurate solution of equation (5.1), then the norm error is not greater than*

$$\frac{|\lambda|^{n+1}L^{n+1}HJ_1^{2n+1}}{1 - |\lambda|^2 L^2 J_1^2}.$$

CHAPTER VI

SEVERAL KINDS OF HIGH ORDER SINGULAR INTEGRALS AND DIFFERENTIAL INTEGRAL EQUATIONS IN REAL CLIFFORD ANALYSIS

In the first section of this chapter, we shall introduce six kinds of high order singular integrals of quasi-Bochner-Martinelli type with one singular point, definitions of their Hadamard principal values, recurrence formulas, calculational formulas and differential formulas. In the second section, after proving the lemma of Hile type, we shall discuss the properties of high order singular integral operators and then prove the Hölder continuity of several kinds of high order singular integrals of quasi-Bochner-Martinelli type on the integral path. In the third section, we shall prove the existence and uniqueness of solutions for three kinds of nonlinear differential integral equations with high order singular integrals of quasi-Bochner-Martinelli type by the method of integral equations. In the fourth section, we shall give the definitions of high order singular integrals with two singular points, and prove the Poincaré-Bertrand permutation formulas for high order singular integrals of quasi-Bochner-Martinelli type in real Clifford analysis by using the differential formulas (see [29]10), [65]).

1 The Hadamard Principal Value and Differential Formulas of High Order Singular Integrals with One Singular Point in Real Clifford Analysis

First of all, we introduce the concept of Hadamard principal value of high order singular integrals with one singular point for functions of one complex variable. Suppose that L is a simple smooth closed curve, f' on L is Hölder continuous, and $f' \in H$. Let $\tau_o \in L$; it is clear that the integral

$$\int_L \frac{f(\tau)}{(\tau - \tau_0)^2} d\tau$$

at $\tau = \tau_0$ possesses a singularity of high order (> 1). In general, it is divergent, even under the sense of Cauchy principal value. For instance, if we define

$$\int_L \frac{f(\tau)}{(\tau-\tau_0)^2} d\tau = \lim_{\eta \to 0} \int_{L-L_\eta} \frac{f(\tau)}{(\tau-\tau_0)^2} d\tau, \tag{1.1}$$

in which $L_\eta = L \cap \{|\tau - \tau_0| \leq \eta\}$, then its limit on the right-hand side of the above equality usually doesn't exist. This is because we know

$$\int_{L-L_\eta} \frac{f(\tau)d\tau}{(\tau-\tau_0)^2} = \int_{L-L_\eta} \frac{f'(\tau)}{(\tau-\tau_0)} d\tau + \left[\frac{f(\tau_2)}{\tau_2-\tau_0} - \frac{f(\tau_1)}{\tau_1-\tau_0}\right] = I_1 + I_2,$$

by means of integration by parts, where $\tau_1, \tau_2 \in L \cap \{|\tau - \tau_0| = \eta\}$. Noting $f' \in H$, if $\eta \to 0$, then

$$I_1 \to \int_L \frac{f'(\tau)}{(\tau-\tau_0)} d\tau,$$

and

$$\begin{aligned}\lim_{\eta\to 0} I_2 &= \lim_{\eta\to 0}\{\frac{f(\tau_2)-f(\tau_0)}{\tau_2-\tau_0} - \frac{f(\tau_1)-f(\tau_0)}{\tau_1-\tau_0} + f(\tau_0)(\frac{1}{\tau_2-\tau_0} - \frac{1}{\tau_1-\tau_0})\} \\ &= f(\tau_0) \lim_{\eta\to 0}(\frac{1}{\tau_2-\tau_0} - \frac{1}{\tau_1-\tau_0}).\end{aligned}$$

This shows that this integral generally doesn't exist (except $f(\tau_0) = 0$) Hence we cannot define this kind of singular integrals by (1.1). Now we first consider the case: $n = 2 > 1$ on the left side of (1.1). It is a singular integral whose singularity is higher than one order. Moreover, we conclude that L and n satisfy conditions such that the integral

$$\int_{L-L_\eta} \frac{f(\tau)}{(\tau-\tau_0)^n} d\tau \tag{1.2}$$

can converge under the sense of Cauchy principal value. Some authors have discussed the problem and acquired some results. But for applications, we shall discuss the high order singular integral (1.2) from another view. This view was first proposed by J. Hadamard for similar singular integrals on the real axis, namely the idea of the so-called finite part of an integral [70]. In 1957, C. Fox generalized this idea to integral (1.2) with the positive integer n [16]. Afterwards Chuanrong Wang discussed the problem in [78], and then Jianke Lu in [45], generalized n to the case of a general positive real number and an integral with many singular points. Using the idea of Hadamard principal value, we can define

$$\int_L \frac{f(\tau)}{(\tau-\tau_0)^2} d\tau = \int_L \frac{f'(\tau)}{\tau-\tau_0} d\tau,$$

i.e.

$$\int_L \left(\frac{-1}{\tau-\tau_0}\right)' f(\tau)d\tau = \int_L -\left(\frac{-1}{\tau-\tau_0}\right) f'(\tau)d\tau,$$

where $-(\frac{-1}{\tau-\tau_0})$ possesses a lower order singularity than $(\frac{-1}{\tau-\tau_0})'$. Thus we define a high order singular integral by induction on an integral having a low order singularity. It is also said that this integral is defined by cutting out the divergent part I_2. This idea is easily generalized to a general high order singular integral (1.2). Provided that we do the integration by parts and cut out the terms with divergent part several times, then we can write the definition as desired. In [46], Jianke Lu straight forwardly defines

$$\int_L \frac{f(\tau)}{(\tau-\tau_0)^{n+1}}d\tau = \frac{1}{n!}\int_L \frac{f^{(n)}(\tau)}{\tau-\tau_0}d\tau, \quad \tau_0 \in L.$$

In 1990, Xiaoqin Wang obtained some results about the Hadamard principal value of high order singular integrals for functions of several complex variables (see [79]).

In this chapter, we discuss the first kind of function in Clifford analysis $f(x) : \mathbf{R^n} \to \mathcal{A}_n(R)$, where the element in $\mathcal{A}_n(R)$ is $u = \sum\limits_A u_A e_A$ ($u_A \in \mathbf{R}$), the element in $\mathbf{R^n}$ is $x = \sum_{k=1}^n x_k e_k$, herein $e_1 = 1$, $\bar{x} = x_1 e_1 - \sum_{k=2}^n x_k e_k$, and denote the operator

$$\bar{\partial}_x = \sum_{k=1}^n e_k \frac{\partial}{\partial x_k}, \ \partial_x = e_1\frac{\partial}{\partial x_1} - \sum_{k=2}^n e_k \frac{\partial}{\partial x_k}.$$

Before giving the induction definition of high order singular integrals, we prove several lemmas.

Lemma 1.1 *Let* $u(x) = \sum\limits_A u_A(x)e_A$, $v(x) = \sum\limits_{i=1}^n v_i(x)e_i$, $x \in \mathbf{R^n}$, $(\ )_{x_j} = \frac{\partial(\ )}{\partial x_j}$. $\frac{\partial u_A}{\partial x_k}$, $(v_i)_{x_k}$ *be continuous. Then*

$$\overline{\partial}_x(uv) = (\overline{\partial}_x u)v + u(\overline{\partial}_x v) + \sum_{j=2}^n (e_j u - u e_j)v_{x_j}, \tag{1.3}$$

$$\partial_x(uv) = (\partial_x u)v + u(\partial_x v) + \sum_{j=2}^n (u e_j - e_j u)v_{x_j}, \tag{1.4}$$

$$\begin{aligned} & u(\overline{\partial}_x\overline{v} + \partial_x v) \\ = \; & -(\overline{\partial}_x u)\overline{v} - (\partial_x u)v + \overline{\partial}_x(u\overline{v}) + \partial_x(uv) \\ & - \sum_{j=2}^{n}(e_j u - ue_j)\overline{v}_{x_j} - \sum_{j=2}^{n}(ue_j - e_j u)v_{x_j}. \end{aligned} \tag{1.5}$$

Proof It is clear that

$$\begin{aligned} \overline{\partial}_x(uv) &= \sum_{j=1}^{n} e_j \sum_A \sum_{i=1}^{n} (u_A v_i)_{x_j} e_A e_i \\ &= \sum_{j=1}^{n} e_j \sum_A (u_A)_{x_j} e_A \sum_{i=1}^{n} v_i e_i + \sum_{j=1}^{n} e_j \sum_A (u_A) e_A \sum_{i=1}^{n} (v_i)_{x_j} e_i \\ &= (\overline{\partial}_x u)v + \sum_{j=1}^{n} e_j u(v_{x_j}) = (\overline{\partial}_x u)v + ue_1 v_{x_1} + \sum_{j=2}^{n} e_j u(v_{x_j}) \\ &= (\overline{\partial}_x u)v + ue_1 v_{x_1} + u\sum_{j=2}^{n} e_j v_{x_j} + \sum_{j=2}^{n}(e_j u - ue_j)v_{x_j} \\ &= (\overline{\partial}_x u)v + u(\overline{\partial}_x v) + \sum_{j=2}^{n}(e_j u - ue_j)v_{x_j}. \end{aligned}$$

Similarly, we can prove (1.4). Substitute $\bar{v}$ into the position of v in (1.3), and add it to (1.4), then we immediately get (1.5).

Corollary 1.2 *If the conditions in Lemma 1.1 are satisfied, then*

(1) If v is independent of x, then $\partial_x(uv) = (\partial_x u)v$.

(2) If $u = u_1 e_1$, then

$$\bar{\partial}_x(uv) = (\bar{\partial}_x u)v + u(\bar{\partial}_x v); \; \partial_x(uv) = (\partial_x u)v + u(\partial_x v),$$

$$u(\bar{\partial}_x\bar{v} + \partial_x v) = -(\bar{\partial}_x u)\bar{v} - (\partial_x u)v + \overline{\partial}_x(u\bar{v}) + \partial_x(uv).$$

Lemma 1.3 *Let $\alpha > 0$, $x, y \in \mathbf{R^n}$, $x \neq y$, $v_1 = \dfrac{x-y}{(-\alpha)|x-y|^{n+\alpha}}$, $v_2 = \dfrac{1}{(2-n-\alpha)|x-y|^{n+\alpha-2}}$. Then*

$$\frac{\partial v_1}{\partial x_j} = \frac{e_j}{(-\alpha)|x-y|^{n+\alpha}} - \frac{(n+\alpha)(x_j-y_j)}{|x-y|^2} v_1, \tag{1.6}$$

$$\frac{\partial \bar{v}_1}{\partial x_j} = \frac{\bar{e}_j}{(-\alpha)|x-y|^{n+\alpha}} - \frac{(n+\alpha)(x_j-y_j)}{|x-y|^2} \bar{v}_1, \tag{1.7}$$

$$\frac{\partial v_2}{\partial x_j} = \frac{x_j - y_j}{|x-y|^{n+\alpha}}, \ 1 \le j \le n, \tag{1.8}$$

where

$$\bar{e}_j = \begin{cases} e_1, & j = 1, \\ -e_j, & 2 \le j \le n, \end{cases} \qquad x - y = \sum_{j=1}^{n} (x_j - y_j) e_j.$$

Proof Noting

$$\begin{aligned} \frac{\partial v_1}{\partial x_j} &= \frac{-1}{\alpha} \sum_{i=1}^{n} [(x_i - y_i)|x-y|^{-n-\alpha}]_{x_j} e_i \\ &= \frac{-1}{\alpha} e_j |x-y|^{-n-\alpha} \\ &\quad + \frac{-1}{\alpha} \sum_{i=1}^{n} (x_i - y_i) e_i \frac{-n-\alpha}{2} \Big(\sum_{i=1}^{n} (x_i - y_i)^2\Big)^{\frac{-n-\alpha-2}{2}} \cdot 2(x_j - y_j) \\ &= \frac{e_j}{-\alpha|x-y|^{n+\alpha}} - (n+\alpha)(x_j - y_j) \frac{v_1}{|x-y|^2}, \end{aligned}$$

and substituting $\overline{v}_1$ into the position of v_1 in (1.6), we get (1.7), and

$$\frac{\partial v_2}{\partial x_j} = \frac{-n-\alpha+2}{2(2-n-\alpha)} \Big(\sum_{i=1}^{n} (x_i - y_i)^2\Big)^{\frac{-n-\alpha}{2}} \cdot 2(x_j - y_j) = \frac{x_j - y_j}{|x-y|^{n+\alpha}},$$

hence (1.8) is valid.

Lemma 1.4 *Let $x, y \in R^n$, $x \ne y$, $\alpha > 0$. Then*

$$\overline{\partial}_x(\bar{x} - \bar{y}) = n, \ \partial_x(x - y) = n; \tag{1.9}$$

$$\begin{cases} \overline{\partial}_x |x-y|^\sigma = \sigma |x-y|^{\sigma-2}(x-y), \\ \partial_x |x-y|^\sigma = \sigma |x-y|^{\sigma-2}(\overline{x} - \overline{y}), \ \sigma > 0; \end{cases} \tag{1.10}$$

$$\overline{\partial}_x\Big(\frac{\bar{x} - \overline{y}}{(-\alpha)|x-y|^{n+\alpha}}\Big) = \partial_x\Big(\frac{x-y}{(-\alpha)|x-y|^{n+\alpha}}\Big) = \frac{1}{|x-y|^{n+\alpha}}; \tag{1.11}$$

$$\begin{cases} \overline{\partial}_x\Big(\dfrac{1}{(2-n-\alpha)|x-y|^{n+\alpha-2}}\Big) = \dfrac{x-y}{|x-y|^{n+\alpha}}, \\ \partial_x\Big(\dfrac{1}{(2-n-\alpha)|x-y|^{n+\alpha-2}}\Big) = \dfrac{\bar{x} - \bar{y}}{|x-y|^{n+\alpha}}. \end{cases} \tag{1.12}$$

Proof It is evident that

$$\bar{\partial}_x(\bar{x}-\bar{y}) = \sum_{j=1}^{n} e_j \frac{\partial}{\partial x_j}[(x_1-y_1)e_1 - \sum_{i=2}^{n}(x_i-y_i)e_i]$$

$$= e_1e_1 + \sum_{j=2}^{n} e_j(-1)e_j = n.$$

Similarly, we can prove $\partial_x(x-y) = n$. In addition, noting that

$$\begin{aligned}\bar{\partial}_x|x-y|^\sigma &= \sum_{j=1}^{n} e_j \frac{\partial}{\partial x_j}(\sum_{i=1}^{n}(x_i-y_i)^2)^{\frac{\sigma}{2}}\\ &= \sum_{j=1}^{n} e_j \frac{\sigma}{2}(\sum_{i=1}^{n}(x_i-y_i)^2)^{\frac{\sigma}{2}-1}\cdot 2(x_j-y_j)\\ &= \sigma|x-y|^{\sigma-2}\sum_{j=1}^{n}(x_j-y_j)e_j\\ &= \sigma|x-y|^{\sigma-2}(x-y),\end{aligned}$$

(1.10) is derived. Moreover we can prove $\partial_x|x-y|^\sigma = \sigma|x-y|^{\sigma-2}(\bar{x}-\bar{y})$. By means of Corollary 1.2 and (1.10), we get

$$\begin{aligned}&\bar{\partial}_x(\frac{\bar{x}-\bar{y}}{(-\alpha)|x-y|^{n+\alpha}})\\ =& [\bar{\partial}_x(\frac{|x-y|^{-n-\alpha}}{-\alpha})](\bar{x}-\bar{y}) + \frac{1}{(-\alpha)|x-y|^{n+\alpha}}\bar{\partial}_x(\overline{x}-\overline{y})\\ =& \frac{n+\alpha}{\alpha}|x-y|^{-n-\alpha-2}(x-y)(\bar{x}-\bar{y}) + \frac{n}{-\alpha|x-y|^{n+\alpha}}\\ =& (\frac{n}{\alpha}+1)\frac{1}{|x-y|^{n+\alpha}} + \frac{n}{-\alpha|x-y|^{n+\alpha}}\\ =& \frac{1}{|x-y|^{n+\alpha}}.\end{aligned}$$

Finally, we can similarly prove $\partial_x(\frac{x-y}{-\alpha|x-y|^{n+\alpha}}) = \frac{1}{|x-y|^{n+\alpha}}$, namely (1.11) is true. On the basis of (1.10), the equality (1.12) can be derived.

Lemma 1.5 *Let* $u(x) = \sum_A u_A(x)e_A$, $v(x) = \sum_{i=1}^{n} v_i(x)e_i$, *and the conditions of Lemma 1.1 be satisfied. Then*

$$\bar{\partial}_x(vu) = (\bar{\partial}_x v)u + v(\bar{\partial}_x u) - 2\sum_{i,j=2,i\neq j}^{n} v_ie_ie_ju_{x_j}, \qquad (1.13)$$

$$\partial_x(vu) = (\partial_x v)u + v(\partial_x u) + 2 \sum_{i,j=2, i\neq j}^{n} v_i e_i e_j u_{x_j}. \tag{1.14}$$

Proof Taking into account

$$\begin{aligned}
\bar{\partial}_x(vu) &= \sum_{j=1}^{n} e_j \sum_{i=1}^{n} \sum_{A} (v_i u_A)_{x_j} e_i e_A \\
&= \sum_{j=1}^{n} e_j v_{x_j} u + \sum_{j=1}^{n} e_j \sum_{i=1}^{n} v_i e_i \sum_{A} (u_A)_{x_j} e_A \\
&= (\bar{\partial}_x v)u + v e_1 u_{x_1} + \sum_{j=2}^{n} e_j v u_{x_j} \\
&= (\bar{\partial}_x v)u + v(e_1 u_{x_1} + \sum_{j=2}^{n} e_j u_{x_j}) + \sum_{j=2}^{n} e_j v u_{x_j} - v \sum_{j=2}^{n} e_j u_{x_j} \\
&= (\bar{\partial}_x v)u + v(\bar{\partial}_x u) + \sum_{j=2}^{n} \sum_{i=1}^{n} v_i e_j e_i u_{x_j} - \sum_{j=2}^{n} \sum_{i=1}^{n} v_i e_i e_j u_{x_j} \\
&= (\bar{\partial}_x v)u + v(\bar{\partial}_x u) - 2 \sum_{i,j=2, i\neq j}^{n} v_i e_i e_j u_{x_j},
\end{aligned}$$

(1.13) is derived. Similarly we can prove (1.14).

Corollary 1.6 *If the conditions in Theorem 1.4 are satisfied, then*

1) If u is independent of x, then $\bar{\partial}_x(vu) = (\bar{\partial}_x v)u$, $\partial_x(vu) = (\partial_x v)u$.

2) When $v = v_1 e_1$, then $\bar{\partial}_x(vu) = (\bar{\partial}_x v)u + v(\bar{\partial}_x u)$, $\partial_x(vu) = (\partial_x v)u + v(\partial_x u)$.

Suppose that D is a connected open set in $\mathbf{R}^n$, Ω is the boundary of D, and $\bar{\partial}_x f(x,y)$, $\partial_x f(x,y)$ $(x, y \in \Omega)$ are Hölder continuous. In (1.4), set

$$v = \frac{1}{(2-n-\alpha)|x-y|^{n+\alpha-2}}, \quad u = f(x,y);$$

by (1.12), (1.8), we have

$$f(x,y)\frac{\overline{x}-\overline{y}}{|x-y|^{n+\alpha}} = \partial_x\left[f(x,y)\frac{1}{(2-n-\alpha)|x-y|^{n+\alpha-2}}\right]$$
$$-\frac{\partial_x f(x,y)}{(2-n-\alpha)|x-y|^{n+\alpha-2}} - \sum_{j=2}^{n}[f(x,y)e_j - e_j f(x,y)]\frac{x_j - y_j}{|x-y|^{n+\alpha}},$$

in which $\dfrac{\overline{x}-\overline{y}}{|x-y|^{n+\alpha}}$ has $\alpha\,(>0)$ order singularity, and $\dfrac{x_j-y_j}{|x-y|^{n+\alpha}}$ also has a high order singularity. By means of the above equality, we know

$\partial_x[\frac{f(x,y)}{(2-n-\alpha)|x-y|^{n+\alpha-2}}]$ also has $\alpha(>0)$ order singularity. However $\frac{1}{|x-y|^{n+\alpha-2}}$ possesses a lower order singularity than $\frac{\overline{x}-\overline{y}}{|x-y|^{n+\alpha}}$. Hence according to the idea of Hadamard's principal value of integrals and using the lower order singular integrals to inductively define the high order singular integrals, we can give the following definitions.

Definition 1.1 The $\alpha\,(>0)$ order singular integral is defined as

$$\int_\Omega \frac{f(x,y)(\bar{x}-\bar{y})}{|x-y|^{n+\alpha}}d\sigma_x = \int_\Omega \frac{\partial_x f(x,y)d\sigma_x}{(n+\alpha-2)|x-y|^{n+\alpha-2}},\ y\in\Omega.$$

Similarly, in (1.3) let $v = \frac{1}{(2-n-\alpha)|x-y|^{n+\alpha-2}}$, $u = f(x,y)$; by means of (1.12), (1.8), we can give the following definition.

Definition 1.2

$$\int_\Omega \frac{f(x,y)(x-y)}{|x-y|^{n+\alpha}}d\sigma_x = \int_\Omega \frac{\bar{\partial}_x f(x,y)d\sigma_x}{(n+\alpha-2)|x-y|^{n+\alpha-2}},\ \alpha>0,\ y\in\Omega.$$

In (1.5), let $u = f(x,y)$, $v = \frac{x-y}{(-\alpha)|x-y|^{n+\alpha}}$; in view of (1.11), (1.6), (1.7), we have

$$\begin{aligned}
&\frac{2f(x,y)}{|x-y|^{n+\alpha}}\\
&= -\left[\frac{(\bar{\partial}_x f(x,y))(\bar{x}-\bar{y})}{(-\alpha)|x-y|^{n+\alpha}} + (\partial_x f(x,y))\frac{x-y}{(-\alpha)|x-y|^{n+\alpha}}\right]\\
&\quad +\bar{\partial}_x\left[\frac{f(x,y)(\bar{x}-\bar{y})}{(-\alpha)|x-y|^{n+\alpha}}\right] + \partial_x\left[\frac{f(x,y)(x-y)}{(-\alpha)|x-y|^{n+\alpha}}\right]\\
&\quad -\sum_{j=2}^n (e_j f(x,y) - f(x,y)e_j)\,\aleph - \sum_{j=2}^n (f(x,y)e_j - e_j f(x,y))\,\overline{\aleph},
\end{aligned}$$

where

$$\aleph = \left(\frac{\overline{e}_j}{(-\alpha)|x-y|^{n+\alpha}} - \frac{(n+\alpha)(x_j-y_j)}{|x-y|^2}\cdot\frac{(\overline{x}-\overline{y})}{(-\alpha)|x-y|^{n+\alpha}}\right).$$

Similarly to the above discussion, we see that in the right-hand side of the above equality, the term $\frac{1}{|x-y|^{n+\alpha}}$ possesses an $\alpha+1$ order singularity and other terms $\frac{\overline{x}-\overline{y}}{(-\alpha)|x-y|^{n+\alpha}}$, $\frac{x-y}{(-\alpha)|x-y|^{n+\alpha}}$ have an $\alpha\,(>0)$ order singularity. So we can inductively give

Definition 1.3 The integral

$$\int_\Omega \frac{f(x,y)d\sigma_x}{|x-y|^{n+\alpha}} = \int_\Omega \frac{(\bar{\partial}_x f)(\bar{x}-\bar{y}) + (\partial_x f)(x-y)}{2\alpha|x-y|^{n+\alpha}} d\sigma_x, \ \alpha > 0, \ y \in \Omega$$

is called the first kind of high order singular integral of quasi-Bochner-Martinelli type with one singular point.

In (1.14), set

$$u = f(x,y), \ v = \frac{1}{(2-n-\alpha)|x-y|^{n+\alpha-2}};$$

by means of the components $v_i = 0\,(2 \leq i \leq n)$ of v and (1.12), we can get

$$\frac{(\bar{x}-\bar{y})f(x,y)}{|x-y|^{n+\alpha}} = \partial_x[\frac{f(x,y)}{(2-n-\alpha)|x-y|^{n+\alpha-2}}] - \frac{\partial_x f(x,y)}{(2-n-\alpha)|x-y|^{n+\alpha-2}}.$$

Similarly to the above discussion, we see that $\dfrac{1}{|x-y|^{n+\alpha-2}}$ has a lower order singularity than the α order of $\dfrac{\bar{x}-\bar{y}}{|x-y|^{n+\alpha}}$, so we can also give

Definition 1.4

$$\int_\Omega \frac{(\bar{x}-\bar{y})f(x,y)d\sigma_x}{|x-y|^{n+\alpha}} = \int_\Omega \frac{\partial_x f(x,y)d\sigma_x}{(n+\alpha-2)|x-y|^{n+\alpha-2}}, \ \alpha > 0, \ y \in \Omega.$$

By means of (1.12), (1.13), we can define other high order singular integrals through a similar method.

Definition 1.5

$$\int_\Omega \frac{(x-y)f(x,y)d\sigma_x}{|x-y|^{n+\alpha}} = \int_\Omega \frac{\bar{\partial}_x f(x,y)d\sigma_x}{(n+\alpha-2)|x-y|^{n+\alpha-2}}, \ \alpha > 0, \ y \in \Omega.$$

In view of Definitions 1.1, 1.2, 1.4 and 1.5, we get

$$\int_\Omega \frac{(\bar{x}-\bar{y})f(x,y)d\sigma_x}{|x-y|^{n+\alpha}} = \int_\Omega \frac{f(x,y)(\bar{x}-\bar{y})}{|x-y|^{n+\alpha}} d\sigma_x, \ \alpha > 0, \ y \in \Omega, \quad (1.15)$$

$$\int_\Omega \frac{(x-y)f(x,y)d\sigma_x}{|x-y|^{n+\alpha}} = \int_\Omega \frac{f(x,y)(x-y)}{|x-y|^{n+\alpha}} d\sigma_x, \ \alpha > 0, \ y \in \Omega. \quad (1.16)$$

The above two high order singular integrals are called the second and third kinds of high order singular integrals of quasi-Bochner-Martinelli type with one singular point respectively.

Definition 1.6 If $f(x,y)$, $(x,y) \in \Omega \times \Omega$ is still Hölder continuous after the action of the operator $\bar{\partial}_x$, ∂_x for $p\,(\le m)$ times, and $0 < \beta < 1$ is its Hölder index, then we say that $f(x,y)$ belongs to $H_x^{(m)}(\beta)$, and write $f \in H_x^{(m)}(\beta)$. When $m = 0$, denote by $f \in H_x^{(0)}(\beta)$ the Hölder continuity of f on $x \in \Omega$, where β is the Hölder index. Similarly we can define $f \in H_y^{(m)}(\beta)$.

In the following, we prove the recurrence formulas of the first, second and third kinds of high order singular integrals of quasi-Bochner-Martinelli type.

Theorem 1.7 *Let $n > 1$, $\alpha > 2m > 0$, $f(x,y) \in H_x^{(2m+2)}(\beta)$, $y \in \Omega$. Then*

$$\int_\Omega \frac{f(x,y)(\bar{x}-\bar{y})d\sigma_x}{|x-y|^{n+\alpha}} = \frac{\mu}{n+\alpha-2m-2}\int_\Omega \frac{\Delta_x^m(\partial_x f(x,y))d\sigma_x}{|x-y|^{n+\alpha-2m-2}}, \quad (1.17)$$

$$\int_\Omega \frac{f(x,y)(x-y)}{|x-y|^{n+\alpha}}d\sigma_x = \frac{\mu}{n+\alpha-2m-2}\int_\Omega \frac{\Delta_x^m(\bar{\partial}_x f(x,y))d\sigma_x}{|x-y|^{n+\alpha-2m-2}}, \quad (1.18)$$

$$\int_\Omega \frac{f(x,y)d\sigma_x}{|x-y|^{n+\alpha}} = \frac{\mu}{\alpha(n+\alpha-2m-2)}\int_\Omega \frac{\Delta_x^{m+1} f(x,y)d\sigma_x}{|x-y|^{n+\alpha-2m-2}}, \quad (1.19)$$

where the operator $\Delta_x = \partial_x\bar{\partial}_x = \bar{\partial}_x\partial_x$,

$$\mu = \frac{(\alpha-2m-2)!!(n+\alpha-2m-2)!!}{(\alpha-2)!!(n+\alpha-2)!!},$$

and $r!!$ expresses the multiplication of the integers from r to the least integer every time decrease 2, and if $-2 < r \le 0$, denote $r!! = 1$, then

$$\mu = \frac{(n+\alpha-2m-2)!!}{(n+\alpha-2)!!(\alpha-2)(\alpha-4)\cdots(\alpha-2m)}. \quad (1.20)$$

Proof On the basis of Definitions 1.1, 1.2 and 1.3, we have

$$\int_\Omega \frac{f(x,y)(\bar{x}-\bar{y})d\sigma_x}{|x-y|^{n+\alpha}} = \int_\Omega \frac{\partial_x f(x,y)d\sigma_x}{(n+\alpha-2)|x-y|^{n+\alpha-2}}$$
$$= \int_\Omega \frac{(\overline{\partial}_x\partial_x f)(\overline{x}-\overline{y}) + (\partial_x\partial_x f)(x-y)}{2(n+\alpha-2)(\alpha-2)|x-y|^{n+\alpha-2}}d\sigma_x$$
$$= \int_\Omega \frac{[\Delta_x\partial_x f(x,y)]d\sigma_x}{(n+\alpha-2)(n+\alpha-4)(\alpha-2)|x-y|^{n+\alpha-4}}$$

$$= \int_\Omega \frac{[(\overline{\partial}_x \Delta_x \partial_x f)(\overline{x}-\overline{y}) + (\partial_x \Delta_x \partial_x f)(x-y)]d\sigma_x}{2(n+\alpha-2)(n+\alpha-4)(\alpha-2)(\alpha-4)|x-y|^{n+\alpha-4}}$$
$$= \int_\Omega \frac{[\Delta_x^2 \partial_x f(x,y)]d\sigma_x}{(n+\alpha-2)(n+\alpha-4)(n+\alpha-6)(\alpha-2)(\alpha-4)|x-y|^{n+\alpha-6}}$$
$$= \frac{(\alpha-6)!!(n+\alpha-6)!!}{(n+\alpha-6)(\alpha-2)!!(n+\alpha-2)!!}\int_\Omega \frac{[\Delta_x^2 \partial_x f(x,y)]d\sigma_x}{|x-y|^{n+\alpha-6}}.$$

Inductively, we can get

$$\int_\Omega \frac{f(x,y)(\bar{x}-\bar{y})d\sigma_x}{|x-y|^{n+\alpha}} = \frac{\mu}{n+\alpha-2m-2}\int_\Omega \frac{(\Delta_x{}^m \partial_x f(x,y))d\sigma_x}{|x-y|^{n+\alpha-2m-2}}.$$

This shows that (1.17) is valid. Similarly, we can prove (1.18). By means of Definitions 1.1, 1.2 and 1.3, we know

$$\int_\Omega \frac{f(x,y)d\sigma_x}{|x-y|^{n+\alpha}} = \frac{1}{2\alpha}\int_\Omega \frac{(\overline{\partial}_x f)(\overline{x}-\overline{y}) + (\partial_x f)(x-y)}{|x-y|^{n+\alpha}} d\sigma_x$$
$$= \frac{1}{\alpha(n+\alpha-2)}\int_\Omega \frac{(\Delta_x f)d\sigma_x}{|x-y|^{n+\alpha-2}}$$
$$= \frac{1}{2\alpha(n+\alpha-2)(\alpha-2)}\int_\Omega \frac{[(\overline{\partial}_x \Delta_x f)(\overline{x}-\overline{y}) + (\partial_x \Delta_x f)(x-y)]}{|x-y|^{n+\alpha-2}} d\sigma_x$$
$$= \int_\Omega \frac{[\Delta_x^2 f(x,y)]d\sigma_x}{\alpha(\alpha-2)(n+\alpha-2)(n+\alpha-4)|x-y|^{n+\alpha-4}}$$
$$= \frac{(\alpha-4)!!(n+\alpha-4)!!}{\alpha(n+\alpha-4)(\alpha-2)!!(n+\alpha-2)!!}\int_\Omega \frac{[\Delta_x^2 f(x,y)]d\sigma_x}{|x-y|^{n+\alpha-4}}.$$

Moreover, it is easy to verify (1.19).

Theorem 1.8 *Let $f(x,y) \in H_x^{(2k+2)}(\beta)$, $y \in \Omega$, $n > 1$, $0 < r < 1$, $\lambda = \dfrac{(n-r-3)!!}{(n+2k-r-1)!!(2k-r-1)!!}$. Then there exist the first, second and third kinds of high order singular integrals of quasi-Bochner-Martinelli type. Moreover, they can be expressed in the forms*

$$\int_\Omega \frac{f(x,y)(\bar{x}-\bar{y})d\sigma_x}{|x-y|^{n+2k+1-r}} = \lambda \int_\Omega \frac{[\Delta_x^k \partial_x f(x,y)]d\sigma_x}{|x-y|^{n-1-r}}, \tag{1.21}$$

$$\int_\Omega \frac{f(x,y)(x-y)d\sigma_x}{|x-y|^{n+2k+1-r}} = \lambda \int_\Omega \frac{[\Delta_x^k \bar{\partial}_x f(x,y)]d\sigma_x}{|x-y|^{n-1-r}}, \tag{1.22}$$

$$\int_\Omega \frac{f(x,y)d\sigma_x}{|x-y|^{n+2k+1-r}} = \frac{\lambda}{2k+1-r}\int_\Omega \frac{[\Delta_x^{k+1} f(x,y)]d\sigma_x}{|x-y|^{n-1-r}}. \tag{1.23}$$

Proof In Theorem 1.7, setting $\alpha = 2k+1-r$, $m=k$, and according to (1.20), we get

$$\frac{\mu}{n+\alpha-2m-2} = \frac{(n-r-1)!!}{(n-r-1)(n+2k-r-1)!!(2k-r-1)!!} = \lambda,$$

hence (1.21), (1.22), (1.23) are correct. The integrals on the right-hand side of (1.21), (1.22),(1.23) converge under the general sense as in Section 4, Chapter II. So there exist the above three kinds of high order singular integrals of quasi-Bochner-Martinelli type, and then (1.21), (1.22), (1.23) are their calculational formulas. This completes the proof.

The values calculated by (1.21) (1.22) (1.23) are called the Hadamard principal value of the first, second and third kinds of high order singular integrals of quasi-Bochner-Martinelli type. In the following, we derive the differential formulas of three kinds of high order singular integrals of quasi-Bochner-Martinelli type.

Theorem 1.9 *Let* $u(y) = \sum\limits_A u_A(y)e_A$, $(u_A(y))_{y_k}$ *be continuous, and* $v_3 = \dfrac{1}{|x-y|^{n-1-r}}$, $0<r<1$, $x,y \in R^n$, $x \neq y$. *Then*

$$\sum_{j=2}^{n}(ue_j - e_ju)\frac{\partial v_3}{\partial y_j} = (n-r-1)\left[\frac{(\bar{x}-\bar{y})u}{|x-y|^{n+1-r}} - \frac{u(\bar{x}-\bar{y})}{|x-y|^{n+1-r}}\right], \quad (1.24)$$

$$\sum_{j=2}^{n}(e_ju - ue_j)\frac{\partial v_3}{\partial y_j} = (n-r-1)\left[\frac{(x-y)u}{|x-y|^{n+1-r}} - \frac{u(x-y)}{|x-y|^{n+1-r}}\right]. \quad (1.25)$$

Proof In accordance with (1.8), we have

$$\frac{\partial v_3}{\partial y_j} = (n-r-1)\frac{(x_j-y_j)}{|x-y|^{n+1-r}},$$

and then

$$\begin{aligned}
&\sum_{j=2}^{n}(ue_j - e_ju)\frac{\partial v_3}{\partial y_j} \\
= \ & (n-r-1)\left[u\sum_{j=2}^{n}\frac{(x_j-y_j)e_j}{|x-y|^{n+1-r}} - \sum_{j=2}^{n}\frac{(x_j-y_j)e_ju}{|x-y|^{n+1-r}}\right] \\
& +(n-1-r)\left[(\frac{-u(x_1-y_1)e_1}{|x-y|^{n+1-r}} + \frac{(x_1-y_1)e_1u}{|x-y|^{n+1-r}}\right] \\
= \ & (n-r-1)\left[\frac{(\bar{x}-\bar{y})u}{|x-y|^{n+1-r}} - \frac{u(\bar{x}-\bar{y})}{|x-y|^{n+1-r}}\right].
\end{aligned}$$

Thus (1.24) is obtained. By means of $[(\bar{x}-\bar{y})+(x-y)]u-u[(\bar{x}-\bar{y})+(x-y)]=0$, we know that (1.25) is true.

Definition 1.7 Let $f(x,y)\in H_x^{(m)}(\beta_1)$, $(x,y)\in\Omega\times\Omega$, and $f(x,y)\in H_y^{(p)}(\beta_2)$, $0<\beta_i<1$, $i=1,2$. Then we say that $f(x,y)$ belongs to $H^{(m,p)}(\beta_1,\beta_2)$, and is denoted by $f(x,y)\in H^{(m,p)}(\beta_1,\beta_2)$.

Theorem 1.10 *Let $f(x,y)\in H^{(m+2k+2,m)}(\beta_1,\beta_2)$, $0<\beta_i<1$, $i=1,2$, $0<r<1$, λ be as stated in Theorem 1.8. Then*

$$\partial_y^m\int_\Omega\frac{f(x,y)(\bar{x}-\bar{y})d\sigma_x}{|x-y|^{n+2k+1-r}}=\lambda\int_\Omega\frac{[(\partial_y+\partial_x)^m\Delta_x^k\partial_x f(x,y)]}{|x-y|^{n-1-r}}d\sigma_x\,,\tag{1.26}$$

$$\partial_y^m\int_\Omega\frac{f(x,y)(x-y)d\sigma_x}{|x-y|^{n+2k+1-r}}=\lambda\int_\Omega\frac{[(\partial_y+\partial_x)^m\Delta_x^k\bar{\partial}_x f(x,y)]d\sigma_x}{|x-y|^{n-1-r}}\,,\tag{1.27}$$

$$\partial_y^m\int_\Omega\frac{f(x,y)d\sigma_x}{|x-y|^{n+2k+1-r}}=\frac{\lambda}{2k+1-r}\int_\Omega\frac{[(\partial_y+\partial_x)^m\Delta_x^{k+1}f(x,y)]d\sigma_x}{|x-y|^{n-1-r}}.\tag{1.28}$$

Proof On the basis of (1.12), we have

$$\partial_y\frac{1}{|x-y|^{n-1-r}}=-(n-1-r)\frac{\overline{y}-\overline{x}}{|x-y|^{n+1-r}}=(n-1-r)\frac{\overline{x}-\overline{y}}{|x-y|^{n+1-r}}.$$

By means of (1.21), (1.4), (1.24), (1.15) and Definition 1.1, we obtain

$$\begin{aligned}
&\partial_y\int_\Omega\frac{f(x,y)(\bar{x}-\bar{y})d\sigma_x}{|x-y|^{n+2k+1-r}}\\
&=\lambda\int_\Omega\frac{[\partial_y\Delta_x^k\partial_x f(x,y)]d\sigma_x}{|x-y|^{n-1-r}}+\lambda(n-1-r)\int_\Omega\frac{[\Delta_x^k\partial_x f(x,y)](\overline{x}-\overline{y})d\sigma_x}{|x-y|^{n+1-r}}\\
&\quad+\lambda(n-1-r)\int_\Omega\frac{(\overline{x}-\overline{y})[\Delta_x^k\partial_x f(x,y)]d\sigma_x}{|x-y|^{n+1-r}}\\
&\quad-\lambda(n-1-r)\int_\Omega\frac{[\Delta_x^k\partial_x f(x,y)](\overline{x}-\overline{y})d\sigma_x}{|x-y|^{n+1-r}}\\
&=\lambda\int_\Omega\frac{(\partial_y\Delta_x^k\partial_x)f(x,y)d\sigma_x}{|x-y|^{n-1-r}}+\lambda(n-1-r)\int_\Omega\frac{[\Delta_x^k\partial_x f(x,y)](\overline{x}-\overline{y})d\sigma_x}{|x-y|^{n+1-r}}\\
&=\lambda\int_\Omega\frac{[\partial_y\Delta_x^k\partial_x f(x,y)]d\sigma_x}{|x-y|^{n-1-r}}+\lambda(n-1-r)\int_\Omega\frac{[\partial_x\Delta_x^k\partial_x f(x,y)]d\sigma_x}{(n-1-r)|x-y|^{n-1-r}}\\
&=\lambda\int_\Omega\frac{(\partial_y+\partial_x)(\Delta_x^k\partial_x f(x,y))}{|x-y|^{n-1-r}}d\sigma_x.
\end{aligned}$$

Inductively, it is easy to see that (1.26) is valid. By means of (1.22), (1.23), we can use the same method to prove (1.27) and (1.28).

Theorem 1.11 *If $f(x,y) \in H^{(m+2k+2,m)}(\beta_1,\beta_2)$, $0<\beta_i<1$, $i=1,2$, $0<r<1$, λ is as stated in Theorem 1.8, then*

$$\bar{\partial}_y^m \int_\Omega \frac{f(x,y)(\bar{x}-\bar{y})d\sigma_x}{|x-y|^{n+2k+1-r}} = \lambda \int_\Omega \frac{[(\bar{\partial}_y+\bar{\partial}_x)^m \Delta_x^k \partial_x f(x,y)]d\sigma_x}{|x-y|^{n-1-r}}, \quad (1.29)$$

$$\bar{\partial}_y^m \int_\Omega \frac{f(x,y)(x-y)d\sigma_x}{|x-y|^{n+2k+1-r}} = \lambda \int_\Omega \frac{[(\bar{\partial}_y+\bar{\partial}_x)^m \Delta_x^k \bar{\partial}_x f(x,y)]d\sigma_x}{|x-y|^{n-1-r}}, \quad (1.30)$$

$$\bar{\partial}_y^m \int_\Omega \frac{f(x,y)d\sigma_x}{|x-y|^{n+2k+1-r}} = \frac{\lambda}{2k+1-r} \int_\Omega \frac{[(\bar{\partial}_y+\bar{\partial}_x)^m \Delta_x^{k+1} f(x,y)]d\sigma_x}{|x-y|^{n-1-r}}. \quad (1.31)$$

Proof In view of (1.12), we have

$$\bar{\partial}_y \frac{1}{|x-y|^{n-1-r}} = \frac{-(n-1-r)(y-x)}{|x-y|^{n+1-r}} = \frac{(n-1-r)(x-y)}{|x-y|^{n+1-r}}.$$

By means of (1.21),(1.3),(1.25),(1.16) and Definition 1.2, we get

$$\begin{aligned}
&\bar{\partial}_y \int_\Omega \frac{f(x,y)(\bar{x}-\bar{y})d\sigma_x}{|x-y|^{n+2k+1-r}} \\
&= \lambda \int_\Omega \frac{[\bar{\partial}_y \Delta_x^k \partial_x f(x,y)]d\sigma_x}{|x-y|^{n-1-r}} + \lambda \int_\Omega \frac{[\Delta_x^k \partial_x f(x,y)](n-1-r)(x-y)d\sigma_x}{|x-y|^{n+1-r}} \\
&\quad + \lambda \int_\Omega \frac{(n-1-r)(x-y)[\Delta_x^k \partial_x f(x,y)]d\sigma_x}{|x-y|^{n+1-r}} \\
&\quad - \lambda \int_\Omega \frac{(n-1-r)[\Delta_x^k \partial_x f(x,y)](x-y)d\sigma_x}{|x-y|^{n+1-r}} \\
&= \lambda \int_\Omega \frac{[\bar{\partial}_y \Delta_x^k \partial_x f(x,y)]d\sigma_x}{|x-y|^{n-1-r}} + \lambda(n-1-r) \int_\Omega \frac{[\Delta_x^k \partial_x f(x,y)](x-y)d\sigma_x}{|x-y|^{n+1-r}} \\
&= \lambda \int_\Omega \frac{[\bar{\partial}_y \Delta_x^k \partial_x f(x,y)]d\sigma_x}{|x-y|^{n-1-r}} + \lambda(n-1-r) \int_\Omega \frac{[\bar{\partial}_x \Delta_x^k \partial_x f(x,y)]d\sigma_x}{(n-1-r)|x-y|^{n-1-r}} \\
&= \lambda \int_\Omega \frac{(\bar{\partial}_y+\bar{\partial}_x)(\Delta_x^k f(x,y))}{|x-y|^{n-1-r}} d\sigma_x.
\end{aligned}$$

Inductively, it is easy to see that (1.29) is true. By means of (1.22), (1.23), we can use the same method to prove (1.30) and (1.31).

Corollary 1.12 *Let $f(x,y) \in H^{(m+p+2k+2,m+p)}(\beta_1,\beta_2)$, $0 < \beta_i < 1$, $i = 1,2$, $0 < r < 1$, λ be as stated in Theorem 1.8. Then*

$$\bar{\partial}_y^m \partial_y^p \int_\Omega \frac{f(x,y)(\bar{x}-\bar{y})d\sigma_x}{|x-y|^{n+2k+1-r}} = \lambda \int_\Omega \frac{[(\bar{\partial}_y+\bar{\partial}_x)^m(\partial_y+\partial_x)^p\Delta_x^k\partial_x f(x,y)]d\sigma_x}{|x-y|^{n-1-r}}, \tag{1.32}$$

$$\bar{\partial}_y^m \partial_y^p \int_\Omega \frac{f(x,y)(x-y)d\sigma_x}{|x-y|^{n+2k+1-r}} = \lambda \int_\Omega \frac{(\bar{\partial}_y+\bar{\partial}_x)^m(\partial_y+\partial_x)^p\Delta_x^k\bar{\partial}_x f(x,y)]d\sigma_x}{|x-y|^{n-1-r}}, \tag{1.33}$$

$$\begin{aligned} &\bar{\partial}_y^m \partial_y^p \int_\Omega \frac{f(x,y)d\sigma_x}{|x-y|^{n+2k+1-r}} \\ = &\frac{\lambda}{2k+1-r}\int_\Omega \frac{[(\bar{\partial}_y+\bar{\partial}_x)^m(\partial_y+\partial_x)^p\Delta_x^{k+1}f(x,y)]d\sigma_x}{|x-y|^{n-1-r}}. \end{aligned} \tag{1.34}$$

Proof It is easy to prove this corollary by using Theorems 1.10 and 1.11.

Theorem 1.13 *If $f(x,y) \in H^{(m+2k+2,p)}(\beta_1,\beta_2)$, $0 < \beta_i < 1$, $i = 1,2$, $t_1, t_2 \in \Omega$, $0 < r < 1$, λ is as stated in Theorem 1.8, then*

$$\bar{\partial}_{t_1}^m \partial_{t_2}^p \int_\Omega \frac{f(x,t_2)(\bar{x}-\bar{t}_1)d\sigma_x}{|x-t_1|^{n+2k+1-r}} = \lambda \int_\Omega \frac{[(\bar{\partial}_x^m\partial_{t_2}^p\Delta_x^k\partial_x f(x,t_2)]d\sigma_x}{|x-t_1|^{n-1-r}}, \tag{1.35}$$

$$\bar{\partial}_{t_1}^m \partial_{t_2}^p \int_\Omega \frac{f(x,t_2)(x-t_1)d\sigma_x}{|x-t_1|^{n+2k+1-r}} = \lambda \int_\Omega \frac{[(\bar{\partial}_x^m\partial_{t_2}^p\Delta_x^k\bar{\partial}_x f(x,t_2)]d\sigma_x}{|x-t_1|^{n-1-r}}, \tag{1.36}$$

$$\bar{\partial}_{t_1}^m \partial_{t_2}^p \int_\Omega \frac{f(x,t_2)d\sigma_x}{|x-t_1|^{n+2k+1-r}} = \frac{\lambda}{2k+1-r} \int_\Omega \frac{[(\bar{\partial}_x^m\partial_{t_2}^p\Delta_x^{k+1} f(x,t_2)]d\sigma_x}{|x-t_1|^{n-1-r}}. \tag{1.37}$$

Proof According to (1.21), Corollaries 1.2, 1.6, (1.12), (1.16) and Definition 1.2, we have

$$\begin{aligned} &\bar{\partial}_{t_1}^m \partial_{t_2}^p \int_\Omega \frac{f(x,t_2)(\bar{x}-\bar{t}_1)d\sigma_x}{|x-t_1|^{n+2k+1-r}} \\ = &\bar{\partial}_{t_1}^m \partial_{t_2}^p \lambda \int_\Omega \frac{[\Delta_x^k\partial_x f(x,t_2)]d\sigma_x}{|x-t_1|^{n-1-r}} \\ = &\lambda\bar{\partial}_{t_1}^m \int_\Omega \frac{[\partial_{t_2}^p\Delta_x^k\partial_x f(x,t_2)]d\sigma_x}{|x-t_1|^{n-1-r}} \end{aligned}$$

$$\begin{aligned}
&= \lambda\bar{\partial}_{t_1}^{m-1}\int_\Omega \frac{(n-1-r)(x-t_1)[\partial_{t_2}^p\Delta_x^k\partial_x f(x,t_2)]d\sigma_x}{|x-t_1|^{n+1-r}} \\
&= \lambda(n-1-r)\bar{\partial}_{t_1}^{m-1}\int_\Omega \frac{\partial_{t_2}^p\Delta_x^k\partial_x f(x,t_2)](x-t_1)d\sigma_x}{|x-t_1|^{n+1-r}} \\
&= \lambda\bar{\partial}_{t_1}^{m-1}\int_\Omega \frac{[(\bar{\partial}_x\partial_{t_2}^p\Delta_x^k\partial_x f(x,t_2)]d\sigma_x}{|x-t_1|^{n-1-r}} \\
&= \lambda\int_\Omega \frac{[(\bar{\partial}_x^m\partial_{t_2}^p\Delta_x^k\partial_x f(x,t_2)]d\sigma_x}{|x-t_1|^{n-1-r}}.
\end{aligned}$$

This shows that (1.35) is correct.

Similarly, By means of (1.22), Corollaries 1.2 and 1.6, (1.12), (1.15) and Definition 1.2, we get (1.36).

At last, by using (1.23), Corollaries 1.2, 1.6, (1.12), (1.16) and Definition 1.2, we can get (1.37).

In the following, we discuss the fourth, fifth and sixth kinds of high order singular integrals of quasi-Bochner-Martinelli type.

Theorem 1.14 *Suppose that* $u(x) = \sum\limits_A u_A(x)e_A$, $v(x) = \sum\limits_{i=1}^n v_i(x)e_i$ *are as those in Lemma 1.1. Then*

$$[(\partial_x-\bar{\partial}_x)(v+\bar{v})]u = (v+\bar{v})[(\bar{\partial}_x-\partial_x)u] + (\partial_x-\bar{\partial}_x)[(v+\bar{v})u]\,, \tag{1.38}$$

$$[\partial_x(v+\bar{v})]u = -(v+\bar{v})\partial_x u + \partial_x[(v+\bar{v})u]\,. \tag{1.39}$$

Proof We substitute v by $\bar{v}$ in Lemma 1.5 and notice when $j \geq 2, \bar{v}_j = -v_j$, then

$$\bar{\partial}_x(\bar{v}u) = (\bar{\partial}_x\bar{v})u + \bar{v}(\bar{\partial}_x u) + 2\sum_{i,j=2,i\neq j}^n v_i e_i e_j u_{x_j}\,, \tag{1.40}$$

and

$$\partial_x(\bar{v}u) = (\partial_x\bar{v})u + \bar{v}(\partial_x u) - 2\sum_{i,j=2,i\neq j}^n v_i e_i e_j u_{x_j}. \tag{1.41}$$

In view of (1.12), (1.14), (1.40), (1.41), we get (1.38). By means of (1.14), (1.41), it is easy to derive (1.39). The proof is finished.

Suppose $\alpha > 0$, and let $v = \frac{\bar{x}-\bar{y}}{(-\alpha)|x-y|^{n+\alpha}}, u = f(x,y)$ in (1.38) of Theorem 1.11. By using Lemma 1.3, we can get

$$[(\partial_x - \bar{\partial}_x)(v+\bar{v})]u = \frac{n+\alpha}{\alpha}\left[\frac{(\bar{x}-\bar{y})^2}{|x-y|^{n+\alpha+2}} - \frac{(x-y)^2}{|x-y|^{n+\alpha+2}}\right]f(x,y), \quad (1.42)$$

$$(v+\bar{v})[(\bar{\partial}_x - \partial_x)u] = \frac{1}{\alpha}\left[\frac{\bar{x}-\bar{y}}{|x-y|^{n+\alpha}} + \frac{x-y}{|x-y|^{n+\alpha}}\right][(\partial_x - \bar{\partial}_x)f(x,y)]. \quad (1.43)$$

The right-hand side of (1.43) has a lower order singularity than the right-hand side of (1.42). Similarly to the discussion of Definition 1.1, due to (1.38), we shall use the integral on the right-hand side of (1.43) to define the following integral.

Definition 1.8

$$\begin{aligned}&\int_\Omega \left[\frac{(\bar{x}-\bar{y})^2}{|x-y|^{n+\alpha-2}} - \frac{(x-y)^2}{|x-y|^{n+\alpha+2}}\right] f(x,y)d\sigma_x \\ =\ &\frac{1}{n+\alpha}\int_\Omega \frac{(\bar{x}-\bar{y})+(x-y)}{|x-y|^{n+\alpha}}[(\partial_x - \bar{\partial}_x)f(x,y)]d\sigma_x,\end{aligned} \quad (1.44)$$

where $\alpha > 0$, $y \in \Omega$.

The singular integral on the left side of (1.44) is called the fourth kind of high order singular integral of quasi-Bochner-Martinelli type with one singular point.

Similarly, let $v = \dfrac{\bar{x}-\bar{y}}{|x-y|^{n+\alpha}}$, $u = f(x,y)$ in (1.39), then we get

$$[\partial_x(v+\bar{v})]u = \left[\frac{2-n-\alpha}{|x-y|^{n+\alpha}} - \frac{(\bar{x}-\bar{y})^2(n+\alpha)}{|x-y|^{n+\alpha+2}}\right] f(x,y), \quad (1.45)$$

$$-(v+\bar{v})\partial_x u = \frac{-[(\bar{x}-\bar{y})+(x-y)]}{|x-y|^{n+\alpha}}\partial_x f(x,y). \quad (1.46)$$

In the terms on the right-hand side of (1.45), (1.46), it is only the term: $\dfrac{(\bar{x}-\bar{y})^2(n+\alpha)}{|x-y|^{n+\alpha+2}}f(x,y)$, whose high order singular integral has not been defined.

Similarly, by means of (1.39), we can inductively define the fifth kind of high order singular integrals of quasi-Bochner-Martinelli type with one singular point.

Definition 1.9

$$\int_{\Omega} \frac{(\bar{x}-\bar{y})^2 f(x,y)d\sigma_x}{|x-y|^{n+\alpha+2}} = \frac{-(n+\alpha-2)}{n+\alpha}\int_{\Omega}\frac{f(x,y)d\sigma_x}{|x-y|^{n+\alpha}} + \frac{1}{n+\alpha}\int_{\Omega}\frac{[(\bar{x}-\bar{y})+(x-y)]\partial_x f(x,y)}{|x-y|^{n+\alpha}}d\sigma_x, \tag{1.47}$$

if $y \in \Omega$, $\alpha > 0$.

Finally according to (1.44), (1.47), it is easy to give the definition of the sixth kind of high order singular integrals of quasi-Bochner-Martinelli type with one singular point.

Definition 1.10

$$\int_{\Omega} \frac{(x-y)^2 f(x,y)d\sigma_x}{|x-y|^{n+\alpha+2}} = \frac{(n+\alpha-2)}{-(n+\alpha)}\int_{\Omega}\frac{f(x,y)d\sigma_x}{|x-y|^{n+\alpha}} + \frac{1}{n+\alpha}\int_{\Omega}\frac{[(\bar{x}-\bar{y})+(x-y)]}{|x-y|^{n+\alpha}}\bar{\partial}_x f(x,y)d\sigma_x, \tag{1.48}$$

for $y \in \Omega$, $\alpha > 0$.

As a supplement to Lemma 1.4, it is easy to prove the following equalities:

$$\bar{\partial}_x\left(\frac{x-y}{|x-y|^{n+\alpha+2}}\right) = \frac{2-n}{|x-y|^{n+\alpha}} - \frac{(n+\alpha)(x-y)^2}{|x-y|^{n+\alpha+2}}, \quad \alpha > 0, \tag{1.49}$$

$$\partial_x\left(\frac{\bar{x}-\bar{y}}{|x-y|^{n+\alpha}}\right) = \frac{2-n}{|x-y|^{n+\alpha}} - \frac{(n+\alpha)(\bar{x}-\bar{y})^2}{|x-y|^{n+\alpha+2}}, \quad \alpha > 0. \tag{1.50}$$

From the second term on the right side of (1.49), (1.50), we see that for investigating high order singular integrals in real Clifford analysis, it is necessary to discuss the fourth, fifth and sixth kinds of high order singular integrals of quasi-Bochner-Martinelli type with one singular point.

By means of Definitions 1.8, 1.9, 1.10 and recurrence formulas, calculational formulas, and differential formulas for the first, second, and third kinds of high order singular integrals of quasi-Bochner-Martinelli type with one singular point, we can get the recurrence formulas, calculational formulas, and differential formulas for the fourth, fifth and sixth kinds of high order singular integrals of quasi-Bochner-Martinelli type. Because of page limitation, we don't prove all formulas, and only prove one of every kind of high order singular integrals; the other proofs are left to readers.

Theorem 1.15 *If $n > 1$, $\alpha > 2m > 0$, $f(x,y) \in H_x^{2m+2}(\beta)$, $y \in \Omega$, μ is as stated in Lemma 1.6, then we have the recurrence formula*

$$\begin{aligned}&\int_\Omega \left[\frac{(\bar{x}-\bar{y})^2}{|x-y|^{n+\alpha+2}} - \frac{(x-y)^2}{|x-y|^{n+\alpha+2}}\right] f(x,y)d\sigma_x \\ =& \frac{\mu}{(n+\alpha)(n+\alpha-2m-2)}\int_\Omega \frac{\Delta_x^m\left[(\partial_x^2-\bar{\partial}_x^2)f(x,y)\right]}{|x-y|^{n+\alpha-2m-2}}d\sigma_x.\end{aligned} \tag{1.51}$$

Proof By Definition 1.8, (1.15), (1.16), (1.17) and (1.18), we have

$$\begin{aligned}&\int_\Omega \left[\frac{(\bar{x}-\bar{y})^2}{|x-y|^{n+\alpha+2}} - \frac{(x-y)^2}{|x-y|^{n+\alpha+2}}\right] f(x,y)d\sigma_x \\ =& \frac{1}{n+\alpha}\int_\Omega \frac{(\bar{x}-\bar{y})+(x-y)}{|x-y|^{n+\alpha}}\left[(\partial_x-\bar{\partial}_x)f(x,y)\right]d\sigma_x \\ =& \frac{\mu}{(n+\alpha)(n+\alpha-2m-2)}\int_\Omega \frac{\Delta_x^m(\partial_x+\bar{\partial}_x)(\partial_x-\bar{\partial}_x)f(x,y)}{|x-y|^{n+\alpha-2m-2}}d\sigma_x, \\ =& \frac{\mu}{(n+\alpha)(n+\alpha-2m-2)}\int_\Omega \frac{\Delta_x^m\left[(\partial_x^2-\bar{\partial}_x^2)f(x,y)\right]d\sigma_x}{|x-y|^{n+\alpha-2m-2}}.\end{aligned}$$

Theorem 1.16 *Suppose that $f(x,y) \in H_x^{(2k+2)}(\beta)$, $y \in \Omega$, λ is as stated in Theorem 1.8, and $n > 1$, $0 < r < 1$. Then*

$$\begin{aligned}&\int_\Omega \frac{(\bar{x}-\bar{y})^2 f(x,y)d\sigma_x}{|x-y|^{n+2k+3-r}} \\ =& \frac{\lambda(2-n)}{(n+2k+1-r)(2k+1-r)}\int_\Omega \frac{\Delta_x^{k+1}f(x,y)}{|x-y|^{n-1-r}}d\sigma_x \\ &+\frac{\lambda}{n+2k+1-r}\int_\Omega \frac{\Delta_x^k\left[\partial_x^2 f(x,y)\right]}{|x-y|^{n-1-r}}d\sigma_x.\end{aligned} \tag{1.52}$$

Proof By Definition 1.9, (1.15), and Theorem 1.8, we get

$$\begin{aligned}&\int_\Omega \frac{(\bar{x}-\bar{y})^2 f(x,y)d\sigma_x}{|x-y|^{n+2k+3-r}} \\ =& \frac{-(n+2k-1-r)}{n+2k+1-r}\int_\Omega \frac{f(x,y)d\sigma_x}{|x-y|^{n+2k+1-r}} \\ &+\frac{1}{n+2k+1-r}\int_\Omega \frac{\left[(\bar{x}-\bar{y})+(x-y)\right]\partial_x f(x,y)d\sigma_x}{|x-y|^{n+2k+1-r}}\end{aligned}$$

$$
\begin{aligned}
&= \frac{-\lambda(n+2k-1-r)}{(n+2k+1-r)(2k+1-r)}\int_\Omega \frac{\left[\Delta_x^{k+1} f(x,y)\right] d\sigma_x}{|x-y|^{n-1-r}} \\
&\quad + \frac{\lambda}{n+2k+1-r}\int_\Omega \frac{\Delta_x^k\left[(\partial_x+\bar\partial_x)\partial_x f(x,y)\right] d\sigma_x}{|x-y|^{n-1-r}} \\
&= \frac{\lambda(2-n)}{(n+2k+1-r)(2k+1-r)}\int_\Omega \frac{\Delta_x^{k+1} f(x,y)}{|x-y|^{n-1-r}} d\sigma_x \\
&\quad + \frac{\lambda}{n+2k+1-r}\int_\Omega \frac{\Delta_x^k\left[\partial_x^2 f(x,y)\right] d\sigma_x}{|x-y|^{n-1-r}}.
\end{aligned}
$$

Theorem 1.17 *Let $f(x,y) \in H^{(m+2k+2,m)}(\beta_1,\beta_2)$, $0<\beta_i<1$, $0<r<1$, λ be the same as that in Theorem 1.8. Then*

$$
\begin{aligned}
&\partial_y^m \int_\Omega \frac{(x-y)^2 f(x,y) d\sigma_x}{|x-y|^{n+2k+3-r}} \\
&= \frac{\lambda(2-n)}{(n+2k+1-r)(2k+1-r)}\int_\Omega \frac{\left[(\partial_y+\partial_x)^m \Delta_x^{k+1} f(x,y)\right] d\sigma_x}{|x-y|^{n-1-r}} \\
&\quad + \frac{\lambda}{n+2k+1-r}\int_\Omega \frac{\left[(\partial_y+\partial_x)^m \Delta_x^k \bar\partial_x^2 f(x,y)\right] d\sigma_x}{|x-y|^{n-1-r}}.
\end{aligned}
\tag{1.53}
$$

Proof By Definitions 1.5, 1.10, and Theorem 1.10, we can get

$$
\begin{aligned}
&\partial_y^m \int_\Omega \frac{(x-y)^2 f(x,y) d\sigma_x}{|x-y|^{n+2k+3-r}} \\
&= \frac{(n+2k-1-r)}{-(n+2k+1-r)}\partial_y^m \int_\Omega \frac{f(x,y) d\sigma_x}{|x-y|^{n+2k+1-r}} \\
&\quad + \frac{1}{n+2k+1-r}\partial_y^m \int_\Omega \frac{\left[(\bar x-\bar y)+(x-y)\right]\bar\partial_x f(x,y) d\sigma_x}{|x-y|^{n+2k+1-r}} \\
&= \frac{(n+2k-1-r)\lambda}{-(n+2k+1-r)(2k+1-r)}\int_\Omega \frac{(\partial_y+\partial_x)^m \Delta_x^{k+1} f(x,y) d\sigma_x}{|x-y|^{n-1-r}} \\
&\quad + \frac{\lambda}{n+2k+1-r}\int_\Omega \frac{\left\{(\partial_y+\partial_x)^m \left[(\Delta_x^{k+1}+\Delta_x^k\bar\partial_x^2) f(x,y)\right]\right\}}{|x-y|^{n-1-r}} d\sigma_x \\
&= \frac{\lambda(2-n)}{(n+2k+1-r)(2k+1-r)}\int_\Omega \frac{\left[(\partial_y+\partial_x)^m \Delta_x^{k+1} f(x,y)\right] d\sigma_x}{|x-y|^{n-1-r}} \\
&\quad + \frac{\lambda}{n+2k+1-r}\int_\Omega \frac{\left[(\partial_y+\partial_x)^m \Delta_x^k \bar\partial_x^2 f(x,y)\right] d\sigma_x}{|x-y|^{n-1-r}}.
\end{aligned}
$$

2 The Hölder Continuity of High Order Singular Integrals in Real Clifford Analysis

By the known inequality

$$|\sigma_1^\mu - \sigma_2^\mu| \le |\sigma_1 - \sigma_2|^\mu, \tag{2.1}$$

where $0 \le \mu \le 1$, $\sigma_i > 0$, $i = 1, 2$ [87], we can prove the following theorem similar to the Hile type lemma (see Section 2, Chapter II).

Theorem 2.1 *Let m be a positive integer, $\alpha > 0$ and $[\alpha]$ be the integral part of α; $x, y, \hat{y}, t, \hat{t} \in \mathbf{R^n}$, $t, \hat{t} \ne 0$, $x \ne y$, $x \ne \hat{y}$. Then*

$$\left|\frac{1}{|t|^m} - \frac{1}{|\hat{t}|^m}\right| \le \sum_{k=0}^{m-1} \left|\frac{\hat{t}}{t}\right|^k \left|\frac{t-\hat{t}}{t}\right| |\hat{t}|^{-m}, \tag{2.2}$$

and

$$\left|\frac{1}{|x-y|^\alpha} - \frac{1}{|x-\hat{y}|^\alpha}\right| \le \sum_{k=0}^{[\alpha]} \left|\frac{x-\hat{y}}{x-y}\right|^{\frac{\alpha k}{[\alpha]+1}} \left|\frac{y-\hat{y}}{x-y}\right|^{\frac{\alpha}{[\alpha]+1}} |x-\hat{y}|^{-\alpha}. \tag{2.3}$$

Proof Noting that

$$\left|\frac{1}{|t|^m} - \frac{1}{\hat{t}^m}\right| = \frac{\left||\hat{t}|^m - |t|^m\right|}{|t|^m|\hat{t}|^m}$$

$$= \frac{\left||\hat{t}| - |t|\right| \sum\limits_{k=0}^{m-1} |\hat{t}|^{m-1-k}|t|^k}{|t|^m|\hat{t}|^m} \le \sum_{k=0}^{m-1} \left|\frac{\hat{t}}{t}\right|^k \left|\frac{t-\hat{t}}{t}\right| |\hat{t}|^{-m},$$

it is clear that (2.2) is true. By means of (2.1), (2.2), we have

$$\left|\frac{1}{|x-y|^\alpha} - \frac{1}{|x-\hat{y}|^\alpha}\right|$$

$$= \left|\frac{1}{(|x-y|^{\frac{\alpha}{[\alpha]+1}})^{[\alpha]+1}} - \frac{1}{(|x-\hat{y}|^{\frac{\alpha}{[\alpha]+1}})^{[\alpha]+1}}\right|$$

$$\le \sum_{k=0}^{[\alpha]} \left|\frac{x-\hat{y}}{x-y}\right|^{\frac{\alpha k}{[\alpha]+1}} \left|\frac{|x-y|^{\frac{\alpha}{[\alpha]+1}} - |x-\hat{y}|^{\frac{\alpha}{[\alpha]+1}}}{|x-y|^{\frac{\alpha}{[\alpha]+1}}}\right| \left(|x-\hat{y}|^{\frac{\alpha}{[\alpha]+1}}\right)^{-[\alpha]-1}$$

$$\le \sum_{k=0}^{[\alpha]} \left|\frac{x-\hat{y}}{x-y}\right|^{\frac{\alpha k}{[\alpha]+1}} \frac{\left||x-y| - |x-\hat{y}|\right|^{\frac{\alpha}{[\alpha]+1}}}{|x-y|^{\frac{\alpha}{[\alpha]+1}}} |x-\hat{y}|^{-\alpha}$$

$$\le \sum_{k=0}^{[\alpha]} \left|\frac{x-\hat{y}}{x-y}\right|^{\frac{\alpha k}{[\alpha]+1}} \left|\frac{y-\hat{y}}{x-y}\right|^{\frac{\alpha}{[\alpha]+1}} |x-\hat{y}|^{-\alpha}.$$

From this section to the last section in this chapter, we suppose that the boundary Ω of the domain D is a smooth, oriented, compact, Liapunov surface, and the orientation of Ω is the induced orientation of D. In view of the definition of Liapunov surface, there exists a positive number $d > 0$, such that for any point $N_0 \in \Omega$, the sphere with the center at N_0 and radius d (or less then d) can be divided Ω into two parts, and the interior part of the sphere is denoted by Ω'. We choose N_0 as the origin of the local generalized sphere coordinate system, such that x_n-axis and the outwards normal direction of Ω at N_0 are identical. If d is small enough and the outward normal line of Ω' at any point $N(\xi_1, \cdots, \xi_n)$ is n_0. Let $r_0 = |N_0N|$. Denote by ρ_0 the length of the project of r_0 on the tangent plane through N_0 and by $(\rho_0, \phi_1, \cdots, \phi_{n-2})$ the local generalized sphere coordinate of N, and let $J = \left|\frac{D(\xi_1,\ldots,\xi_{n-1})}{D(\rho_0,\varphi_1,\ldots,\varphi_{n-2})}\right|$ be the Jacobian determinant of coordinate transformation. In Section 2, Chapter I, we have obtained

$$\cos(n_0, x_n) \geq \frac{1}{2}, \quad |J| \leq {\rho_0}^{n-2}. \tag{2.4}$$

For any $\Phi \in H^{(\bar{n}+2k+2,\bar{n})}(\beta_1, \beta_2)$, $0 < \beta_i < 1$, $i = 1, 2$, the norm of Φ is defined as

$$\|\Phi\|_\beta = \sum_{0\leq m+p\leq\bar{n}} C(\bar{\partial}_x^{m+2k+2}\partial_y{}^p\Phi,\Omega\times\Omega) + \sum_{0\leq m+p\leq\bar{n}} H(\bar{\partial}_x^{m+2k+2}\partial_y{}^p\Phi,\Omega\times\Omega).$$

For convenience, we denote $\hat{\Phi} = \bar{\partial}_x^{m+2k+2}\partial_y^p\Phi$, where

$$C(\hat{\Phi}, \Omega\times\Omega) = \max_{(x,y)\in\Omega\times\Omega} |\hat{\Phi}(x,y)|,$$

$$H(\hat{\Phi},\Omega\times\Omega) = \sup_{(x_1,y_1),(x_2,y_2)\in\Omega\times\Omega} \frac{|\hat{\Phi}(x_1,y_1) - \hat{\Phi}(x_2,y_2)|}{|x_1-x_2|^{\beta_1} + |y_1-y_2|^{\beta_2}},$$

in which $(x_1, y_1) \neq (x_2, y_2)$. This definition is different from the definition of $H(f, \partial\Omega\times\partial\Omega\beta)$ in Section 4, Chapter II, but it is the same as that in Tongde Zhong's paper in 1980 [88]2), and it is easy to prove $H^{(\bar{n}+2k+2,\,\bar{n})}(\beta_1, \beta_2)$ is a compact Banach space, and its norm possesses the property

$$\|F+G\|_\beta \leq \|F\|_\beta + \|G\|_\beta, \ \|FG\|_\beta \leq 2^{n-1}\|F\|_\beta\|G\|_\beta,$$

where F, $G \in H^{(\bar{n}+2k+2,\bar{n})}(\beta_1, \beta_2)$.

Theorem 2.2 *If the operator on Ω is defined as*

$$(P\varphi)(y) = \int_\Omega \frac{\varphi(x,y)d\sigma_x}{|x-y|^{n-1-r}},$$

here $0 < r < 1$, $\varphi(x,y) \in H^{(0,0)}(\beta_1,\beta_2)$, $0 < \beta_i < 1$, $i = 1,2$, $\omega = \min(\beta_1,\beta_2)$, $r > \beta_1$, $\beta_1 < \frac{n-1-r}{n-1}$, $y \in \Omega$, *then* $P\varphi \in H_y^{(0)}(\omega)$ *on* Ω.

Proof We introduce the operator

$$(\theta_1\varphi)(y) = \int_\Omega \frac{\varphi(x,y)-\varphi(y,y)}{|x-y|^{n-1-r}}d\sigma_x, \quad (\theta_2\varphi)(y) = \int_\Omega \frac{\varphi(y,y)d\sigma_x}{|x-y|^{n-1-r}}.$$

Obviously $(P\varphi)(y) = (\theta_1\varphi)(y) + (\theta_2\varphi)(y)$. Firstly, we discuss the Hölder continuity of $(\theta_1\varphi)(y)$. From Section 2, Chapter 1, we know $d\sigma_x = m(x)ds_x$, $m(x) = \sum_{j=1}^{n} e_j\cos(m,e_j)$ is the outward normal direction of x on Ω. For any $y_1, y_2 \in \Omega$, we denote $\delta = |y_1 - y_2|$. Let $3\delta < d$, here d is as stated before. We make a sphere with the center at y_1 and radius 3δ. Denote by Ω_1 the interior part of Ω and by Ω_2 the left part. It is easy to see that

$$\begin{aligned}
&|(\theta_1\varphi)(y_1) - (\theta_1\varphi)(y_2)|\\
\le\ & |\textstyle\int_{\Omega_1}\dfrac{\varphi(x,y_1)-\varphi(y_1,y_1)}{|x-y_1|^{n-1-r}}d\sigma_x| + \left|\displaystyle\int_{\Omega_1}\dfrac{\varphi(x,y_2)-\varphi(y_2,y_2)}{|x-y_2|^{n-1-r}}d\sigma_x\right|\\
+\ & \left|\textstyle\int_{\Omega_2}\left(\dfrac{1}{|x-y_1|^{n-1-r}} - \dfrac{1}{|x-y_2|^{n-1-r}}\right)(\varphi(x,y_1)-\varphi(y_1,y_1))d\sigma_x\right|\\
+\ & \left|\textstyle\int_{\Omega_2}\dfrac{[(\varphi(x,y_1)-\varphi(x,y_2)) + (\varphi(y_2,y_2)-\varphi(y_1,y_1))]}{|x-y_2|^{n-1-r}}d\sigma_x\right|\\
=\ & L_1 + L_2 + L_3 + L_4.
\end{aligned}$$

The formula (2.4) about $N_0 \in \Omega$ discussed at the beginning of the section is used as y_1, and the projective domain of Ω_1 on the tangent plane of y_1 is denoted by π_1, from which we can get

$$\begin{aligned}
L_1 &\le G_1\|\varphi\|_\beta \int_{\Omega_1}\frac{ds_x}{|x-y_1|^{n-1-r-\beta_1}}\\
&= G_1\|\varphi\|_\beta \textstyle\int_{\Omega_1}\dfrac{cos(n_0,x_n)ds_x}{|x-y_1|^{n-1-r-\beta_1}cos(n_0,x_n)}\\
&\le G_1\|\varphi\|_\beta \displaystyle\int_{\pi_1}\frac{d\xi_1\cdots d\xi_{n-1}}{(|x-y_1|^{n-1-r-\beta_1})\frac{1}{2}} \qquad (2.5)\\
&\le G_2\|\varphi\|_\beta \textstyle\int_0^{3\delta}\dfrac{(\rho_0)^{n-2}d\rho_0}{{\rho_0}^{n-1-r-\beta_1}} \le G_3\|\varphi\|_\beta\delta^r\delta^{\beta_1}\\
&\le G_4\|\varphi\|_\beta\delta^{\beta_1} = G_4\|\varphi\|_\beta\delta^\omega\delta^{\beta_1-\omega}\\
&\le G_5\|\varphi\|_\beta\delta^\omega = G_5\|\varphi\|_\beta|y_1-y_2|^\omega.
\end{aligned}$$

Similarly, we have

$$L_2 \le G_6\|\varphi\|_\beta|y_1-y_2|^\omega, \qquad (2.6)$$

where $G_i(i = 1, ..., 6)$ are positive constants independent of y_1, y_2. In the following, $G_i(i \geq 1)$ are denoted as positive constants with the similar property. Next, we estimate L_3. By means of Theorem 2.1, the inequality

$$\begin{aligned} & \left| \frac{1}{|x-y_1|^{n-1-r}} - \frac{1}{|x-y_2|^{n-1-r}} \right| \\ \leq & \sum_{k=0}^{n-2} \left| \frac{x-y_2}{x-y_1} \right|^{\frac{k(n-1-r)}{n-1}} \left| \frac{y_1-y_2}{x-y_1} \right|^{\frac{n-1-r}{n-1}} | x-y_2 |^{-n+1-r} \end{aligned}$$

is derived. For any $x \in \Omega_2$, we have $|x-y_1| \geq 3\delta = 3|y_1-y_2|$, $|x-y_2| \geq 2\delta$, hence $\frac{1}{2} \leq \left|\frac{x-y_2}{x-y_1}\right| \leq 2$, $|x-y_2| \geq \frac{1}{2}|x-y_1|$. Moreover, noting that $(n-1-r)/(n-1) > \beta_1$, we have

$$\left| \frac{y_1-y_2}{x-y_1} \right|^{\frac{n-1-r}{n-1}} \leq \left| \frac{y_1-y_2}{x-y_1} \right|^{\beta_1} = (\frac{\delta}{|x-y_1|})^{\beta_1},$$

and then

$$\begin{aligned} L_3 \leq G_7 \int_{\Omega_2} \frac{\delta^{\beta_1} \|\varphi\|_\beta |x-y_1|^{\beta_1} ds_x}{|x-y_1|^{n-1-r+\beta_1}} = G_7 |y_1-y_2|^{\beta_1} \int_{\Omega_2} \frac{\|\varphi\|_\beta ds_x}{|x-y_1|^{n-1-r}} \\ \leq G_8 \|\varphi\|_\beta |y_1-y_2|^{\beta_1} \leq G_9 \|\varphi\|_\beta |y_1-y_2|^{\omega}. \end{aligned} \tag{2.7}$$

By virtue of

$$\begin{aligned} & |(\varphi(x,y_1)-\varphi(x,y_2)) + (\varphi(y_2,y_2)-\varphi(y_1,y_1))| \\ \leq & \|\varphi\|_\beta |y_1-y_2|^{\beta_2} + \|\varphi\|_\beta \left[|y_1-y_2|^{\beta_1} + |y_1-y_2|^{\beta_2} \right] \\ \leq & G_{10} \|\varphi\|_\beta |y_1-y_2|^{\omega}, \end{aligned}$$

we know

$$L_4 \leq G_{11} \|\varphi\|_\beta |y_1-y_2|^{\omega}. \tag{2.8}$$

Next we discuss the Hölder continuity of $(\theta_2\varphi)(y)$. It is not difficult to see that

$$\begin{aligned} & |(\theta_2\varphi)(y_1) - (\theta_2\varphi)(y_2)| \\ \leq & \left| (\varphi(y_1,y_1)-\varphi(y_2,y_2)) \int_\Omega \frac{d\sigma_x}{|x-y_1|^{n-1-r}} \right|, \\ + & |\varphi(y_2,y_2)| \left| \int_\Omega \left(\frac{1}{|x-y_1|^{n-1-r}} - \frac{1}{|x-y_2|^{n-1-r}} \right) d\sigma_x \right| \end{aligned}$$

$$\begin{aligned}
&\leq G_{12}\|\varphi\|_\beta(|y_1-y_2|^{\beta_1}+|y_1-y_2|^{\beta_2})+\|\varphi\|_\beta\left|\int_{\Omega_1}\frac{d\sigma_x}{|x-y_1|^{n-1-r}}\right| \\
&+\ \|\varphi\|_\beta\left|\int_{\Omega_1}\frac{d\sigma_x}{|x-y_2|^{n-1-r}}\right| \\
&+\ \|\varphi\|_\beta\left|\int_{\Omega_2}\left(\frac{1}{|x-y_1|^{n-1-r}}-\frac{1}{|x-y_2|^{n-1-r}}\right)d\sigma_x\right| \\
&=\ L_5+L_6+L_7+L_8.
\end{aligned}$$

Firstly, it is easy to see that

$$L_5 \leq G_{13}\|\varphi\|_\beta|y_1-y_2|^\omega. \tag{2.9}$$

Secondly, similarly to the deduction of (2.5), and according to the condition of $r>\beta_1\geq\omega$, we get

$$L_6 \leq G_{14}\|\varphi\|_\beta\delta^r \leq G_{15}\|\varphi\|_\beta\delta^\omega = G_{15}\|\varphi\|_\beta|y_1-y_2|^\omega. \tag{2.10}$$

Moreover, we have

$$L_7 \leq G_{16}\|\varphi\|_\beta|y_1-y_2|^\omega. \tag{2.11}$$

Finally, similarly to the deduction of (2.7), and noting that $r-\beta_1>0$, we get

$$L_8 \leq G_{17}\|\varphi\|_\beta|y_1-y_2|^{\beta_1}\int_{\Omega_2}\frac{d\sigma_x}{|x-y_1|^{n-1-(r-\beta_1)}} \leq G_{18}\|\varphi\|_\beta|y_1-y_2|^\omega. \tag{2.12}$$

In view of (2.5) – (2.12), the inequality

$$|(P\varphi)(y_1)-(P\varphi)(y_2)| \leq G_{19}\|\varphi\|_\beta|y_1-y_2|^\omega \leq G_{20}|y_1-y_2|^\omega, \tag{2.13}$$

is derived, namely $P\varphi\in H_y^0(\omega)$. The proof of this theorem is finished.

Now, we verify the hölder continuity for every kind of high order singular integrals of quasi-Bochner-Martinelli type on an integral path.

Theorem 2.3 *Let $f(x,y)\in H^{(2k+2,0)}(\beta_1,\beta_2)$, $0<r<1$, $r>\beta_1$, $n-1>r/(1-\beta_1)$, $\omega=\min(\beta_1,\beta_2)$, $y\in\Omega$, $0<\beta_i<1$, $i=1,2$. Then the six kinds of high order singular integrals of quasi-Bochner-Martinelli type as stated before all belong to $H_y^0(\omega)$ on Ω.*

Proof On the basis of Theorem 2.2 and the calculation formulas of every kind of high order singular integrals of quasi-Bochner-Martinelli type, we can prove this theorem.

3 Nonlinear Differential Integral Equations including Three Kinds of High Order Singular Integrals of Quasi-Bochner-Martinelli Type in Real Clifford Analysis

From the Definitions 1.8, 1.9, 1.10, we know that the fourth, fifth, and sixth kinds of high order singular integrals of quasi-Bochner-Martinelli type with one singular point can be expressed by the first, second, and third kinds of high order singular integrals of quasi-Bochner-Martinelli type with one singular point. So when we consider the nonlinear differential equations including high order singular integrals of quasi-Bochner-Martinelli type in real Clifford analysis, it is sufficient to discuss the equations with the first, second, and third kinds of high order singular integrals of quasi-Bochner-Martinelli type. In the field of differential integral equations, Y. Hino investigated the linear Volterra differential integral equation in 1990 (see [24]).

Due to the enlightenment from the Volterra differential integral equation, in this section we shall prove the existence and uniqueness of solutions for some nonlinear differential integral equations with the first, second, and third kinds of high order singular integrals of quasi-Bochner-Martinelli type by using the results in Sections 1 and 2, the method of integral equations and the Schauder fixed-point theorem.

In this section, Ω is the same as that in the above section. Now we introduce three high order singular integral operators over $\Omega \times \Omega$, i.e.

$$\begin{aligned}
(S_1\varphi)(t_1,t_2) &= \int_\Omega \frac{\varphi(x,t_2)(\bar{x}-\bar{t}_1)d\sigma_x}{|x-t_1|^{n+2k+1-r_1}},\\
(S_2\varphi)(t_1,t_2) &= \int_\Omega \frac{\varphi(x,t_2)(x-t_1)d\sigma_x}{|x-t_1|^{n+2k+1-r_2}},\\
(S_3\varphi)(t_1,t_2) &= \int_\Omega \frac{\varphi(x,t_2)d\sigma_x}{|x-t_1|^{n+2k+1-r_3}},
\end{aligned}$$

where $0 < r_i < 1$, $i = 1,2,3$, $(t_1,t_2) \in \Omega \times \Omega$, and the nonlinear differential integral equation including $S_i\varphi\,(1 \le i \le 3)$ is

$$\begin{aligned}
W\varphi &= \sum_{i=1}^{3}(a_i(t_1,t_2)\sum_{0\le k_i+m_i\le n_i}\bar{\partial}_{t_1}^{k_i}\partial_{t_2}^{m_i}S_i\varphi)\\
&\quad +g(t_1,t_2)f[t_1,\,t_2,\,\bar{\partial}_{t_1}^{p_1}\partial_{t_2}^{q_1}S_1\varphi,\,\bar{\partial}_{t_1}^{p_2}\partial_{t_2}^{q_2}S_2\varphi,\,\bar{\partial}_{t_1}^{p_3}\partial_{t_2}^{q_3}S_3\varphi]\\
&= \varphi(t_1,t_2),\ \ (t_1,t_2) \in \Omega\times\Omega,
\end{aligned} \tag{3.1}$$

here k_i, m_i, n_i, p_i, q_i are non-negative integers.

Theorem 3.1 *If the operator*

$$(P\Psi)(t_1,t_2)=\int_\Omega \frac{\Psi(x,t_2)d\sigma_x}{|x-t_1|^{n-1-r}},\quad 0<r<1,$$

is given, then for any $\Psi(x,y)\in H^{(0,0)}_{(\beta_1,\beta_2)}$, $0<\beta_i<1$, $i=1,2$, $n-1>r/(1-\beta_1)$, *there exists a positive constant* J_1 *independent of* Ψ, *such that*

$$\|P\Psi\|_\beta\le J\|\Psi\|_\beta. \tag{3.2}$$

Proof From Section 2, Chapter II, it is clear that $d\sigma_x=m(x)ds_x$. Now we introduce the singular integral operator

$$\begin{aligned}(Q\Psi)(t_1,t_2)&=\int_\Omega \frac{(\Psi(t_1,t_2)-\Psi(x,t_2))d\sigma_x}{|x-t_1|^{n-1-r}}\\ &=\Psi(t_1,t_2)\int_\Omega \frac{d\sigma_x}{|x-t_1|^{n-1-r}}-P\Psi.\end{aligned}\tag{3.3}$$

Firstly, we estimate

$$\begin{aligned}|(Q\Psi)(t_1,t_2)|&\le G_1\|\Psi\|_\beta\int_\Omega \frac{|t_1-x|^{\beta_1}|m(x)|ds_x}{|x-t_1|^{n-1-r}}\\ &\le G_1\|\Psi\|_\beta\int_\Omega \frac{ds_x}{|x-t_1|^{n-1-r-\beta_1}}\le G_2\|\Psi\|_\beta,\end{aligned}\tag{3.4}$$

where G_1,G_2 are positive constants independent of Ψ.

In order to consider $H(Q\Psi,\Omega\times\Omega,\beta)$, denote $\sum=\Omega\times\Omega$, and for any (t_1,t_2), $(\hat{t_1},\hat{t_2})\in\sum$, set $\delta=|t_1-\hat{t_1}|$. Let $3\delta<d$, and d be as stated in Section 2. We make a sphere with the center at t_1 and radius 3δ. Denote by Ω_1 the interior part and by Ω_2 the left part, hence

$$\begin{aligned}&|(Q\Psi)(t_1,t_2)-(Q\Psi)(\hat{t_1},\hat{t_2})|\\ \le&\left|\int_{\Omega_1}\frac{(\Psi(t_1,t_2)-\Psi(x,t_2))d\sigma_x}{|x-t_1|^{n-1-r}}\right|+\left|\int_{\Omega_1}\frac{(\Psi(\hat{t_1},\hat{t_2})-\Psi(x,\hat{t_2}))d\sigma_x}{|x-\hat{t_1}|^{n-1-r}}\right|\\ +&\left|\int_{\Omega_2}\left(\frac{1}{|x-t_1|^{n-1-r}}-\frac{1}{|x-\hat{t_1}|^{n-1-r}}\right)(\Psi(t_1,t_2)-\Psi(x,t_2))d\sigma_x\right|\\ +&\left|\int_{\Omega_2}\frac{[(\Psi(x,\hat{t_2})-\Psi(x,t_2))+(\Psi(t_1,t_2)-\Psi(\hat{t_1},\hat{t_2}))]}{|x-\hat{t_1}|^{n-1-r}}d\sigma_x\right|\\ =&\ L_1+L_2+L_3+L_4.\end{aligned}\tag{3.5}$$

The formula (2.4) about $N_0 \in \Omega$ in Section 2 is used to t_1, and the projective domain of Ω on the tangent plane of t_1 is denoted by π_1, thus we can obtain

$$\begin{aligned} L_1 &\le G_3\|\Psi\|_\beta \int_{\Omega_1} \frac{ds_x}{|x-t_1|^{n-1-r-\beta_1}} \\ &= G_3\|\Psi\|_\beta \int_{\Omega_1} \frac{cos(n_0,x_n)ds_x}{|x-t_1|^{n-1-r-\beta_1} cos(n_0,x_n)} \\ &\le G_3\|\Psi\|_\beta \int_{\pi_1} \frac{d\xi_1 ... d\xi_{n-1}}{|x-t_1|^{n-1-r-\beta_1}\frac{1}{2}} \\ &\le G_4\|\Psi\|_\beta \int_0^{3\delta} \frac{\rho_0^{n-2} d\rho_0}{\rho_0^{n-1-r-\beta_1}} \\ &\le G_5\|\Psi\|_\beta \delta^r \delta^{\beta_1} \le G_6\|\Psi\|_\beta |t_1-\hat{t_1}|^{\beta_1} \\ &\le G_6\|\Psi\|_\beta [|t_1-\hat{t_1}|^{\beta_1} + |t_2-\hat{t_2}|^{\beta_2}], \end{aligned} \tag{3.6}$$

where $G_i (i=3,...,6)$ are positive constants independent of $t_i, \hat{t_i}, i=1,2$. Similarly, we have

$$L_2 \le G_7\|\Psi\|_\beta \,[|t_1-\hat{t_1}|^{\beta_1} + |t_2-\hat{t_2}|^{\beta_2}]. \tag{3.7}$$

Secondly, we estimate L_3. By means of Theorem 2.1, the inequality

$$\begin{aligned} &\left| \frac{1}{|x-t_1|^{n-1-r}} - \frac{1}{|x-\hat{t_1}|^{n-1-r}} \right| \\ \le &\sum_{k=0}^{n-2} \left|\frac{x-\hat{t_1}}{x-t_1}\right|^{\frac{k(n-1-r)}{n-1}} \left|\frac{t_1-\hat{t_1}}{x-t_1}\right|^{\frac{n-1-r}{n-1}} |x-\hat{t_1}|^{-n+1+r} \end{aligned}$$

is derived. For any $x \in \Omega_2$, we get $|x-t_1| \ge 3\delta$, $|x-\hat{t_1}| \ge 2\delta$, hence

$$\frac{1}{2} \le \left|\frac{x-\hat{t_1}}{x-t_1}\right| \le 2, \; |x-\hat{t_1}| \ge \frac{1}{2}|x-t_1|.$$

Moreover we have

$$\left|\frac{t_1-\hat{t_1}}{x-t_1}\right|^{\frac{n-1-r}{n-1}} \le \left|\frac{t_1-\hat{t_1}}{x-t_1}\right|^{\beta_1} = (\frac{\delta}{|x-t_1|})^{\beta_1},$$

and then

$$\begin{aligned} L_3 &\le G_8 \int_{\Omega_2} \frac{\delta^{\beta_1}|x-t_1|^{\beta_1}\|\Psi\|_\beta ds_x}{|x-t_1|^{n-1-r+\beta_1}} = G_8|t_1-\hat{t_1}|^{\beta_1} \int_{\Omega_2} \frac{\|\Psi\|_\beta ds_x}{|x-t_1|^{n-1-r}} \\ &\le G_9\|\Psi\|_\beta [|t_1-\hat{t_1}|^{\beta_1} + |t_2-\hat{t_2}|^{\beta_2}]. \end{aligned} \tag{3.8}$$

In view of

$$\begin{aligned}
&|(\Psi(x,\hat{t}_2)-\Psi(x,t_2))+(\Psi(t_1,t_2)-\Psi(\hat{t}_1,\hat{t}_2))|\\
\le\ & [G_{10}|t_2-\hat{t}_2|^{\beta_2}+G_{11}(|t_1-\hat{t}_1|^{\beta_1}+|t_2-\hat{t}_2|^{\beta_2})]\|\Psi\|_\beta\\
\le\ & G_{12}\|\Psi\|_\beta[|t_1-\hat{t}_1|^{\beta_1}+|t_2-\hat{t}_2|^{\beta_2}],
\end{aligned}$$

it is easy to see that

$$L_4\le G_{13}\|\Psi\|_\beta[|t_1-\hat{t}_1|^{\beta_1}+|t_2-\hat{t}_2|^{\beta_2}]. \tag{3.9}$$

In view of (3.5) – (3.9), we know when $3|t_1-\hat{t}_1|<d$,

$$|(Q\Psi)(t_1,t_2)-(Q\Psi)(\hat{t}_1,\hat{t}_2)|\le G_{14}\|\Psi\|_\beta[|t_1-\hat{t}_1|^{\beta_1}+|t_2-\hat{t}_2|^{\beta_2}]. \tag{3.10}$$

It is clear that when $3|t_1-\hat{t}_1|\ge d$, the above estimation is correct. From (3.4), (3.10), it follows that $\|Q\Psi\|_\beta\le G_{15}\|\Psi\|_\beta$. So in view of (3.3), we have

$$\|P\Psi\|_\beta\le\|Q\Psi\|_\beta+G_{16}\|\Psi\|_\beta\le J_1\|\Psi\|_\beta,$$

here $G_i(7\le i\le 16)$ are positive constants independent of $t_i,\hat{t}_i,(i=1,2)$, $G_{16}=\sup\limits_{t_1\in\Omega}\int_\Omega\frac{d\sigma_x}{|x-t_1|^{n-1-r}}$, J_1 is a positive constant independent of Ψ. This completes the proof of Theorem 3.1.

Corollary 3.2 *Suppose* $\varphi(t_1,t_2)\in H^{(\bar{n}+2k+2,\bar{n})}(\beta_1,\beta_2)\subset H^{(0,0)}(\beta_1,\beta_2)$, $0\le k_i+m_i\le n_i\le\bar{n}$, $n-1>r_i/(1-\beta_1)$, $i=1,2,3$, *where* $\bar{n},k,k_i,m_i,n_i$ *are non-negative integers. Then* $\|\bar{\partial}_{t_1}^{k_i}\partial_{t_2}^{m_i}S_i\varphi(t_1,t_2)\|_\beta\le J_2\|\Psi\|_\beta$, *where the norm is the norm of the element in space* $H^{(0,0)}(\beta_1,\beta_2)$.

Proof It is easy to prove this corollary by means of Theorems 1.13 and 3.1.

The set of functions which have arbitrary order generalized derivative $\bar{\partial}_{t_1}$, ∂_{t_2} on $\Omega\times\Omega$ is denoted by $D_\infty(\Omega\times\Omega)$. It is clear that $D_\infty(\Omega\times\Omega)\subset H^{(0,0)}(\beta_1,\beta_2)$.

Theorem 3.3 *Let* $g(t_1,t_2),f(t_1,t_2,\Phi^1(t_1,t_2),\Phi^2(t_1,t_2),\Phi^3(t_1,t_2))$ *and* $a_i(t_1,t_2)\ (i=1,2,3)$ *in the nonlinear differential integral equations (3.1) belong to* $D_\infty(\Omega\times\Omega)$, *where* $\Phi^i(t_1,t_2)=(\bar{\partial}_{t_1}^{p_i}\partial_{t_2}^{q_i}S_i\varphi)(t_1,t_2),(t_1,t_2)\in\Omega\times\Omega$, $p_i+q_i\le n_i\le\bar{n}$, $s_i\phi$, $0<r_i<1$, $n-1>r_i/(1-\beta_1)\ (i=1,2,3)$ *are as stated in Corollary 3.2, and* $f(0,0,0,0,0)=0$. *Then when* $0<r=2^{n-1}J_2(\bar{n}+1)^2\sum\limits_{i=1}^{n}\|a_i\|_\beta<1$, $\|g\|_\beta<\sigma$, $0<\sigma\le\frac{M(1-r)}{2^{n-1}(G_{19}+G_{20}M)}$, *the nonlinear differential integral equation (3.1) has a solution* $\varphi\in D_\infty(\Omega\times\Omega)$; *and when* $f\equiv 1$, $0<r<1$, *the solution is unique, where* M *is a* *given positive number such that* $\|\varphi\|_\beta \le M$, *and* G_{19}, G_{20} *are positive constants.*

Proof According to the condition $f \in D_\infty(\Omega \times \Omega)$, for any (t_1, t_2), $(\hat{t}_1, \hat{t}_2) \in \Omega \times \Omega$ and any $\Phi_i^j \in A_n(R)$, $(i = 1, 2)$, $1 \le j \le 3$, we have

$$\begin{aligned} & |f(t_1, t_2, \Phi_1^{(1)}, \Phi_1^{(2)}, \Phi_1^{(3)}) - f(\hat{t}_1, \hat{t}_2, \Phi_2^{(1)}, \Phi_2^{(2)}, \Phi_2^{(3)})| \\ & \le G_{17}[|t_1 - t_2|^{\beta_1} + |\hat{t_1} - \hat{t_2}|^{\beta_2}] + G_{18} \sum_{j=1}^{3} |\Phi_1^{(j)} - \Phi_2^{(j)}|, \end{aligned} \tag{3.11}$$

in which G_{17}, G_{18} are positive constants independent of t_i, $\hat{t}_i$, $\Phi_i^{(j)}$, $i = 1, 2$, $1 \le j \le 3$. Next, we consider the subset

$$T = \left\{ \varphi(t_1, t_2) \middle| \begin{array}{l} \varphi \in D_\infty(\Omega \times \Omega) \subset H^{(0,0)}(\beta_1, \beta_2), \\ (t_1, t_2) \in \Omega \times \Omega, \ \|\varphi\|_\beta \le M, \ M > 0 \end{array} \right\}$$

of $H^{(0,0)}(\beta_1, \beta_2)$, in which the norm is defined in the space $H^{(0,0)}(\beta_1, \beta_2)$. For any $\varphi \in T$, by means of (3.1), Theorem 1.13 and the above conditions, we know $W_\varphi \in D_\infty(\Omega \times \Omega)$. In view of (3.1) and Corollary 3.1, the estimate

$$\begin{aligned} \|W\varphi\|_\beta & \le 2^{n-1} \sum_{i=1}^{3} \|a_i\|_\beta \sum_{0 \le k_i + m_i \le n_i} \|\bar{\partial}_{t_1}^{k_i} \partial_{t_2}^{m_i} S_i \varphi\|_\beta + 2^{n-1} \|g\|_\beta \|f\|_\beta \\ & \le 2^{n-1} \sum_{i=1}^{3} \|a_i\|_\beta (\bar{n} + 1)^2 J_2 \|\varphi\|_\beta + 2^{n-1} \|g\|_\beta \|f\|_\beta \end{aligned} \tag{3.12}$$

is concluded. In accordance with $f(0, 0, 0, 0, 0) = 0$, (3.11) and Corollary 3.2, we have

$$\begin{aligned} |f| & \le G_{17}[|t_1 - 0|^{\beta_1} + |t_2 - 0|^{\beta_2}] + G_{18} \sum_{i=1}^{3} |\bar{\partial}_{t_1}^{p_i} \partial_{t_2}^{q_i} S_i \varphi| \\ & \le G_{19} + 3G_{18} J_2 \|\varphi\|_\beta. \end{aligned} \tag{3.13}$$

Noting (3.11), we have

$$\begin{aligned} & |f(t_1, t_2, \bar{\partial}_{t_1}^{p_1} \partial_{t_2}^{q_1} S_1 \varphi(t_1, t_2), ..., \bar{\partial}_{t_1}^{p_3} \partial_{t_2}^{q_3} S_3 \varphi(t_1, t_2)) \\ & -f(\hat{t}_1, \hat{t}_2, \bar{\partial}_{t_1}^{p_1} \partial_{t_2}^{q_1} S_1 \varphi(\hat{t}_1, \hat{t}_2), ..., \bar{\partial}_{t_1}^{p_3} \partial_{t_2}^{q_3} S_3 \varphi(\hat{t}_1, \hat{t}_2))| \\ \le \ & G_{17}[|t_1 - \hat{t}_1|^{\beta_1} + |t_2 - \hat{t}_2|^{\beta_2}] + 3G_{18} J_2 \|\varphi\|_\beta [|t_1 - \hat{t}_1|^{\beta_1} + |t_2 - \hat{t}_2|^{\beta_2}] \\ \le \ & [G_{17} + 3G_{18} J_2 \|\varphi\|_\beta] \ [|t_1 - \hat{t}_1|^{\beta_1} + |t_2 - \hat{t}_2|^{\beta_2}]. \end{aligned} \tag{3.14}$$

From (3.13),(3.14), it follows that

$$\|f\|_\beta \le G_{19} + G_{20}\|\varphi\|_\beta. \tag{3.15}$$

By means of (3.12), (3.13) and the conditions, we get

$$\begin{aligned}\|W\varphi\|_\beta &\le 2^{n-1}\sum_{i=1}^{3}\|a_i\|_\beta(\bar{n}+1)^2 J_2 M + 2^{n-1}\delta(G_{19}+G_{20}M)\\ &\le Mr + 2^{n-1}\delta(G_{19}+G_{20}M) \le M.\end{aligned}$$

This shows that W maps the set T into itself.

Next, we prove that W is a continuous mapping. we are free to choose a sequence $\varphi^{(n)} \in T(n = 1, 2, ...)$, such that $\{\varphi^{(n)}(t_1, t_2)\}$ uniformly converges to $\varphi(t_1, t_2) \in T$, $(t_1, t_2) \in \Omega \times \Omega$. For arbitrarily given positive number ε, when n is large enough, $\|\varphi^{(n)} - \varphi\|_\beta$ can be small enough, hence by (3.11), (3.12) and Corollary 3.2, we see that when n is large enough, the following inequality holds:

$$\begin{aligned}&\|W\varphi^{(n)}(t_1,t_2) - W\varphi(t_1,t_2)\|_\beta\\ \le\ & 2^{n-1}\sum_{i=1}^{3}\|a_i\|_\beta(\bar{n}+1)^2 J_2\|\varphi^{(n)}-\varphi\|_\beta\\ &+2^{n-1}\|g\|_\beta(3G_{18}J_2\|\varphi^{(n)}-\varphi\|_\beta)<\varepsilon.\end{aligned}$$

This shows that W is a continuous mapping, which maps T into itself. According to the Ascoli-Arzela theorem, we see that T is a compact set of space $H^{0,0}(\beta_1, \beta_2)$. Hence the continuous mapping W maps the closed convex set T in $H^{(0,0)}(\beta_1, \beta_2)$ onto itself. Moreover $W(T)$ is also a compact set in $H^{(0,0)}(\beta_1, \beta_2)$. By the Schauder fixed-point theorem, there exists a function $\varphi_0 \in D_\infty(\Omega\times\Omega)$ satisfying equation (3.1); here we mention that though the conditions added to a_i, g, f are stronger, the solution found still satisfies the above condition. At last when $f \equiv 1$, similarly to Section 4, Chapter II, by using the contraction mapping theorem, we can verify the uniqueness of the solution for equation (3.1).

4 A Kind of High Order Singular Integrals of Quasi-Bochner-Martinelli Type With two Singular Points and Poincaré-Bertrand Permutation Formulas in Real Clifford Analysis

From the enlightenment of Sheng Gong's paper about singular integrals of several complex variables [20], in this section, we first discuss high order singular integrals of quasi-Bochner-Martinelli type with two singular points.

Denote still by Ω the boundary of the connected open set D in $\mathbf{R}^n$. If $x \neq t\,(x, t \in \Omega)$, by means of the separability of D, we assume that D^x, D^t satisfy the conditions $D^x \cap D^t = \emptyset$, $\overline{D^x} \cup \overline{D^t} = \overline{D}$, Ω^x, Ω^t are the boundaries of D^x, D^t respectively, $x \in \Omega^x$, $t \in \Omega^t$, and the orientations of Ω^x, Ω^t are harmonious with the orientation of Ω. Moreover, the orientations of Ω, Ω^x, Ω^t are induced orientations of D, D^x, D^t respectively. In addition, suppose that $\overline{D^x} \bigcap \overline{D^t} = \Sigma$, and for any $y \in \Sigma$, $|y - x| = |y - t|$ holds. The integral kernels with singular points x, t are denoted by $K^x(x, y)$, $K^t(x, y)$ respectively. By the additive property of general integrals, the singular integral with two singular points can be defined as follows.

Definition 4.1 Let $\varphi(x, y) \in H^{(m,p)}(\beta_1, \beta_2)$, $0 < \beta_i < 1$, $i = 1, 2$; the singular integral with two singular points on Ω is defined as

$$\begin{aligned}
&\int_\Omega K^t(y, t)\varphi(x, y)K^x(x, y)d\sigma_y \\
= &\int_{\Omega^t} K^t(y, t)\varphi(x, y)K^x(x, y)d\sigma_y + \int_{\Omega^x} K^t(y, t)\varphi(x, y)K^x(x, y)d\sigma_y,
\end{aligned}$$

where m, p are determined by the orders of singularity of kernels $K^x(x, y)$, $K^t(y, t)$ respectively.

In the following, we first discuss the singular integral which can exchange the integral order.

Theorem 4.1 *Let* $\varphi(x, y) \in H^{(0,0)}(\beta_1, \beta_2)$, $0 < \beta_i < 1$, $r > 0$, $h > 0$, $n - 1 > r + h$, $x, t \in \Omega$. *Then*

$$\begin{aligned}
&\int_\Omega \left[\int_\Omega \frac{\varphi(x, y)}{|y - t|^{n-1-r}|x - y|^{n-1-h}} d\sigma_y\right] d\sigma_x \\
&= \int_\Omega \left[\frac{1}{|y - t|^{n-1-r}} \int_\Omega \frac{\varphi(x, y) d\sigma_x}{|x - y|^{n-1-h}}\right] d\sigma_y.
\end{aligned} \tag{4.1}$$

Proof According to the Hadamard theorem [19], when $(n-1) > r+h$, we have

$$\left|\int_\Omega \frac{\varphi(x, y)d\sigma_y}{|y - t|^{n-1-r}|x - y|^{n-1-h}}\right| \leq \frac{G_{21}}{|x - t|^{(n-1)-(r+h)}}.$$

This shows that the above integral with two singular points has a weak singularity, and then the left side of (4.1) is the integral in a general sense. By means of the Fubini theorem [19], we know that this integral order can be exchanged. Hence (4.1) holds.

Now, we deduce the Poincaré-Bertrand permutation formula for the first and second kinds of high order singular integrals of quasi-Bochner-Martinelli type.

Theorem 4.2. *Let $\varphi(x,y) \in H^{(1,1)}(\beta_1,\beta_2)$, $0<\beta_i<1$, $r\,,h>\beta_i$, $i=1,2$, $0<r<1$, $n-1>h+r$, $x,t\in\Omega$. Then*

$$\begin{aligned}
&\int_\Omega \left[\frac{\bar{y}-\bar{t}}{|y-t|^{n+1-r}}\int_\Omega \frac{\varphi(x,y)d\sigma_x}{|x-y|^{n-1-h}}\right]d\sigma_y \\
=&\int_\Omega \left[\int_\Omega \frac{(\bar{y}-\bar{t})\varphi(x,y)d\sigma_y}{|y-t|^{n+1-r}|x-y|^{n-1-h}}\right]d\sigma_x \\
&+\int_\Omega \left[\int_{\Omega^t} \frac{1}{(n-1-r)|y-t|^{n-1-r}}\partial_x\left(\frac{\varphi(x,y)}{|x-y|^{n-1-h}}\right)d\sigma_y\right]d\sigma_x \\
&+\int_\Omega \left[\int_{\Omega^x} \frac{1}{(n-1-r)|x-y|^{n-1-h}}(\partial_x+\partial_y)\left(\frac{\varphi(x,y)}{|y-t|^{n-1-r}}\right)d\sigma_y\right]d\sigma_x.
\end{aligned} \tag{4.2}$$

Proof Under the conditions of Definition 1.1, Theorem 1.8, Corollary 1.2, Lemma 1.4 and Theorem 4.1, we can get

$$\begin{aligned}
&\int_\Omega \frac{\bar{y}-\bar{t}}{|y-t|^{n+1-r}}\int_\Omega \frac{\varphi(x,y)d\sigma_x}{|x-y|^{n-1-h}}d\sigma_y \\
=&\int_\Omega \left[\frac{1}{(n-1-r)|y-t|^{n-1-r}}\int_\Omega \partial_y\left(\frac{\varphi(x,y)}{|x-y|^{n-1-h}}\right)d\sigma_x\right]d\sigma_y \\
=&\int_\Omega \left[\frac{1}{(n-1-r)|y-t|^{n-1-r}}\left(\int_\Omega \frac{\partial_y\varphi(x,y)d\sigma_x}{|x-y|^{n-1-h}}\right)\right]d\sigma_y \\
&+\int_\Omega \left[\frac{1}{(n-1-r)|y-t|^{n-1-r}}\int_\Omega \frac{\varphi(x,y)(\bar{x}-\bar{y})(n-1-h)}{|x-y|^{n+1-h}}d\sigma_x\right]d\sigma_y \qquad (4.3)\\
=&\int_\Omega \left[\int_\Omega \frac{\partial_y\varphi(x,y)d\sigma_y}{(n-1-r)|y-t|^{n-1-r}|x-y|^{n-1-h}}\right]d\sigma_x \\
&+\frac{n-1-h}{n-1-r}\int_\Omega \left[\frac{1}{|y-t|^{n-1-r}}\int_\Omega \frac{\partial_x\varphi(x,y)d\sigma_x}{(n-1-h)|x-y|^{n-1-h}}\right]d\sigma_y \\
=&\frac{1}{n-1-r}\int_\Omega \left[\int_\Omega \frac{(\partial_y+\partial_x)\varphi(x,y)d\sigma_y}{|y-t|^{n-1-r}|x-y|^{n-1-h}}\right]d\sigma_x.
\end{aligned}$$

On the basis of Definitions 4.1 and 1.1, Corollary 1.2, Lemma 1.4 and

(4.3), we can derive

$$\begin{aligned}
&\int_{\Omega}\left[\int_{\Omega}\frac{(\bar{y}-\bar{t})\varphi(x,y)d\sigma_y}{|y-t|^{n+1-r}|x-y|^{n-1-h}}\right]d\sigma_x\\
=&\int_{\Omega}\left[\left(\int_{\Omega^t}+\int_{\Omega^x}\right)\left(\frac{(\bar{y}-\bar{t})\varphi(x,y)d\sigma_y}{|y-t|^{n+1-r}|x-y|^{n-1-h}}\right)\right]d\sigma_x\\
=&\int_{\Omega}\left[\int_{\Omega^t}\frac{1}{(n-1-r)|y-t|^{n-1-r}}\partial_y\left(\frac{\varphi(x,y)}{|x-y|^{n-1-h}}\right)d\sigma_y\right]d\sigma_x\\
&+\int_{\Omega}\left[\int_{\Omega^x}\frac{(\bar{y}-\bar{t})\varphi(x,y)d\sigma_y}{|x-y|^{n-1-h}|y-t|^{n+1-r}}\right]d\sigma_x\\
=&\int_{\Omega}\left[\int_{\Omega^t}\frac{\partial_y\varphi(x,y)d\sigma_y}{(n-1-r)|y-t|^{n-1-r}|x-y|^{n-1-h}}\right]d\sigma_x\\
&+\int_{\Omega}\left[\int_{\Omega^t}\frac{\varphi(x,y)(n-1-h)(\bar{x}-\bar{y})d\sigma_y}{(n-1-r)|y-t|^{n-1-r}|x-y|^{n+1-h}}\right]d\sigma_x\\
&+\int_{\Omega}\left[\int_{\Omega^x}\frac{1}{|x-y|^{n-1-h}(1+r-n)}\partial_y\left(\frac{1}{|y-t|^{n-1-r}}\right)\varphi(x,y)d\sigma_y\right]d\sigma_x\\
&+\int_{\Omega}\left[\int_{\Omega^x}\frac{\partial_y\varphi(x,y)d\sigma_y}{(n-1-r)|y-t|^{n-1-r}|x-y|^{n-1-h}}\right]d\sigma_x\\
&-\int_{\Omega}\left[\int_{\Omega^x}\frac{\partial_y\varphi(x,y)d\sigma_y}{(n-1-r)|y-t|^{n-1-r}|x-y|^{n-1-h}}\right]d\sigma_x\\
=&\int_{\Omega}\left[\int_{\Omega}\frac{\partial_y\varphi(x,y)d\sigma_y}{(n-1-r)|y-t|^{n-1-r}|x-y|^{n-1-h}}\right]d\sigma_x\\
&+\int_{\Omega}\left[\int_{\Omega}\frac{\partial_x\varphi(x,y)d\sigma_y}{(n-1-r)|y-t|^{n-1-r}|x-y|^{n-1-h}}\right]d\sigma_x\\
&-\int_{\Omega}\left[\int_{\Omega^t}\frac{\partial_x\varphi(x,y)d\sigma_y}{(n-1-r)|y-t|^{n-1-r}|x-y|^{n-1-h}}\right]d\sigma_x\\
&+\int_{\Omega}\left[\int_{\Omega^t}\frac{\varphi(x,y)(n-1-h)(\bar{x}-\bar{y})d\sigma_y}{(n-1-r)|y-t|^{n-1-r}|x-y|^{n+1-h}}\right]d\sigma_x\\
&+\int_{\Omega}\left[\int_{\Omega^x}\frac{1}{(1+r-n)|x-y|^{n-1-h}}\partial_y\left(\frac{1}{|y-t|^{n-1-r}}\right)\varphi(x,y)d\sigma_y\right]d\sigma_x\\
&-\int_{\Omega}\left[\int_{\Omega^x}\frac{\partial_x\varphi(x,y)d\sigma_y}{(n-1-r)|y-t|^{n-1-r}|x-y|^{n-1-h}}\right]d\sigma_x\\
&-\int_{\Omega}\left[\int_{\Omega^x}\frac{\partial_y\varphi(x,y)d\sigma_y}{(n-1-r)|y-t|^{n-1-r}|x-y|^{n-1-h}}\right]d\sigma_x\\
=&\int_{\Omega}\left[\int_{\Omega}\frac{(\partial_y+\partial_x)\varphi(x,y)d\sigma_y}{(n-1-r)|y-t|^{n-1-r}|x-y|^{n-1-h}}\right]d\sigma_x\\
&-\int_{\Omega}\left[\int_{\Omega^t}\frac{\partial_x\varphi(x,y)d\sigma_y}{(n-1-r)|y-t|^{n-1-r}|x-y|^{n-1-h}}\right]d\sigma_x\\
&+\int_{\Omega}\left[\int_{\Omega^t}\left(\frac{\varphi(x,y)\cdot(-1)}{(n-1-r)|y-t|^{n-1-r}}\right)\partial_x\left(\frac{1}{|x-y|^{n-1-h}}\right)d\sigma_y\right]d\sigma_x
\end{aligned}$$

$$
\begin{aligned}
&-\int_\Omega\left[\int_{\Omega^x}\left(\frac{1}{(n-1-r)|x-y|^{n-1-h}}\right)\partial_y\left(\frac{1}{|y-t|^{n-1-r}}\right)\varphi(x,y)d\sigma_y\right]d\sigma_x\\
&-\int_\Omega\left[\int_{\Omega^x}\frac{(\partial_x+\partial_y)\varphi(x,y)d\sigma_y}{(n-1-r)|y-t|^{n-1-r}|x-y|^{n-1-h}}\right]d\sigma_x\\
=\;&\int_\Omega[\int_\Omega\frac{(\partial_x+\partial_y)\varphi(x,y)d\sigma_y}{(n-1-r)|y-t|^{n-1-r}|x-y|^{n-1-h}}]d\sigma_x\\
&-\int_\Omega[\int_{\Omega^t}\frac{1}{(n-1-r)|y-t|^{n-1-r}}\partial_x(\frac{\varphi(x,y)}{|x-y|^{n-1-h}})d\sigma_y]d\sigma_x\\
&-\int_\Omega[\int_{\Omega^x}\frac{1}{(n-1-r)|x-y|^{n-1-h}}\partial_y(\frac{1}{|y-t|^{n-1-r}})\varphi(x,y)d\sigma_y]d\sigma_x\\
&-\int_\Omega[\int_{\Omega^x}\frac{(\partial_x+\partial_y)\varphi(x,y)d\sigma_y}{(n-1-r)|y-t|^{n-1-r}|x-y|^{n-1-h}}]d\sigma_x\\
=\;&\int_\Omega[\frac{\bar y-\bar t}{|y-t|^{n+1-r}}\int_\Omega\frac{\varphi(x,y)d\sigma_x}{|x-y|^{n-1-h}}]d\sigma_y\\
&-\int_\Omega[\int_{\Omega^t}\frac{1}{(n-1-r)|y-t|^{n-1-r}}\partial_x(\frac{\varphi(x,y)}{|x-y|^{n-1-h}})d\sigma_y]d\sigma_x\\
&-\int_\Omega[\int_{\Omega^x}(\frac{1}{(n-1-r)|x-y|^{n-1-h}}\partial_y(\frac{1}{|y-t|^{n-1-r}})\varphi(x,y)d\sigma_y]d\sigma_x\\
&-\int_\Omega[\int_{\Omega^x}\frac{(\partial_y+\partial_x)\varphi(x,y)d\sigma_y}{(n-1-r)|y-t|^{n-1-r}|x-y|^{n-1-h}}]d\sigma_x.
\end{aligned}
\tag{4.4}
$$

In view of (4.4) and Corollary 1.2, we have

$$
\begin{aligned}
&\int_\Omega\left[\frac{\bar y-\bar t}{|y-t|^{n+1-r}}\int_\Omega\frac{\varphi(x,y)d\sigma_x}{|x-y|^{n-1-h}}\right]d\sigma_y\\
=\;&\int_\Omega\left[\int_\Omega\frac{(\bar y-\bar t)\varphi(x,y)d\sigma_y}{|y-t|^{n+1-r}|x-y|^{n-1-h}}\right]d\sigma_x\\
&+\int_\Omega\left[\int_{\Omega^t}\frac{1}{(n-1-r)|y-t|^{n-1-r}}\partial_x\left(\frac{\varphi(x,y)}{|x-y|^{n-1-h}}\right)d\sigma_y\right]d\sigma_x\\
&+\int_\Omega\left[\int_{\Omega^x}\frac{1}{(n-1-r)|x-y|^{n-1-h}}\partial_y(\frac{\varphi(x,y)}{|y-t|^{n-1-r}})d\sigma_y\right]d\sigma_x\\
&+\int_\Omega\left[\int_{\Omega^x}\frac{\partial_x\varphi(x,y)d\sigma_y}{(n-1-r)|x-y|^{n-1-h}|y-t|^{n-1-r}}\right]d\sigma_x\\
=\;&\int_\Omega\left[\int_\Omega\frac{(\bar y-\bar t)\varphi(x,y)d\sigma_y}{|y-t|^{n+1-r}|x-y|^{n-1-h}}\right]d\sigma_x\\
&+\int_\Omega\left[\int_{\Omega^t}\frac{1}{(n-1-r)|y-t|^{n-1-r}}\partial_x\left(\frac{\varphi(x,y)}{|x-y|^{n-1-h}}\right)d\sigma_y\right]d\sigma_x
\end{aligned}
$$

$$
\begin{aligned}
&+\int_{\Omega}\left[\int_{\Omega^x}\frac{1}{(n-1-r)|x-y|^{n-1-h}}\partial_y\left(\frac{\varphi(x,y)}{|y-t|^{n-1-r}}\right)d\sigma_y\right]d\sigma_x\\
&+\int_{\Omega}\left[\int_{\Omega^x}\frac{1}{(n-1-r)|x-y|^{n-1-h}}\partial_x\left(\frac{\varphi(x,y)}{|y-t|^{n-1-r}}\right)d\sigma_y\right]d\sigma_x.
\end{aligned}
\tag{4.5}
$$

Merge the last two terms in (4.5), and we get (4.2). The proof of this theorem is completed.

In short, the method of the proof in Theorem 4.2 is as follows: Firstly, by using calculational formulas, differential formulas in Section 1 and the Fubini theorem (see [6],[7][19]), we establish the relation between the singular integral before permutation, the singular integral after permutation with high order quasi-Bochner-Martinelli type kernels and the singular integrals with two singular points to be exchanged in integral order respectively. Next we can get the Poincaré-Bertrand permutation formulas. The above differential formulas play an important pole in the proof. This method is different from that used to prove the Poincaré-Bertrand permutation formulas of non-high order singular integrals in Section 2, Chapter V.

If $W(k_i, r_i)\,(1 \le i \le 6)$ are expressed the general first−sixth kinds of high order quasi-Bochner-Martinelli kernels in Theorems 1.8 and 1.16, then we can use the same method in the proof of Theorem 4.2 to prove the Poincaré-Bertrand permutation formulas of the general high order singular integrals with the kernels $W(k_i, r_i)$ and $W(k_j, r_j)\,(1 \le i, j \le 6)$. Because of page limitation we do not prove them one by one. The proofs are left as exercises.

Concluding Remark It is well known that the representation in circular cylinder domains for several complex variables is worse than the representation in the unit disk for functions of one complex variable. Hence in theory of several complex variables, the regular functions determined by Cauchy integral formulas in several domains possess different representations, which were an important research subject by many scholars. In 1961, F. Norguet enumerated 66 main papers about Cauchy integral formulas of several complex variables in the appendix of his paper [56]. In [56], the main Cauchy integral formulas are classified as five kinds, for example, the Cauchy integral formulas in four kinds of canonical domains, Cauchy integral formulas of Bochner-Martinelli type and so on. These Cauchy integral formulas are generalized to the Cauchy integral formula of a complex variable from different angles and views. We are enlightened by the variability of Cauchy integral formulas in several complex variables. Because of the non-commutativity property in Clifford algebra, the expressions of integral operators adopted in this chapter are different from the Cauchy integral expression

$$\frac{1}{\omega_n}\int_{\Omega}\frac{\overline{x}-\overline{y}}{|x-y|^n}d\sigma_x f(x),\ \Omega=\partial D\subset R^n,$$

used by other scholars (see [6],[19]). In this chapter, we use integral expressions, that be got in Chapter I, such that many calculational formulas, recurrence formulas and differential formulas of singular integrals as stated in Section 1 are succinct and regular. Furthermore, we can give another method to prove the Poincaré-Bertrand permutation formula of high order singular integrals by using differential formulas. It is also interesting to investigate various integral expressions of generalized or doubly regular functions in Clifford analysis.

CHAPTER VII

RELATION BETWEEN CLIFFORD ANALYSIS AND ELLIPTIC EQUATIONS

In this chapter, we first introduce the solvability of some oblique derivative problems for uniformly and degenerate elliptic equations of second order, and then discuss the existence of solutions of some boundary value problems for some degenerate elliptic systems of first order in Clifford analysis by using the above results for elliptic equations.

1 Oblique Derivative Problems for Uniformly Elliptic Equations of Second Order

1.1 Formulation of oblique derivative problem for nonlinear elliptic equations

Let Q be a bounded domain in $\mathbf{R}^N$ and the boundary $\partial Q \in C^2_\mu\,(0 < \mu < 1)$, herein $N\,(> 1)$ is a positive integer. We consider the nonlinear elliptic equation of second order

$$F(x, u, D_x u, D_x^2 u) = 0 \text{ in } Q, \tag{1.1}$$

namely

$$Lu = \sum_{i,j=1}^{N} a_{ij} u_{x_i x_j} + \sum_{i=1}^{N} b_i u_{x_i} + cu = f \text{ in } Q, \tag{1.2}$$

where $D_x u = (u_{x_i})$, $D_x^2 u = (u_{x_i x_j})$, and

$$a_{ij} = \int_0^1 F_{\tau r_{ij}}(x, u, p, \tau r)d\tau,\ b_i = \int_0^1 F_{\tau p_i}(x, u, \tau p, 0)d\tau,$$

$$c = \int_0^1 F_{\tau u}(x, \tau u, 0, 0)d\tau,\ f = -F(x, 0, 0, 0),$$

$$r = D_x^2 u,\ p = D_x u,\ r_{ij} = \frac{\partial u}{\partial x_i \partial x_j},\ p_i = \frac{\partial u}{\partial x_i}.$$

Suppose that (1.1) (or (1.2)) satisfies Condition C, i.e. for arbitrary functions $u_1(x)$, $u_2(x) \in C^1_\beta(\overline{Q}) \cap W^2_2(Q)$, $F(x,u,D_xu,D^2_xu)$ satisfy the following conditions:

$$\begin{aligned} &F(x,u_1,D_xu_1,D^2_xu_1) - F(x,u_2,D_xu_2,D^2_xu_2) \\ &\quad = \sum_{i,j=1}^{N} \tilde{a}_{ij} u_{x_ix_j} + \sum_{i=1}^{N} \tilde{b}_i u_{x_i} + \tilde{c}u, \end{aligned} \tag{1.3}$$

where $0 < \beta < 1, u = u_1 - u_2$ and

$$\begin{aligned} &\tilde{a}_{ij} = \int_0^1 F_{u_{x_ix_j}}(x,\tilde{u},\tilde{p},\tilde{r})d\tau,\ \tilde{b}_i = \int_0^1 F_{u_{x_i}}(x,\tilde{u},\tilde{p},\tilde{r})d\tau, \\ &\tilde{c} = \int_0^1 F_u(x,\tilde{u},\tilde{p},\tilde{r})d\tau,\ \tilde{u} = u_2 + \tau(u_1 - u_2), \\ &\tilde{p} = D_x[u_2 + \tau(u_1 - u_2)],\ \tilde{r} = D^2_x[u_2 + \tau(u_1 - u_2)], \end{aligned}$$

and $\tilde{a}_{ij}, \tilde{b}_i, \tilde{c}, f$ satisfy the conditions

$$q_0 \sum_{j=1}^{N} |\xi_j|^2 \leq \sum_{i,j=1}^{N} \tilde{a}_{ij}\xi_i\xi_j \leq q_0^{-1} \sum_{j=1}^{N} |\xi_j|^2, 0 < q_0 < 1, \tag{1.4}$$

$$\begin{aligned} &\sup_Q \sum_{i,j=1}^{N} \tilde{a}^2_{ij} / \inf_Q [\sum_{i=1}^{N} \tilde{a}_{ii}]^2 \leq q_1 < \frac{2N-1}{2N^2-2N-1},\ L_p(f,Q) \leq k_1, \\ &|\tilde{a}_{ij}|, |\tilde{b}_i| \leq k_0 \text{ in } Q,\ i,j = 1,...,N, -k_0 \leq \inf_Q \tilde{c} \leq \sup_Q \tilde{c} < 0, \end{aligned} \tag{1.5}$$

in which $k_0, k_1, p\,(> N+2)$ are non-negative constants. Moreover, for almost every point $x \in Q$ and $D^2_xu \in \mathbf{R}^{N(N+1)/2}$, the functions $\tilde{a}_{ij}(x,u,D_xu,D^2_xu)$, $\tilde{b}_i(x,u,D_xu)$, $\tilde{c}(x,u)$ are continuous in $u \in \mathbf{R}$ and $D_xu \in \mathbf{R}^N$.

The so-called oblique derivative problem (Problem O) is to find a continuously differentiable solution $u = u(x) \in C^1_\beta(\overline{Q}) \cap W^2_2(Q)$ satisfying the boundary conditions:

$$\begin{aligned} &lu = d\frac{\partial u}{\partial \nu} + bu = g(x),\ \ x \in \partial Q, \ \text{ i.e.} \\ &lu = \sum_{j=1}^{N} d_j \frac{\partial u}{\partial x_j} + bu = g(x),\ \ x \in \partial Q, \end{aligned} \tag{1.6}$$

in which $d_j(x), b(x), g(x)$ satisfy the conditions

$$
\begin{gathered}
C^1_\alpha[b(x),\partial Q]\le k_0,\, C^1_\alpha[g(x),\overline{Q}]\le k_2,\\
\cos(\nu,n)\ge q_0>0,\ b(x)\ge 0 \ \text{on}\ \partial Q,\\
\sum_{j=1}^N d_j\cos(\nu,x_j)\ge q_0>0,\ \ C^1_\alpha[d_j(x),\partial Q]\le k_0, j=1,...,,N,
\end{gathered}
\tag{1.7}
$$

in which n is the unit outward normal on ∂Q, $\alpha\,(0<\alpha<1), k_0, k_2$, $q_0\,(0<q_0<1)$ are non-negative constants. In particular, if Problem O has the conditions $d=1, \nu=n$, $b=0$ on ∂Q in (1.6), then Problem O is the Neumann boundary value problem, which will be called Problem N.

In the following, we give a priori estimates of solutions for Problem O. Then, by using the method of parameter extension and the Leray-Schauder theorem, we prove the existence and uniqueness of solutions for Problem O.

1.2 A priori estimates of solutions for Problem O for (1.2)

We first prove the following theorem.

Theorem 1.1 *If the equation (1.2) satisfies Condition C, then the solution of Problem O is unique.*

Proof Let $u_1(x), u_2(x)$ be two solutions of Problem O, it is easily seen that $u=u_1-u_2$ is a solution of the boundary value problem

$$\sum_{i,j=1}^N \tilde a_{ij}u_{x_ix_j}+\sum_{i=1}^N \tilde b_i u_{x_i}+\tilde c u=0 \ \text{in}\ Q, \tag{1.8}$$

$$lu(x)=0 \ \text{i.e.}\ d\frac{\partial u}{\partial\nu}+bu=0,\ \ x\in\partial Q, \tag{1.9}$$

where $\tilde a_{ij}, \tilde b_i, \tilde c$ are as stated in (1.3). By the maximum principle of solutions for (1.8), $u(x)$ attains its maximum in $\overline{Q}$ at a point $P_0\in\partial Q$, and $lu|_{P_0}>0$; this contradicts (1.9), hence $u(x)=0$ in $\overline{Q}$, i.e. $u^1(x)=u^2(x)$, $x\in Q$.

In the following, we shall give the estimates of $C^1(\bar Q)$ and $C^1_\beta(\overline{Q})$ of solutions $u(x)$ of Problem O.

Theorem 1.2 *Under the same condition as in Theorem 1.1, any solution $u(x)$ of Problem O for (1.2) satisfies the estimate*

$$||u||_{C^1(\bar{Q})} = ||u||_{C(\bar{Q})} + \sum_{i=1}^{N} ||u_{x_i}||_{C(\overline{Q})} \le M_1, \tag{1.10}$$

where M_1 is a non-negative constant only dependent on q, p, α, k, Q, i.e. $M_1 = M_1(q, p, \alpha,\ k, Q), q = (q_0, q_1), k = (k_0, k_1, k_2)$.

Proof Suppose that (1.10) is not true, there exist sequences of functions $\{a_{ij}^m\}, \{b_i^m\}, \{c^m\}, \{f^m\}$ and $\{a^m(x)\}, \{b^m(x)\}, \{g^m(x)\}$, which satisfy Condition C and the conditions in (1.7), and $\{a_{ij}^m\}$, $\{b_i^m\}$, $\{c^m\}, \{f^m\}$ weakly converge to $a_{ij}^0, b_i^0, c^0, f^0$, and $\{a^m\}, \{b^m\}, \{g^m\}$ uniformly converge to a^0, b^0, g^0 on ∂Q respectively. Furthermore the boundary value problem

$$\sum_{i,j=1}^{N} a_{ij}^m u_{x_i x_j} + \sum_{i=1}^{N} b_i^m u_{x_i} + c^m u = f^m \text{ in } Q, \tag{1.11}$$

$$lu^m(x) = g^m(x), \text{ i.e. } a^m \frac{\partial u^m}{\partial \nu} + b^m u^m = g^m(x),\ x \in \partial Q \tag{1.12}$$

has a solution $u^m(x)$, such that $||u^m||_{C^1(\bar{Q})} = A_m (m = 1, 2, ...)$ is unbounded (there is no harm in assuming that $A_m \ge 1$, and $\lim_{m\to\infty} A_m = +\infty$). It is easy to see that $U^m = u^m/A_m$ is a solution of the initial-boundary value problem

$$\tilde{L}U^m = \sum_{i,j=1}^{N} a_{ij}^m U_{x_i x_j}^m = B^m, B^m = -\sum_{i=1}^{N} b_i^m U_{x_i}^m - c^m U^m + f^m/A_m, \tag{1.13}$$

$$lU^m(x) = \frac{g^m}{A_m}, \text{ i.e. } a^m \frac{\partial U^m}{\partial n} + b^m U^m = \frac{g^m}{A_m},\ x \in \partial Q. \tag{1.14}$$

Noting that $\sum_{i=1}^{N} b_i^m U_{x_i}^m + c^m U^m$ in (1.13) is bounded, by using the result in Theorem 1.3 below, we can obtain the estimate

$$||U^m||_{C^1_\beta(\bar{Q})} = ||U^m||_{C_\beta(\bar{Q})} + \sum_{i=1}^{N} ||U_{x_i}^m||_{C_\beta(\bar{Q})} \le M_2, \tag{1.15}$$

$$||U^m||_{W_2^2(Q)} \le M_3 = M_3(q, p, \alpha, k, Q),\ m = 1, 2, ... \tag{1.16}$$

where $\beta (0 < \beta \le \alpha)$, $M_j = M_j(q, p, \alpha, k, Q) (j = 2, 3)$ are non-negative constants. Hence from $\{U^m\}, \{U_{x_i}^m\}$, we can choose a subsequence $\{U^{m_k}\}$ such that $\{U^{m_k}\}$, $\{U_{x_i}^{m_k}\}$ uniformly converge to $U^0, U_{x_i}^0$ in $\overline{Q}$

respectively, $\{U^{m_k}_{x_ix_j}\}$ weak converges to $U^0_{x_ix_j}$ in Q, and U^0 is a solution of the boundary value problem:

$$\sum_{i,j=1}^{N} a^0_{ij}\tilde{U}^0_{x_ix_j} + \sum_{i=1}^{N} b^0_i\tilde{U}^0_{x_i} + c^0U^0 = 0, \tag{1.17}$$

$$lU^0(x) = 0, \text{ i.e. } a\frac{\partial U^0}{\partial \nu} + bU^0 = 0, \;\; x \in \partial Q. \tag{1.18}$$

According to Theorem 1.1, we know that $U^0(x) = 0$, $x \in \overline{Q}$. However, from $||U^m||_{C^1(\bar{Q})} = 1$, we can derive that there exists a point $x^* \in \bar{Q}$, such that $|U^0(x^*)| + \sum_{i=1}^N | U^0_{x_i}(x^*)| > 0$. This contradiction proves that (1.10) is true.

Theorem 1.3 *Under the same condition as in Theorem 1.1, any solution $u(x)$ of Problem O satisfies the estimates*

$$||u||_{C^1_\beta(\bar{Q})} \le M_4 = M_4(q, p, \alpha, k, Q), \tag{1.19}$$

$$||u||_{W^2_2(Q)} \le M_5 = M_5(q, p, \alpha, k, Q), \tag{1.20}$$

where $\beta(0 < \beta \le \alpha)$, M_4, M_5 are non-negative constants.

Proof First of all, we find a solution $\hat{u}(x)$ of the equation

$$\Delta\hat{u} - \hat{u} = 0 \tag{1.21}$$

with the boundary condition (1.6), which satisfies the estimate

$$||\hat{u}||_{C^2(\bar{Q})} \le M_6 = M_6(q, p, \alpha, k, Q). \tag{1.22}$$

(see [38]). Thus the function

$$\tilde{u}(x) = u(x) - \hat{u}(x) \tag{1.23}$$

is a solution of the equation

$$L\tilde{u} = \sum_{i,j=1}^{N} a_{ij}\tilde{u}_{x_ix_j} + \sum_{i=1}^{N} b_i\tilde{u}_{x_i} + c\tilde{u} = \tilde{f}, \tag{1.24}$$

$$l\tilde{u}(x) = 0, \;\; x \in \partial Q, \tag{1.25}$$

where $\tilde{f} = f - L\hat{u}$. Introduce a local coordinate system on the neighborhood G of a surface $S_1 \in \partial Q$:

$$x_i = h_i(\xi_1, ..., \xi_{N-1})\xi_N + g_i(\xi_1, ..., \xi_{N-1}),\; i = 1, ..., N, \tag{1.26}$$

where $\xi_N = 0$ is just the surface $S_1 : x_i = g_i(\xi_1, ..., \xi_{N-1})(i = 1, ..., N)$, and

$$h_i(\xi) = \left.\frac{d_i(x)}{d(x)}\right|_{x_i=g_i(\xi)}, \ i = 1, ..., N, d^2(x) = \sum_{i=1}^{N} d_i^2(x).$$

Then the boundary condition (1.25) can be reduced to the form

$$\frac{\partial \tilde{u}}{\partial \xi_N} + \tilde{b} = 0 \text{ on } \xi_N = 0, \tag{1.27}$$

where $\tilde{u} = \tilde{u}[x(\xi)]$, $\tilde{b} = b[x(\xi)]$. Secondly, we find a solution $v(x)$ of Problem N for the equation (1.21) with the boundary condition

$$\frac{\partial v}{\partial \xi_N} = \tilde{b} \text{ on } \xi_N = 0, \tag{1.28}$$

which satisfies the estimate

$$||v||_{C^2(\bar{Q})} \leq M_7 = M_7(q, p, \alpha, k, Q) < \infty, \tag{1.29}$$

(see [38]) and the function

$$V(x) = \tilde{u}(x)e^{v(x)} \tag{1.30}$$

is a solution of the boundary value problem in the form

$$\sum_{i,j=1}^{N} \tilde{a}_{ij} V_{\xi_i\xi_j} + \sum_{i=1}^{N} \tilde{b}_i V_{x_i} + \tilde{c}V = \tilde{f}, \tag{1.31}$$

$$\frac{\partial V}{\partial \xi_N} = 0, \ \ \xi_N = 0. \tag{1.32}$$

On the basis of Theorem 1.4 below, we can derive the estimates of $V(\xi)$, i.e.

$$||V||_{C^1_\beta(\bar{Q})} \leq M_8 = M_8(q, p, \alpha, k, Q), \tag{1.33}$$

$$||V||_{W^2_2(Q)} \leq M_9 = M_9(q, p, \alpha, k, Q), \tag{1.34}$$

where $\beta\,(0 < \beta \leq \alpha)$, M_8, M_9 are non-negative constants. Combining (1.22), (1.29), (1.33) and (1.34), the estimates (1.19) and (1.20) are obtained.

Now, we shall give some estimates of solutions of Problem N for (1.2).

Theorem 1.4 *Suppose that the equation (1.2) satisfies Condition C. Then any solution $u(x)$ of Problem N satisfies the estimates*

$$C^1_\beta[u, \bar{Q}] \leq M_{10} = M_{10}(q, p, \alpha, k, Q), \tag{1.35}$$

$$||u||_{W_2^2(Q)} \le M_{11} = M_{11}(q,p,\alpha,k,Q), \tag{1.36}$$

where M_{10}, M_{11} are non-negative constants, and β is a constant as stated in (1.19).

Proof First of all, choosing that x^* is an arbitrary point in Q and ε is a small positive number, we construct a function $J(x) \in C^2_\alpha(\bar{Q})$ such that

$$J(x) = \begin{cases} 1, & x \in \Omega_\varepsilon, \\ 0, & x \in Q \backslash \partial Q_{2\varepsilon}, \end{cases} \quad 0 \le J(x) \le 1, \ x \in \partial Q_{2\varepsilon} \backslash Q_\varepsilon,$$

in which $Q_\varepsilon = \{|x - x^*| \le \varepsilon\}$ and $J(x)$ satisfies the estimate

$$C^2[J(x), \overline{Q}] \le M_{12} = M_{12}(\varepsilon, Q). \tag{1.37}$$

Denote $U(x) = J(x)u(x)$; obviously $U = U(x)$ is a solution of the boundary value problem

$$\sum_{i,j=1}^{N} a_{ij}U_{x_ix_j} + \sum_{i=1}^{N} b_iU_{x_i} + cU = f^*, \ \ x \in \bar{Q}, \tag{1.38}$$

$$lU(x) = 0, \ \ x \in \partial Q, \tag{1.39}$$

in which $d^* = JLu + \sum_{i,j=1}^N a_{ij}[J_{x_i}u_{x_j} + J_{x_ix_j}u] + \sum_{i=1}^N b_iJ_{x_i}u$. By the method of inner estimate in [1],[11], we can obtain

$$C^1_\beta[U, \bar{Q}] \le M_{13}, \ C^1_\beta[u, Q_\varepsilon] \le M_{14}, \tag{1.40}$$

where $M_j = M_j(q,p,\alpha,k,Q_\varepsilon)$, $j = 13, 14$. Combining (1.37) and (1.40), we obtain the estimate

$$C^1_\beta[u, Q_\varepsilon] \le M_{15} = M_{15}(q,\alpha,k,Q,p). \tag{1.41}$$

Next, we choose any point $x^* \in S_2 = \partial Q$ and a small positive number d such that $S_3 = S_2 \cap \{|x - x^*| \le d\}$. Then we can find a solution $\hat{u}(x)$ of (1.21) on $\overline{Q}$, such that $\hat{u}(x)$ satisfies the boundary condition

$$\frac{\partial \hat{u}}{\partial n} = g(x), \ \ x \in S_3,$$

which satisfies the estimate (1.22). Thus $\tilde{u}(x) = u(x) - \hat{u}(x)$ is a solution of the equation as stated in (1.24), and $\tilde{u}(x)$ satisfies the boundary condition

$$\frac{\partial \tilde{u}}{\partial n} = 0, \ \ x \in S_3. \tag{1.42}$$

We can define a non-singular transformation of a second order continuously differentiable function $\zeta = \zeta(x)$, such that S_3 maps onto S_4 on the plane $\zeta_N = 0$, the domain Q onto the domain Q_1 in the half space $\zeta_N < 0$, and the equation (1.24) and boundary condition (1.42) are reduced to the equation and boundary condition as follows:

$$\sum_{i,j=1}^{N} A_{ij}\tilde{u}_{\zeta_i\zeta_j} + \sum_{i=1}^{N} B_i\tilde{u}_{\zeta_i} + C\tilde{u} = D \text{ in } Q_1 \tag{1.43}$$

$$\frac{\partial \tilde{u}}{\partial \xi_N} = 0 \text{ on } S_4. \tag{1.44}$$

Now, we extend the function $\tilde{u}$ to a symmetric domain Q_2 of Q_1 about S_4, i.e. we define a function

$$U = \begin{cases} \tilde{u}(\zeta), & \zeta \in Q_1, \\ \tilde{u}(\zeta^*), & \zeta^* \in Q_2, \end{cases} \tag{1.45}$$

where $\zeta^* = (\zeta_1, ..., \zeta_{N-1}, -\zeta_N)$, and $U(\zeta)$ is a solution of the equation

$$\sum_{i,j=1}^{N} \tilde{A}_{ij}U_{\zeta_i\zeta_j} + \sum_{i=1}^{N} \tilde{B}_iU_{\zeta_i} + \tilde{C}U = \tilde{D} \text{ in } Q_1 \cup Q_2, \tag{1.46}$$

where

$$\tilde{A}_{ij} = \begin{cases} A_{ij}(\zeta), \\ (-1)^k A_{ij}(\zeta^*), \end{cases} \quad k = \begin{cases} 1, i \neq j, \ i \text{ or } j = N, \\ 0, \text{ other cases}, \end{cases}$$

$$\tilde{B}_i = \begin{cases} B_i(\zeta), \\ (-1)^k B_i(\zeta^*), \end{cases} \quad k = \begin{cases} 1, \ i = N, \\ 0, \ i \neq N, \end{cases}$$

$$\tilde{C} = \begin{cases} C(\zeta), \\ C(\zeta^*), \end{cases} \quad \tilde{D} = \begin{cases} D(\zeta), & \zeta \in Q_1, \\ D(\zeta^*), & \zeta \in Q_2. \end{cases}$$

By using a similar method in the proof of (1.41), we can derive that $U(\zeta)$ and $u(x)$ satisfy the estimate

$$C^1_\beta[U, Q_1 \cup Q_2] \leq M_{16}, \ \ C^1_\beta[u, Q_1] \leq M_{17}, \tag{1.47}$$

where $M_j = M_j(q, p, \alpha, k, Q), j = 16, 17$. Combining (1.41) and (1.47), the estimates (1.35) and (1.36) are obtained.

1.3 Solvability of oblique derivative problem for elliptic equations

We first consider a special equation of (1.2), namely

$$\Delta u = g_m(x, u, D_x u, D_x^2 u),$$
$$g_m = \Delta u - \sum_{i,j=1}^{N} a_{ijm} u_{x_i x_j} - \sum_{i=1}^{N} b_{im} u - c_m u + f_m \text{ in } Q, \tag{1.48}$$

where $\Delta u = \sum_{i=1}^{N} \partial^2 u/\partial x_i^2$, $\Lambda = (2N-1)\inf_Q \sum_{i=1}^{N} a_{ii}/(2N^2-2N-1)$, and the coefficients

$$a_{ijm} = \begin{cases} a_{ij}/\Lambda, \\ \delta_{ij}/\Lambda, \end{cases} \quad b_{im} = \begin{cases} b_i/\Lambda, \\ 0, \end{cases} \quad i, j = 1, ..., N,$$

$$c_m = \begin{cases} c/\Lambda, \\ 0, \end{cases} \quad f_m = \begin{cases} f/\Lambda & \text{in } Q_m, \\ 0 & \text{in } \mathbf{R}^N \backslash Q_m, \end{cases}$$

where $Q_m = \{(x) \in Q \,|\, \operatorname{dist}(x, \partial Q) \geq 1/m\}$, m is a positive integer, $\delta_{ii} = 1$, $\delta_{ij} = 0\,(i \neq j,\, i, j = 1, ..., N)$. In particular, the linear case of equation (1.48) can be written as

$$\Delta u = g_m(x, u, D_x u, D_x^2 u), \quad g_m = \sum_{i,j=1}^{N} [\delta_{ij} - a_{ijm}(x)] u_{x_i x_j}$$
$$- \sum_{i=1}^{N} b_{im}(x) u_{x_i} - c_m(x) u + f_m(x) \text{ in } Q. \tag{1.49}$$

In the following, we will give a representation of solutions of Problem O for equation (1.48).

Theorem 1.5 *Under the same condition as in Theorem 1.1, if $u(x)$ is any solution of Problem O for equation (1.48), then $u(x)$ can be expressed in the form*

$$u(x) = U(x) + V(x) = U(x) + v_0(x) + v(x),$$
$$v(x) = \tilde{H}\rho = \int_{Q_0} G(x-\zeta)\rho(\zeta)d\zeta, \tag{1.50}$$
$$G = \begin{cases} |x-\zeta|^{2-N}/(N(2-N)\omega_N), N > 2, \\ \log|x-\zeta|/2\pi,\ N = 2, \end{cases}$$

where $\omega_N = 2\pi^{N/2}/(N\Gamma(N/2))$ *is the volume of a unit ball in* $\mathbf{R}^N$, $\rho(x) = \Delta u = g_m$ *and* $V(x)$ *is a solution of Problem* D_0 *for (1.48) in* $Q_0 = \{|x| < R\}$ *with the boundary condition* $V(x) = 0$ *on* ∂Q_0*; here* R *is an appropriately large number, such that* $Q_0 \supset \overline{Q}$, *and* $U(x)$ *is a solution of Problem* $\tilde{O}$ *for* $\Delta U = 0$ *in* Q *with the boundary condition (1.58) below, which satisfy the estimates*

$$\begin{gathered} C^1_\beta[U, \overline{Q}] + ||U||_{W^2_2(Q)} \le M_{18}, \\ C^1_\beta[V, \overline{Q_0}] + ||V||_{W^2_2(Q_0)} \le M_{19}, \end{gathered} \tag{1.51}$$

where $\beta(0 < \beta \le \alpha)$, $M_j = M_j(q, p, \alpha, k, Q_m)\,(j$ *=18,19) are non-negative constants,* $q = (q_0, q_1)$, $k = (k_0, k_1, k_2)$.

Proof It is easy to see that the solution $u(x)$ of Problem O for equation (1.48) can be expressed by the form (1.50). Noting that $a_{ijm} = 0\,(i \ne j)$, $b_{im} = 0$, $c_m = 0$, $f_m(x) = 0$ in $\mathbf{R}^N \backslash Q_m$ and $V(x)$ is a solution of Problem D_0 for (1.48) in Q_0, we can obtain that $V(x)$ in $\hat{Q}_{2m} = \overline{Q} \backslash Q_{2m}$ satisfies the estimate

$$C^2[V(x), \hat{Q}_{2m}] \le M_{20} = M_{20}(q, p, \alpha, k, Q_m).$$

On the basis of Theorem 1.3, we can see that $U(x)$ satisfies the first estimate in (1.51), and then $V(x)$ satisfies the second estimate in (1.51).

Theorem 1.6 *If equation (1.2) satisfies Condition C, then Problem O for (1.48) has a solution* $u(x)$.

Proof In order to prove the existence of solutions of Problem O for the nonlinear equation (1.48) by using the Leray-Schauder theorem, we introduce an equation with the parameter $h \in [0, 1]$,

$$\Delta u = h g_m(x, u, D_x u, D^2_x u) \text{ in } Q. \tag{1.52}$$

Denote by B_M a bounded open set in the Banach space $B = \hat{W}^2_2(Q) = C^1_\beta(\overline{Q}) \cap W^2_2(Q) (0 < \beta \le \alpha)$, the elements of which are real functions $V(x)$ satisfying the inequalities

$$||V||_{\hat{W}^2_2(Q)} = C^1_\beta[V, \bar{Q}] + ||V||_{W^2_2(Q)} < M_{21} = M_{19} + 1, \tag{1.53}$$

in which M_{19} is a non-negative constant as stated in (1.51). We choose any function $\tilde{V}(x) \in \overline{B_M}$ and substitute it into the appropriate positions in the right-hand side of (1.52), and then we define an integral $\tilde{v}(x) = \tilde{H}\rho$ as

$$\tilde{v}(x) = \tilde{H}\tilde{\rho}, \quad \tilde{\rho}(x) = \Delta \tilde{V}. \tag{1.54}$$

Next we find a solution $\tilde{v}_0(x)$ of the boundary value problem in Q_0:

$$\Delta\tilde{v}_0 = 0 \text{ on } Q_0, \tag{1.55}$$

$$\tilde{v}_0(x) = -\tilde{v}(x) \text{ on } \partial Q_0, \tag{1.56}$$

and denote the solution $\hat{V}(x) = \tilde{v}(x) + \tilde{v}_0(x)$ of the corresponding Problem D_0 in Q_0. Moreover on the basis of the result in [38], we can find a solution $\tilde{U}(x)$ of the corresponding Problem $\tilde{O}$ in Q,

$$\Delta\tilde{U} = 0 \text{ on } Q, \tag{1.57}$$

$$\frac{\partial\tilde{U}}{\partial\nu} + b(x)\tilde{U} = g(x) - \frac{\partial\hat{V}}{\partial\nu} + b(x)\hat{V} \text{ on } \partial Q. \tag{1.58}$$

Now we discuss the equation

$$\Delta V = hg_m(x, \tilde{u}, D_x\tilde{u}, D_x^2\tilde{U} + D_x^2 V),\ \ 0 \le h \le 1, \tag{1.59}$$

where $\tilde{u} = \tilde{U} + \hat{V}$. By Condition C, applying the principle of contracting mapping, we can find a unique solution $V(x)$ of Problem D_0 for equation (1.59) in Q_0 satisfying the boundary condition

$$V(x) = 0 \text{ on } \partial Q_0. \tag{1.60}$$

Denote $u(x) = U(x) + V(x)$, where the relation between U and V is the same as that between $\tilde{U}$ and $\tilde{V}$, and by $V = S(\tilde{V}, h)$, $u = S_1(\tilde{V}, h)$ $(0 \le h \le 1)$ the mappings from $\tilde{V}$ onto V and u respectively. Furthermore, if $V(x)$ is a solution of Problem D_0 in Q_0 for the equation

$$\Delta V = hg_m(x, u, D_x u, D_x^2 U + D_x^2 V)),\ \ 0 \le h \le 1, \tag{1.61}$$

where $u = S_1(V, h)$, then from Theorem 1.3, the solution $V(x)$ of Problem D_0 for (1.61) satisfies the estimate (1.53), consequently $V(x) \in B_M$. Set $B_0 = B_M \times [0, 1]$. In the following, we shall verify that the mapping $V = S(\tilde{V}, h)$ satisfies the three conditions of the Leray-Schauder theorem:

1) For every $h \in [0, 1]$, $V = S(\tilde{V}, h)$ continuously maps the Banach space B into itself, and is completely continuous on B_M. Besides, for every function $\tilde{V}(x) \in \overline{B_M}$, $S(\tilde{V}, h)$ is uniformly continuous with respect to $h \in [0, 1]$.

In fact, we arbitrarily choose $\tilde{V}_l(x) \in \overline{B_M}$ $(l = 1, 2, ...)$; it is clear that from $\{\tilde{V}_l(x)\}$ there exists a subsequence $\{\tilde{V}_{l_k}(x)\}$ such that $\{\tilde{V}_{l_k}(x)\}$, $\{\tilde{V}_{l_k x_i}(x)\}$ $(i = 1, ..., N)$ and corresponding functions $\{\tilde{U}_{l_k}(x)\}$,

$\{\tilde{U}_{l_k x_i}(x)\}$, $\{\tilde{u}_{l_k}(x)\}$, $\{\tilde{u}_{l_k x_i}(x)\}$ $(i = 1, ..., N)$ uniformly converge to $\tilde{V}_0(x)$, $\tilde{V}_{0x_i}(x)$, $\tilde{U}_0(x)$, $\tilde{U}_{0x_i}(x)$, $\tilde{u}_0(x)$, $\tilde{u}_{0x_i}(x)$ $(i = 1, ..., N)$ in $\overline{Q_0}$, $\overline{Q}$ respectively, in which $\tilde{u}_{l_k} = S_1(\tilde{V}_{l_k}, h)$, $\tilde{u}_0 = S_1(\tilde{V}_0, h)$. We can find a solution $V_0(x)$ of Problem D_0 for the equation

$$\Delta V_0 = h g_m(x, \tilde{u}_0, D_x\tilde{u}_0, D_x^2\tilde{U}_0 + D_x^2 V_0), \quad 0 \le h \le 1 \text{ in } Q_0. \tag{1.62}$$

From $V_{l_k} = S(\tilde{V}_{l_k}, h)$ and $V_0 = S(\tilde{V}_0, h)$, we have

$$\Delta(V_{l_k} - V_0) = h[g_m(x, \tilde{u}_{l_k}, D_x\tilde{u}_{l_k}, D_x^2\tilde{U}_{l_k} + D_x^2 V_{l_k})$$

$$-g_m(x, \tilde{u}_{l_k}, D_x\tilde{u}_{l_k}, D_x^2\tilde{U}_{l_k} + D_x^2 V_0) + C_{l_k}(x)], \quad 0 \le h \le 1,$$

where

$$C_{l_k}(x) = g_m(x, \tilde{u}_{l_k}, D_x\tilde{u}_{l_k}, D_x^2\tilde{U}_{l_k} + D_x^2 V_0)$$

$$-g_m(x, \tilde{u}_0, D_x\tilde{u}^0, D_x^2\tilde{U}_0 + D_x^2 V_0), \quad x \in Q_0.$$

According to a similar method to deriving (2.43), Chapter II in [81], we can prove that

$$L_2[C_{l_k}(x), \overline{Q_0}] \to 0 \text{ as } k \to \infty. \tag{1.63}$$

Moreover according to Theorem 1.3, we can derive that

$$||V_{l_k} - V_0||_{\hat{W}_2^2(Q_0)} \le M_{22} L_2[C_{l_k}, \overline{Q_0}],$$

where $M_{22} = M_{22}(q, p, \alpha, k_0, Q_m)$ is a non-negative constant, hence $||V_{l_k} - V_0||_{\hat{W}_2^2(Q_0)} \to 0$ as $k \to \infty$. Thus from $\{V_{l_k}(x) - V_0(x)\}$, there exists a subsequence (for convenience we denote the subsequence again by $\{V_{l_k}(x) - V_0(x)\}$) such that $||V_{l_k}(x) - V_0(x)||_{\hat{W}_2^2(Q_0)}$ $= C_\beta^1[V_{l_k}(x) - V_0(x,t), \overline{Q_0}] + ||V_{l_k}(x) - V_0(x)||_{W_2^2(Q_0)} \to 0$ as $k \to \infty$. From this we can obtain that the corresponding subsequence $\{u_{l_k}(x) - u_0(x)\} = \{S_1(V_{l_k}, h) - S_1(V_0, h)\}$ possesses the property: $||u_{l_k}(x) - u_0(x)||_{\hat{W}_2^2(Q)} \to 0$ as $k \to \infty$. This shows the complete continuity of $V = S(\tilde{V}, h)$ $(0 \le h \le 1)$ in $\overline{B_M}$. By using a similar method, we can prove that $V = S(\tilde{V}, h)$ $(0 \le h \le 1)$ continuously maps $\overline{B_M}$ into B, and $V = S(\tilde{V}, h)$ is uniformly continuous with respect to $h \in [0, 1]$ for $\tilde{V} \in \overline{B_M}$.

2) For $h = 0$, from (1.53) and (1.59), it is clear that $V = S(\tilde{V}, 0) \in B_M$.

3) From Theorem 1.3 and (1.53), we see that $V = S(\tilde{V}, h)(0 \le h \le 1)$ does not have a solution $u(x)$ on the boundary $\partial B_M = \overline{B_M} \backslash B_M$.

Hence by the Leray-Schauder theorem (see [40]), we know that Problem D_0 for equation (1.59) with $h=1$ has a solution $V(x) \in B_M$, and then Problem P of equation (1.52) with $h=1$, i.e. (1.48) has a solution $u(x)=S_1(V,h)=U(x)+V(x)=U(x)+v_0(x)+v(x) \in B$.

Theorem 1.7 *Under the same conditions as in Theorem 1.1, Problem O for the equation (1.2) has a solution.*

Proof By Theorem 1.3 and Theorem 1.6, Problem O for equation (1.48) possesses a solution $u_m(x)$, and the solution $u_m(x)$ of Problem O for (1.48) satisfies the estimates (1.19) and (1.20), where $m=1,2,\ldots$. Thus, we can choose a subsequence $\{u_{m_k}(x)\}$, such that $\{u_{m_k}(x)\}$, $\{u_{m_k x_i}(x)\}$ $(i=1,\ldots,N)$ in $\overline{Q}$ uniformly converge to $u_0(x)$, $u_{0x_i}(x)$ $(i=1,\ldots,N)$ respectively. Obviously, $u_0(x)$ satisfies the boundary conditions of Problem O. On the basis of principle of compactness of solutions for equation (1.48), we can see that $u_0(x)$ is a solution of Problem O for (1.2).

2 Boundary Value Problems of Degenerate Elliptic Equations of Second Order

2.1 Formulation of the Oblique Derivative Problem for Degenerate Elliptic Equations

Let $G \in C^2_\alpha(0<\alpha<1)$ be a bounded domain in the upper-half space $x_N>0$, whose boundary is $\partial G = S_1 \cup S_2$, in which S_1 is located on $x_N=0$ and S_2 is located in $x_N>0$. Denote $\Gamma = \overline{S_2} \cap \{x_N=0\}$. We consider the degenerate elliptic equation of second order

$$Lu=\sum_{i,j=1}^{N} a_{ij}(x)u_{x_i x_j}+\sum_{i=1}^{N} b_i(x)u_{x_i}+\frac{h(x)}{x_N^k}u_{x_N}+c(x)u=f(x) \text{ in } G. \quad (2.1)$$

Suppose that the equation (2.1) satisfies the following conditions, i.e. **Condition** C:

1) There exists a positive number $q_0\,(<1)$, such that for $x=(x_1,\ldots,x_N)\in G$, the following equality holds:

$$\sum_{i,j=1}^{N} a_{ij}(x)\xi_i\xi_j \geq q_0\sum_{i=1}^{N}\xi_i^2; \quad (2.2)$$

there is no harm in assuming that $a_{NN}=1$ and $(a_{ij}(x))$ is symmetrical.

2) Coefficients of (2.1) satisfy the conditions

$$C_\alpha[a_{ij}(x),\overline{G}], C_\alpha[b_i(x),\overline{G}], C_\alpha[c(x),\overline{G}] \le k_0, i,j=1,...,N,$$
$$C_\alpha[f(x),\overline{G}] \le k_1,\ C^2[h(x),\overline{G}] \le k_0,\ c(x)\le 0,\ k>0, \tag{2.3}$$

where $\alpha(0<\alpha<1), k_j(j=0,1)$ are non-negative constants.

Problem Q Find a bounded solution $u(x)$ of equation (2.1) in $\overline{G}$ satisfying the oblique derivative boundary condition

$$lu = \frac{\partial u}{\partial \nu} + b(x)u = g(x) \text{ on } S_2, \tag{2.4}$$

where ν is a vector on every point of S_2, and $b(x), g(x)$ satisfy the conditions

$$C^1_\alpha[\cos(\nu,n),S_2], C^1_\alpha[b(x),S_2] \le k_0, C^1_\alpha[g(x),S_2] \le k_2,$$
$$\cos(\nu,n)>0,\ b(x)<0 \text{ on } S_2, \tag{2.5}$$

in which n is the inner normal vector on S_2. The contents of this section are mainly chosen from [42].

2.2 Unique Solvability of the Oblique Derivative Problem for Degenerate Elliptic Equations

Lemma 2.1 *If the boundary ∂Q of the domain Q belongs to C^2_α and the operator*

$$Lu = \sum_{i,j=1}^N a_{ij}(x)u_{x_ix_j} + \sum_{i=1}^N b_i(x)u_{x_i} + c(x)u$$

satisfies the uniformly elliptic condition, $a_{ij}(x), b_i(x), c(x) \in C_\alpha(\overline{Q})$, and the coefficients of boundary operator

$$lu = \sum_{i=1}^N d_i(x)u_{x_i} + b(x)u$$

satisfy the conditions

$$\sum_{i=1}^N d_i(x)\cos(n,x_i)|_{\partial Q} \ge q_0 > 0, C^1_\alpha[d_i(x),\partial Q] \le k_0, C^1_\alpha[b(x),\partial Q] \le k_2, \tag{2.6}$$

then any function $u(x)$ $(\in C^2_\alpha(\overline{Q}))$ *satisfies the estimate*

$$C^2_\alpha[u,\overline{Q}] \le M_1\{C_\alpha[Lu,\overline{Q}] + C^1_\alpha[Bu,\partial Q] + c[u,\overline{Q}]\},$$

where the non-negative constant M_1 *is only dependent on the coefficients of* L, l *and* ∂Q (see [38]).

We use the parameter expression of S_2 as follows:

$$x_i = x_i(\xi_1, ..., \xi_{N-1}), i = 1, ..., N,$$

in which $x_i(\xi_1, ..., \xi_{N-1}) \in C^2_\alpha$ and define by

$$X_i = \frac{1}{\Omega}\frac{\partial(x_{i+1}, ..., x_N, x_1, ..., x_{i-1})}{\partial(\xi_1, ..., \xi_{N-1})},$$

the direct cosine of an inner normal line of S_2, where

$$\Omega = \left[\sum_{i=1}^{N}\left(\frac{\partial(x_{i+1}, ..., x_N, x_1, ..., x_{i-1})}{\partial(\xi_1, ..., \xi_{N-1})}\right)^2\right]^{1/2} > 0.$$

Theorem 2.2 *Suppose that equation (2.1) satisfies Condition C and* $C^1_\alpha[a_{ij},\overline{G}] \le k_0$, $c < 0$ *in* $\overline{G}$. *Then Problem Q of (2.1) has a bounded solution* $u(x) \in C^2_\alpha(G \cup S_2)$.

Proof Choose a decreasing sequence of positive numbers $\{\varepsilon_m\}$ such that $\varepsilon_m = 0$ as $m \to \infty$, and define a sequence of domains $\{G_m\}$ satisfying the conditions:

1) $G_1 \subset G_2 \subset ...;$

2) $\cup G_m = G;$

3) G_m and its boundary are in $x_N > 0$, and $G_m \cap \{x_N \ge \varepsilon_m\} = G \cap \{x_N \ge \varepsilon_m\}$;

4) The boundary S_m of G_m belongs to C^2_α.

Introduce the boundary condition

$$\frac{\partial u_m}{\partial \nu_m} + b_m(x)u_m = g_m(x) \text{ on } S_m,$$

where we choose $\nu_m, b_m(x), g_m(x)$, such that when $x_m > 2\varepsilon_m, \cos(\nu_m, n) > 0$; herein n is the inner normal vector of S_m and $\nu_m = \nu$, $b_m(x) = b(x)$, $g_m(x) = g(x)$. Next for $S_m \cap \{x_N \le \varepsilon_m\}$, set

$$\frac{\partial u_m}{\partial \nu_m} = \frac{\partial u_m}{\partial l} = \sum_{i,j=1}^{N}\frac{1}{B}a_{ij}(x)X_j\frac{\partial u_m}{\partial x_i},\ B = \left[\sum_{i,j=1}^{N}(\sum_{i,j=1}^{N}a_{ij}(x)X_j)^2\right]^{1/2},$$

in which l is the secondary normal vector and its directional cosine is $Y_i = \sum_{j=1}^{N} a_{ij}(x)X_j/B$. It is clear that $\cos(n,l) > 0$, and $b_m(x) < -1$, $C^1_\alpha[b_m, S_m \cap \{x_N \le \varepsilon_m\}]$, $C^1_\alpha[g_m, S_m \cap \{x_N \le \varepsilon_m\}] \le k_3$, where k_3 is a positive constant. Moreover for $\{3\varepsilon_m/2 < x_N \le 2\varepsilon_m\}$, let $\nu = \nu$, and $C^1_\alpha[b_m, S_m \cap \{3\varepsilon_m/2 < x_N \le 2\varepsilon_m\}]$, $C^1_\alpha[g_m, S_m \cap \{3\varepsilon_m/2 < x_N \le 2\varepsilon_m\}] \le k_3$. Finally for $S_m \cap \{\varepsilon_m < x_N \le 3\varepsilon_m/2\}$, $\partial u_m/\partial\nu_m$ is defined to be the relation

$$\frac{\partial u_m}{\partial l} = k(\xi_1, ..., \xi_{N-1})\frac{\partial u_m}{\partial \nu_m} + \sum_{i=1}^{N-1} h_i(\xi_1, ..., \xi_{N-1})\frac{\partial u_m}{\partial \xi_i},$$

where $k(\xi_1, ..., \xi_{N-1}) > 0$, $h_i(\xi_1, ..., \xi_{N-1}) \in C^1_\alpha$, $b_m < -1$, and $b_m(x)$, $g_m(x)$ satisfy the condition as before.

Noting that

$$\frac{\partial}{\partial \nu} = \sum_{i=1}^{N} \cos(\nu, x_i)\frac{\partial}{\partial x_i} = \sum_{i=1}^{N} \frac{\cos(\nu, n)}{\cos(n, l)} \cos(l, x_i)\frac{\partial}{\partial x_i}$$

$$+\sum_{i=1}^{N}\left[\cos(\nu, x_i) - \frac{\cos(\nu, n)}{\cos(n, l)}\cos(l, x_i)\right]\frac{\partial}{\partial x_i} \quad \text{on} \quad \frac{3\varepsilon_m}{2} < x_N < 2\varepsilon_m,$$

we have

$$\frac{\partial}{\partial l} = \frac{\cos(n,l)}{\cos(\nu,n)}\frac{\partial}{\partial \nu}$$
$$+\sum_{i=1}^{N}\left[\cos(l, x_i) - \frac{\cos(n, l)}{\cos(\nu, n)}\cos(\nu, x_i)\right]\frac{\partial}{\partial x_i} \quad \text{on} \quad \frac{3\varepsilon_m}{2} < x_N < 2\varepsilon_m. \tag{2.7}$$

Denote $\alpha_i = \cos(l, x_i) - \cos(n, l)\cos(\nu, x_i)/\cos(\nu, n)$, then it is obvious

$$\sum_{i=1}^{N} \alpha_i X_i = 0.$$

Besides for any differentiable function $v(x_1, ..., x_N)$, we have

$$\frac{\partial v}{\partial \xi_i} = \sum_{j=1}^{N} \frac{\partial v}{\partial x_j}\frac{\partial x_j}{\partial \xi_i}, \; i = 1, ..., N-1, \;\text{ i.e. } \sum_{j=1}^{N} \frac{\partial v}{\partial x_j}\frac{\partial x_j}{\partial \xi_i} - \frac{\partial v}{\partial \xi_i} = 0.$$

There is no harm in assuming that $\partial(x_1, ..., x_{N-1})/\partial(\xi_1, ..., \xi_{N-1}) \neq 0$, we obtain

$$\frac{\partial v}{\partial x_j} = \frac{\partial v}{\partial x_N}\frac{\partial(x_{j+1}, ..., x_N, x_1, ..., x_{j-1})}{\partial(\xi_1, ..., \xi_{N-1})}$$
$$+\frac{\partial(x_{j+1}, ..., x_{N-1}, -v, x_1, ..., x_{j-1})}{\partial(\xi_1, ..., \xi_{N-1})} \Bigg/ \frac{\partial(x_1, ..., x_{N-1})}{\partial(\xi_1, ..., \xi_{N-1})},$$
$$j = 1, ..., N-1,$$

thus

$$\sum_{i=1}^{N} \alpha_i \frac{\partial v}{\partial x_i} = \sum_{i=1}^{N-1} \alpha_i \frac{\partial(x_{i+1}, ..., x_N, x_1, ..., x_{i-1})}{\partial(\xi_1, ..., \xi_{N-1})} \frac{\partial v}{\partial x_N} \Big/ \frac{\partial(x_1, ..., x_{N-1})}{\partial(\xi_1, ..., \xi_{N-1})}$$

$$+ \sum_{i=1}^{N-1} \alpha_i \frac{\partial(x_{i+1}, ..., x_{N-1}, -v, x_1, ..., x_{i-1})}{\partial(\xi_1, ..., \xi_{N-1})} \Big/ \frac{\partial(x_1, ..., x_{N-1})}{\partial(\xi_1, ..., \xi_{N-1})}$$

$$+ \alpha_N \frac{\partial v}{\partial x_N} = \frac{1}{\Omega X_N} \sum_{i=1}^{N-1} \alpha_i \frac{\partial(x_{i+1}, ..., x_{N-1}, -v, x_1, ..., x_{N-1})}{\partial(\xi_1, ..., \xi_{N-1})}$$

$$+ \left(\sum_{i=1}^{N-1} \frac{\alpha_i X_i}{X_N} + \alpha_N \right) \frac{\partial v}{\partial x_N}.$$

Besides we can find $N-1$ functions $\tilde{h}(\xi_1, ..., \xi_{N-1}) \in C_\alpha^1$, such that

$$\sum_{i=1}^{N} \alpha_i \frac{\partial v}{\partial x_i} = \sum_{i=1}^{N-1} \tilde{h}(\xi_1, ..., \xi_{N-1}) \frac{\partial v}{\partial \xi_i},$$

thus (2.7) can be written as

$$\frac{\partial}{\partial l} = k'(\xi_1, ..., \xi_{N-1}) \frac{\partial}{\partial \nu} + \sum_{i=1}^{N-1} h_i'(\xi_1, ..., \xi_{N-1}) \frac{\partial}{\partial \xi_i},$$

where $k'(\xi_1, ..., \xi_{N-1}) > 0, h_i'(\xi_1, ..., \xi_{N-1}) \in C_\alpha^1$. Especially $k'(\xi_1, ..., \xi_{N-1}) = 1$, $h_i'(\xi_1, ..., \xi_{N-1}) = 0 \, (i = 1, ..., N-1)$ in $x_n \le \varepsilon_N$. Hence we have

$$\frac{\partial}{\partial l} = K(\xi_1, ..., \xi_{N-1}) \frac{\partial}{\partial \nu_m} + \sum_{i=1}^{N-1} H_i(\xi_1, ..., \xi_{N-1}) \frac{\partial}{\partial \xi_i}, i = 1, ..., N-1.$$

Now we consider the boundary value problem (Problem Q_m)

$$Lu_m = f(x) \text{ in } G_m, \quad \frac{\partial u_m}{\partial \nu_m} + b_m u_m = g_m(x) \text{ on } S_m. \tag{2.8}$$

By the result in [38] and Lemma 2.1, we know that there exists a solution $u_m(x)$ of Problem Q_m, and the solution $u_m(x)$ satisfies the estimate

$$C[u_m(x), \overline{G_m}] \le M_1, C_\alpha^2[u_m(x), \overline{G_m}] \le M_2, \tag{2.9}$$

where $M_1 = M_1(q_0, \alpha, k, G), M_2 = M_2(q_0, \alpha, k, G_m)$ are two positive constants, $k = (k_0, k_1, k_2)$. In the following, we shall prove that $u_m(x)$,

$u_{mx_i}(x)$, $u_{mx_ix_j}$ $(i,j = 1,...,N)$ uniformly converge to the functions $u_0(x)$, $u_{0x_i}(x)$, $u_{0x_ix_j}$ $(i,j = 1,...,N)$ in $G \cap \{x_N \geq 3\eta > 0\}$ respectively. In fact, for arbitrary small positive number η, we can assume that $\psi(x) \in C^2_\alpha(\overline{G})$, $\psi(x) = 1$ for $x_N \geq 2\eta$, $\psi(x) = 0$ for $x_N \leq \eta$, and $\partial\psi/\partial\nu = 0$ on S_2. Let $v_m = u_m(x)\psi(x)$, then we have

$$\begin{aligned} Lv_m &= \psi(x)f(x) + u_m(x)L\psi \\ &+2\sum_{i,j=1}^{N} a_{ij}(x)\frac{\partial u_m}{\partial x_i}\frac{\partial \psi}{\partial x_j} - c(x)\psi(x)u_m(x) = F_m(x). \end{aligned} \tag{2.10}$$

Denote by G_η a bounded domain in G with the boundary $S_\eta \in C^2_\alpha$, and $S_\eta \cap \{x_N \geq \eta\} = S_2 \cap \{x_N \geq \eta\}$, $\overline{G_\eta} \subset \{x_N > 0\}$. It is clear that there exists a positive integer m such that $G_\eta \subset G_m$, when $2\varepsilon_m \leq \eta$, and

$$\begin{aligned} \frac{\partial v_m}{\partial \nu} + b(x)v_m(x) &= \psi(x)g_m(x) + b(x)u_m(x)\frac{\partial\psi(x)}{\partial\nu} \\ &= \psi(x)g(x) \text{ for } x_N \geq \eta; \end{aligned} \tag{2.11}$$

moreover for $x_N < \eta$, we have $\psi(x) \equiv 0$, $v_m(x) \equiv 0$, hence

$$\frac{\partial v_m}{\partial \nu} + b(x)v_m(x) = \psi(x)g(x) \text{ for } x_N < \eta.$$

Thus

$$lv_m = \frac{\partial v_m}{\partial \nu} + b(x)v_m(x) = \psi(x)g(x) \text{ on } S_\eta. \tag{2.12}$$

By the results in [38], Section 1 and Lemma 2.1, the boundary value problem (2.10),(2.12) has a bounded solution $v_m(x)$, and the solution satisfies the estimate

$$C^2_\alpha[u_m(x), \overline{G_\eta}] \leq M_3 = M_3(q_0, \alpha, k, G_\eta). \tag{2.13}$$

Hence $v_m(x)$, $v_{mx_i}(x)$, $v_{mx_ix_j}$ $(i,j = 1,...,N)$ are uniformly bounded and equicontinuous. According to the Ascoli-Arzela theorem, from $\{v_m(x)\}$ we can choose a subsequence $\{v_{m_k}(x)\}$, such that $\{v_{m_k}\}$, $\{v_{m_kx_i}\}$, $\{v_{m_kx_ix_j}\}$ uniformly converge to $u(x)$, u_{x_i}, $u_{x_ix_j}$ in G_η and $u(x) \in C^2_\alpha(G_\eta)$. Hence $u(x)$ satisfies the equation and boundary condition

$$Lu = f(x) \text{ in } x_N \geq 3\eta, \quad \frac{\partial u}{\partial \nu} + b(x)u = g(x) \text{ on } S_2 \cap \{x_N \geq 3\eta\}.$$

This prove that there exists a bounded solution of the above problem in $G \cap \{x_N \geq 3\eta\}$. Noting the arbitrariness of η, we verify that the boundary value problem (2.1),(2.4), i.e. Problem Q, has a bounded solution $u(x) \in C^2_\alpha(G \cup S_2)$.

Corollary 2.3 *If the coefficient $c(x) \le 0$ of equation (2.1), and the boundary condition*

$$lu = \frac{\partial u}{\partial \nu} + b(x)u = \sum_{i=1}^{N} d_i(x)u_{x_i} + b(x)u = g(x) \text{ on } S_2 \qquad (2.14)$$

satisfies

$$C^1_\alpha[d_i(x), S_2] \le k_0, \sum_{i=1}^{N} d_i(x)\cos(n, x_i)|_{S_2} \ge q_0 > 0,$$

$$\cos(\nu, n) > 0, \ b(x) \le 0 \text{ on } S_2,$$

and one of $b_i(x)(i = 1, ..., N)$ for instance $b_j(x) > 0(1 \le j \le N)$, then the result in Theorem 2.2 is still valid.

Proof Introduce a transformation

$$u(x) = (C - e^{\mu x_j})v(x), \ 1 \le j \le N;$$

equation (2.1) is reduced to

$$\sum_{i,j=1}^{N} a_{ij}(x)v_{x_i x_j} + \sum_{i=1}^{N-1} \tilde{b}_i(x)v_{x_i} + \frac{h(x)}{x_N^k}v_{x_N} + \tilde{c}(x)v = \tilde{f}(x) \text{ in } G,$$

where $\tilde{c}(x) = c(x) - (a_{11}(x)\mu^2 + b_j(x)\mu)e^{\mu x_j}/(C - e^{\mu x_j})$. Noting $a_{11} > q_0$, we can choose positive constants C and μ such that $\tilde{c}(x) < 0$ in $\overline{G}$ and $(C - e^{\mu x_j}) > 1$, then the boundary condition (2.14) is transformed into

$$\sum_{i=1}^{N} \tilde{d}_i(x)v_{x_i} + \tilde{b}(x)v = \tilde{g}(x) \text{ on } S_2,$$

in which $\tilde{b}(x) = b(x) - \mu b_j(x)e^{\mu x_j}/(C - e^{\mu x_j}) < 0$.

Theorem 2.4 *If the conditions in Theorem 2.2 and $b(x) \le 0$ on S_2 hold, then when $g(x) \equiv 0$, equation (2.1) has a bounded solution $u(x) \in C^2_\alpha(G \cup S_2)$ satisfying the boundary condition (2.4) on S_2.*

The proof is the same with that of Theorem 2.2.

Theorem 2.5 *Let the conditions in Theorem 2.2 hold, and $\cos(\nu, x_N) > 0$ on Γ. Moreover one of the following conditions holds:*

a) $k = 1$, $h(x_1, ..., x_{N-1}, 0) \geq 1$, *and* $h(x)$ *about* x_N *is even;*

b) $k > 1$, $h(x_1, ..., x_{N-1}, 0) > 0$.

Then Problem Q of (2.1) has at most a bounded solution $u(x) \in C^2_\alpha(G \cup S_2)$.

Proof It is sufficient to prove the bounded solution $u(x) \equiv 0$ of the boundary value problem: $Lu = 0$ and $\partial u/\partial \nu + bu = 0$ on S_2. For this we construct a barrier function $W(x)$ as follows:

1) $W(x) \geq c_0 > 0$ in $\overline{G}$, here c_0 is a constant;

2) When $x_N \to 0$, $W(x) \to \infty$ uniformly holds;

3) $LW < 0$ in G;

4) $\dfrac{\partial W}{\partial \nu} + bW < M_1^2$ on S_2.

If there exists the function $W(x)$, by Theorem 2.2 we can establish a bounded function $w(x)$ satisfying $Lw = 0$ and $\partial w/\partial \nu + bw > \partial W/\partial \mathbf{l} + bW$. Let $V(x) = W(x) - w(x)$. Then $LV < 0$, $\partial V/\partial \nu + bV < 0$, and from the boundedness of $w(x)$, it follows that $\lim_{X_N \to 0} V(x) = +\infty$. On the basis of the result in [58], we can derive $\varepsilon V \pm u_0 \geq 0$, where ε is a positive number. In fact, $L(\varepsilon V \pm u_0) = \varepsilon LV < 0$ in G, $\partial(\varepsilon V \pm u_0)/\partial \nu + b(\varepsilon V \pm u_0) \leq 0$ on S_2, and $V \to \infty$ as $X_N \to 0$, hence $\varepsilon V \pm u_0$ cannot take the negative minimum in G. Due to the arbitrariness of ε, we obtain $u_0 \equiv 0$.

Now we make the barrier function in case a) as follows:

$$W(x) = \ln x_N - (x_1 - \tilde{a})^J + K, \tag{2.15}$$

where $\tilde{a}$ is chosen such that $x_1 - \tilde{a} > 1$ for $x = (x_1, ..., x_N) \in G$, and J, K are undetermined constants. For a sufficiently small x_N, we can choose that a positive number A is large enough, such that

$$\frac{1 - h(x)}{(x_N)^2}$$
$$= [1 - h(x_1, ..., x_{N-1}, 0) - \tfrac{\partial h(x_1,...,x_{N-1},0)x_N}{\partial x_N} + O(x^2_{x_N})]x_N^{-2} \leq A;$$

the above inequality is also valid for any x_N in G. From (2.2), $a_{11} \geq q_0$, we can select a J large enough such that

$$LW = -a_{11}(x)J(J-1)(x_1-\tilde{a})^{J-2}$$
$$-b_1(x)J(x_1-\tilde{a})^{J-1}+[1-h(x)]x_N^{-2}+c(x)W<0,$$

and then choose a sufficiently large K such that $W(x) \geq c_0 > 0$ in G. Moreover $\partial W/\partial\nu = -(\partial x_N/\partial\nu)/x_N + O(1)$. According to the hypothesis: $\cos(\nu, x_N) > 0$ near to Γ, hence $\partial W/\partial\nu + bW < M_1^2$ on S_2; this shows that the function $W(x)$ in (2.15) satisfies all conditions of the barrier function.

Finally we consider the case b); we choose the function

$$W(x) = x_N^{-\beta} - (x_1-\tilde{a})^J + K, \ \ 0<\beta<1, \tag{2.16}$$

and for a sufficiently small x_N, it is clear that

$$\beta[\beta+1-h(x)x_N^{-k+1}]x^{-\beta-2} < -\frac{\beta}{2}h(x_1,...,x_{N-1},0)x_N^{-k-\beta-1} < 0,$$

where J, K are chosen as stated in case a). Moreover from $\cos(\nu, x_N) > 0$ on Γ, we have $\partial W/\partial\nu + b(x)W < M_1^2$. Thus the function $W(x)$ in (2.16) satisfies all conditions of the barrier function. This completes the proof.

Theorem 2.6 *Under the same conditions as in Theorems 2.4 and 2.5, there exists at most a bounded solution $u(x) \in C_\alpha^2(G\cup S_2)$ of (2.1) satisfying the boundary condition*

$$\frac{\partial u}{\partial l} + bu = 0 \ \text{ on } \ S_2.$$

3 The Schwarz Formulas and Dirichlet Problem in Half-space and in a Ball

First of all, we introduce the regular function in real Clifford analysis. The Clifford algebra $\mathcal{A}_n(R)$ over the space $\mathbf{R}^n$ is defined as follows: Let $e_1 = 1, e_2, \cdots, e_n$ be the standard orthogonal basis in $\mathbf{R}^n$, and denote by e_s the general basis element of $\mathcal{A}_n(R)$, where s is any subset of $\{1, 2, 3, ..., n\}$, i.e.

$$\begin{cases} e_1 = 1, \ e_j^2 = -1, \ 2 \leq j \leq n, \\ e_je_k = -e_ke_j, \ 2 \leq j < k \leq n, \\ e_s = e_{j_1}e_{j_2}\cdots e_{j_s}, \ 2 \leq j_1 < j_2 < \cdots < j_s \leq n, \\ s = \{j_1, j_2, ..., j_s\}. \end{cases}$$

It is not difficult to see that $\mathcal{A}_n(R)$ is a 2^{n-1}-dimensional Clifford algebra space. An arbitrary element of the Clifford algebra $\mathcal{A}_n(R)$ can be written as

$$x = x_1e_1 + \cdots + x_ne_n + \cdots + x_{x_{j_1\ldots\gamma_s}}e_{j_1}\ldots e_{\gamma_s} + \cdots + x_{2\ldots n}e_2\ldots e_n = \sum_s x_se_s,$$

in which $x_1, ..., x_n, ..., x_{j_1\ldots j_s}, ..., x_{2\ldots n} \in \mathbf{R}$.

Let $x = \sum x_se_s$, $y = \sum y_se_s \in \mathcal{A}_n(R)$, where $x_s, y_s \in \mathbf{R}$. Then $xy = \sum_{s,k} x_sy_ke_se_k$, in which e_se_k is a basis element of $\mathcal{A}_n(R)$ and $x_sy_k \in \mathbf{R}$. For $x = x_1e_1 + x_2e_2 + \cdots + x_ne_n \in \mathbf{R}^n$, its conjugate element is defined by $\bar{x} = x_1e_1 - x_2e_2 - \cdots - x_ne_n$. It is obvious that

$$x\bar{x} = \bar{x}x = |x|^2.$$

If $x \neq 0$, then x is invertible, and $x^{-1} = \bar{x}/|x|^2$.

Denote by D a connected open set in $\mathbf{R}$, and by

$$F_D^{(r)} = \{f | f : D \to \mathcal{A}_n(\mathbf{R}),\ f(x) = \sum_s f_s(x)e_s,\ f_s \in C^r(D)\}$$

the set of continuously differentiable functions up to degree r, the values of which belong to $\mathcal{A}_n(R)$. Define the differential operators

$$\bar{\partial} = e_1\frac{\partial}{\partial x_1} + e_2\frac{\partial}{\partial x_2} + \cdots + e_n\frac{\partial}{\partial x_n},\quad \partial = e_1\frac{\partial}{\partial x_1} - \cdots - e_n\frac{\partial}{\partial x_n};$$

then we have

$$\bar{\partial}\partial = \partial\bar{\partial} = \sum_{j=1}^{n} \frac{\partial^2}{\partial x_j^2} = \Delta.$$

If $f(x) \in F_D^{(r)}(r \geq 1)$ and $\bar{\partial} f = 0$, then $f(x)$ is called a (left-)regular function, and then $f_s(x)$ is harmonic in D. In particular if $n = 2$, the regular function $f(x)$ in a domain D in $\mathbf{R}^2$ is an analytic function, and if $n = 3$, the regular function $f(x) = \sum_s f_s(x)e_s = f_1(x) + f_2(x)e_2 + f_3(x)e_3 + f_{23}(x)e_2e_3$ in a domain in $\mathbf{R}^3$ is a solution (f_1, f_2, f_3, f_{23}) of the system of partial differential equations of first order

$$\begin{pmatrix} \frac{\partial}{\partial x_1} & -\frac{\partial}{\partial x_2} & -\frac{\partial}{\partial x_3} & 0 \\ \frac{\partial}{\partial x_2} & \frac{\partial}{\partial x_1} & 0 & \frac{\partial}{\partial x_3} \\ \frac{\partial}{\partial x_3} & 0 & \frac{\partial}{\partial x_1} & -\frac{\partial}{\partial x_2} \\ 0 & -\frac{\partial}{\partial x_3} & \frac{\partial}{\partial x_2} & \frac{\partial}{\partial x_1} \end{pmatrix} \begin{pmatrix} f_1 \\ f_2 \\ f_3 \\ f_{23} \end{pmatrix} = \begin{pmatrix} 0 \\ 0 \\ 0 \\ 0 \end{pmatrix}. \tag{3.0}$$

We can see that it is an analytic function in the 3-dimensional domain. Hence, the general regular functions possess some properties similar to those of analytic functions in the planar domain.

In the following, we first discuss the 3-dimensional space $\mathbf{R}^3$ and the corresponding Clifford algebra $\mathcal{A}_3(R)$ over $\mathbf{R}^3$. In this case, $x = x_1 + x_2e_2 + x_3e_3 + x_{23}e_{23} \in \mathcal{A}_3(R)$, and we define by

$$\mathrm{Re}\, x = x_1 + x_2e_2,\ \mathrm{Im}\, x = x_3 - x_{23}e_2, \tag{3.1}$$

the real part and imaginary part of x. It is evident that

$$x = \mathrm{Re}\, x + e_3\mathrm{Im}\, x,\ \mathrm{Re}\, x = \frac{1}{2}(x + \tilde{x}),\ \mathrm{Im}\, x = -e_3\frac{x - \tilde{x}}{2}, \tag{3.2}$$

where $\tilde{x} = \mathrm{Re}\, x - e_3\mathrm{Im}\, x$. Let $y = \mathrm{Re}\, y + e_3\mathrm{Im}\, y \in \mathcal{A}_3(R)$, then we can verify that $\widetilde{xy} = \tilde{x}\tilde{y}$.

3.1 Schwarz Formula and Dirichlet Problem for the Halfspace

Theorem 3.1 *Suppose that $u(y)$ is a Hölder continuous function in the plane $E = \{x_3 = 0\}$, and $u(y) = 0$ if $|y|$ is large enough. Then there exists a regular function $f(x)$ in the upper halfspace D, such that*

$$\mathrm{Re} f^+(y) = u(y),\ y \in E. \tag{3.3}$$

Then the function $f(x)$ can be expressed as

$$f(x) = \frac{1}{2\pi}\iint_E \frac{x_3 + e_3[(x_1 - y_1) + e_2(x_2 - y_2)]}{|y - x|^3}u(y)dS_y + e_3g, \tag{3.4}$$

where we assume that $\lim_{|x|\to\infty} f(x) = f(\infty) = e_3(a_3 - a_{23}e_2)$, a_3, a_{23} are real constants, and then $e_3g = f(\infty)$. This is also a representation of the solution of the Dirichlet problem for regular functions in the halfspace D.

Proof We assume that $f(x)$ is a desired regular function, where x is any point in D. Let Σ_R be the upper half sphere with the center at the origin and radius R, and denote by D_R a domain with the boundary Σ_R and $E_R = \{|x| < R,\ x \in E\}$. We choose R so large that $x \in D_R$ and $|y - x| \geq R/2$, $y \in E_R$. By the Cauchy integral formula we have

$$\begin{aligned} f(x) = &\frac{1}{4\pi}\int_{\Sigma_R} \frac{\bar{y} - \bar{x}}{|y - x|^3}n(y)f(y)dS_y \\ &+\frac{1}{4\pi}\int_{E_R} \frac{\bar{y} - \bar{x}}{|y - x|^3}n(y)f(y)dS_y = I_1 + I_2. \end{aligned} \tag{3.5}$$

Moreover from

$$\left|\frac{\bar{y}-\bar{x}}{|y-x|^3}-\frac{\bar{y}}{|y|^3}\right|\leq |x|\left(\frac{1}{|y-x||y|^2}+\frac{1}{|y-x|^2|y|}\right)$$
$$\leq |x|(\frac{2}{R^3}+\frac{4}{R^3})=\frac{6|x|}{R^3}, \tag{3.6}$$

it follows that

$$\left|\frac{1}{4\pi}\int_{\Sigma_R}\left[\frac{\bar{y}-\bar{x}}{|y-x|^3}-\frac{\bar{y}}{|y|^3}\right]n(y)f(y)dS_y\right|$$
$$\leq\frac{1}{4\pi}\int_{\Sigma_R}\frac{6|x|}{R^3}|f(y)|dS_y\leq\frac{M|x|}{4\pi R^3}\int_{\Sigma_R}dS_y \tag{3.7}$$
$$=\frac{M|x|}{2R}\to 0 \text{ as } R\to\infty.$$

For arbitrarily given positive constant ε, there exists a large positive constant R_0, such that $|f(y)-f(\infty)|<\varepsilon$ if $|y|\geq R_0$. Thus

$$\left|\int_{\Sigma_R}\frac{\bar{y}}{|y|^3}n(y)[f(y)-f(\infty)]dS_y\right|\leq\frac{4}{\pi}\varepsilon,$$

which implies that

$$\int_{\Sigma_R}\frac{\bar{y}}{|y|^3}n(y)[f(y)-f(\infty)]dS_y\to 0 \text{ as } R\to\infty. \tag{3.8}$$

Noting that

$$I_1=\frac{1}{4\pi}\int_{\Sigma_R}\frac{\bar{y}-\bar{x}}{|y-x|^3}n(y)f(y)dS_y=\frac{1}{4\pi}\int_{\Sigma_R}\left[\frac{\bar{y}-\bar{x}}{|y-x|^3}-\frac{\bar{y}}{|y|^3}\right]n(y)f(y)dS_y$$
$$+\frac{1}{4\pi}\int_{\Sigma_R}\frac{\bar{y}}{|y|^3}n(y)[f(y)-f(\infty)]dS_y+\frac{1}{4\pi}\int_{\Sigma_R}\frac{\bar{y}}{|y|^3}n(y)f(\infty)dS_y,$$

and

$$\frac{1}{4\pi}\int_{\Sigma_R}\frac{\bar{y}}{|y|^3}n(y)f(\infty)dS_y=\frac{1}{2}f(\infty),$$

and applying (3.7)and (3.8),

$$\lim_{R\to\infty}I_1=\frac{1}{2}f(\infty) \tag{3.9}$$

follows.

As for the integral I_2, taking $n(y) = -e_3$, $y = y_1 + e_2 y_2$ on E_R into account, we have

$$\begin{aligned} I_2 &= \frac{1}{4\pi}\int_{E_R} \frac{\bar{y}-\bar{x}}{|y-x|^3} n(y) f(y) dS_y \\ &= \frac{1}{4\pi}\int_{E_R} \frac{(y_1-x_1)-e_2(x_2-y_2)+e_3x_3}{|y-x|^3}(-e_3) f(y) dS_y \qquad (3.10) \\ &= \frac{1}{4\pi}\int_{E_R} \frac{x_3+e_3[(x_1-y_1)+e_2(x_2-y_2)]}{|y-x|^3} n(y) f(y) dS_y. \end{aligned}$$

Letting R tend to ∞,

$$f(x) = \frac{1}{4\pi}\int_{E_R} \frac{x_3+e_3[(x_1-y_1)+e_2(x_2-y_2)]}{|y-x|^3} f(y) dS_y + \frac{1}{2} f(\infty) \qquad (3.11)$$

can be derived. If $\operatorname{Im} x < 0$, then according to Cauchy's theorem for regular functions, we can similarly obtain

$$0 = \frac{1}{4\pi}\int_{E_R} \frac{-x_3+e_3[(x_1-y_1)+e_2(x_2-y_2)]}{|y-x|^3} f(y) dS_y + \frac{1}{2} f(\infty).$$

From the above formula, it is easy to derive

$$0 = \frac{1}{4\pi}\int_{E} \frac{x_3+e_3[(x_1-y_1)+e_2(x_2-y_2)}{|y-x|^3} \tilde{f}(y) dS_y - \frac{1}{2}\tilde{f}(\infty). \qquad (3.12)$$

Adding (3.11) and (3.12), and noting that $\tilde{f}(\infty) = -f(\infty)$, we get

$$\begin{aligned} f(x) &= \frac{1}{4\pi}\int_{E} \frac{x_3+e_3[(x_1-y_1)+e_2(x_2-y_2)]}{|y-x|^3} [f(y)+\tilde{f}(y)] dS_y + f(\infty) \\ &= \frac{1}{2\pi}\int_{E} \frac{x_3+e_3[(x_1-y_1)+e_2(x_2-y_2)]}{|y-x|^3} u(y) dS_y + f(\infty). \end{aligned}$$

It remains to verify that $f(x)$ is just the desired function. It is sufficient to prove that

$$\operatorname{Re} f^+(x^0) = u(x^0), \qquad (3.13)$$

where x^0 is any point on E. We rewrite (3.4) in the form

$$\begin{aligned} f(x) &= \frac{1}{2\pi}\int_{E_R} \frac{x_3+e_3[(x_1-y_1)+e_2(x_2-y_2)]}{|y-x|^3} u(y) dS_y \\ &+ \frac{1}{2\pi}\int_{E\backslash E_R} \frac{x_3+e_3[(x_1-y_1)+e_2(x_2-y_2)]}{|y-x|^3} u(y) dS_y + f(\infty), \end{aligned}$$

where $E_R = \{\,|y| < R,\, y \in E\}$ and Σ_R are as stated before. Due to $u(y) = 0$ if $|y|$ is sufficiently large, and for piecewise smooth surface $\Sigma_R \cup E_R$ and sectionally Hölder-continuous function $u(y)$, the Plemelj formula is still true. Hence

$$g^+(x^0) - g^-(x^0) = 2u(x^0), \tag{3.14}$$

where

$$g^+(x) = \frac{1}{2\pi}\int_{E_R\cup\Sigma_R} \frac{x_3 + e_3[(x_1 - y_1) + e_2(x_2 - y_2)]}{|y-x|^3} u(y)dS_y,\ x \in D_R.$$

Setting $\tilde{\zeta} = x$, $\zeta \notin \bar{D}_R$,

$$\begin{aligned}\tilde{g}^+(x) &= \frac{1}{2\pi}\int_{E_R\cup\Sigma_R} \frac{x_3 - e_3[(x_1 - y_1) + e_2(x_2 - y_2)]}{|y-x|^3} u(y)dS_y \\ = \tilde{g}^+(\tilde{\zeta}) &= \frac{1}{2\pi}\int_{E_R\cup\Sigma_R} \frac{-\zeta_3 - e_3[(x_1 - y_1) + e_2(x_2 - y_2)]}{|y-\zeta|^3} u(y)dS_y\end{aligned}$$

can be obtained. Letting x tend to x^0, we know that $\tilde{g}^+(x^0) = -g^-(x^0)$. From (3.14), it follows that

$$\begin{gathered} g^+(x^0) - g^-(x^0) = g^+(x^0) + \tilde{g}^+(x^0) = 2u(x^0), \text{ i.e.} \\ \operatorname{Re} g^+(x^0) = u(x^0). \end{gathered} \tag{3.15}$$

Noting that

$$\begin{aligned}&\lim_{x\to x^0} \frac{1}{2\pi}\iint_{E\setminus E_R} \frac{x_3 + e_3[(x_1 - y_1) + e_2(x_2 - y_2)]}{|y-x|^3} u(y)dS_y + f(\infty) \\ &\quad = \frac{1}{2\pi}\iint_{E\setminus E_R} \frac{e_3[(x_1^0 - y_1) + e_2(x_2^0 - y_2)]}{|y-x^0|^3} u(y)dS_y + f(\infty),\end{aligned}$$

it is easily seen that

$$\operatorname{Re}\left[\frac{1}{2\pi}\int_{E\setminus E_R} \frac{e_3[(x_1^0 - y_1) + e_2(x_2^0 - y_2)]}{|y-x^0|^3} u(y)dS_y + f(\infty)\right] = 0.$$

Thus we obtain (3.13).

Finally, on account of $u(y) = 0$ if $|y|$ is large enough,

$$\lim_{|x|\to\infty} f(x) = f(\infty)$$

can be derived. This shows that the above regular function $f(x)$ is unique.

Remark By using a similar method, we can also get an expression for the solution of the Dirichlet problem for regular functions in the halfspace D with weaker conditions, namely we suppose $u(y)$ is a bounded continuous function in the plane $E = \{x_3 = 0\}$, and the solution $f(x)$ of the Dirichlet problem satisfies the condition $|\mathrm{Re} f(x)| = O(1/|x|)$ as $|x|$ tends to ∞. Then $f(x)$ may be expressed in the form (3.4), where $g(x_1, x_2) = g_1(x_1, x_2) + e_2 g_2(x_1, x_2)$ is an arbitrary function satisfying $g_{x_1} = e_2 g_{x_2}$ (see [3]).

3.2 Schwarz Formula and Dirichlet Problem for a Ball

Next, we shall give the Schwarz formula for regular functions in a ball $G = \{|x| < R,\ 0 < R < \infty\}$ in $\mathbf{R}^3$. We need the following lemmas.

Lemma 3.2 *A function $f(x)$ with values in the Clifford algebra $\mathcal{A}_3(R)$ is regular in G if and if only $\mathrm{Re}\, f(x)$ and $\mathrm{Im}\, f(x)$ satisfy the system of first order equations*

$$2\frac{\partial}{\partial x_*}\mathrm{Re}\, f = \frac{\partial}{\partial x_3}\mathrm{Im}\, f,\ x \in G, \tag{3.16}$$

$$\frac{\partial}{\partial x_3}\mathrm{Re}\, f = -2\frac{\partial}{\partial x^*}\mathrm{Im}\, f,\ x \in G, \tag{3.17}$$

where $x^* = x_1 + x_2 e_2$, $x_* = x_1 - x_2 e_2$, and

$$\frac{\partial}{\partial x^*} = \frac{1}{2}\left(\frac{\partial}{\partial x_1} - e_2\frac{\partial}{\partial x_2}\right),\ \frac{\partial}{\partial x_*} = \frac{1}{2}\left(\frac{\partial}{\partial x_1} + e_2\frac{\partial}{\partial x_2}\right).$$

Proof It is clear that

$$2\frac{\partial}{\partial x^*}(e_3 \mathrm{Im}\, f) = 2e_3\frac{\partial}{\partial x_*}\mathrm{Im}\, f.$$

Hence

$$\bar{\partial} f = \left(2\frac{\partial}{\partial x_*} + e_3\frac{\partial}{\partial x_3}\right)(\mathrm{Re}\, f + e_3 \mathrm{Im}\, f)$$

$$= \left(2\frac{\partial}{\partial x_*}\mathrm{Re}\, f - \frac{\partial}{\partial x_3}\mathrm{Im}\, f\right) + e_3\left(\frac{\partial}{\partial x_3}\mathrm{Re}\, f + 2\frac{\partial}{\partial x^*}\mathrm{Im}\, f\right). \tag{3.18}$$

If $f(x)$ is regular in G, i.e. $\bar{\partial} f = 0$, from (3.18) follows (3.16) and (3.17). The inverse statement is also true.

Suppose that $f(x)$ is a regular function and $f(x) \in C^2(G)$. Then

$$\bar{\partial}\partial\,\mathrm{Re}\, f = \partial\bar{\partial}\,\mathrm{Re}\, f = \Delta \mathrm{Re}\, f = 0,\ \Delta \mathrm{Im}\, f = 0.$$

So Re f and Im f are called harmonic functions in Clifford analysis, and Im f is called the conjugate harmonic function of Re f.

Lemma 3.3 *Let $f(x) = u_1(x) + e_2u_2(x)$, $u_1(x)$ and $u_2(x)$ be real harmonic functions in G and $u_1(x), u_2(x) \in C^1(\overline{G})$. Then the conjugate harmonic function $v(x)$ of $u(x)$ is given by the formula*

$$v(x) = \int_0^{x_3} 2\frac{\partial}{\partial x_*}u dx_3 - \frac{1}{2}\tilde{T}(\frac{\partial}{\partial x_3}u(x_1, x_2, 0)) + g(x_1, x_2), \tag{3.19}$$

where $g(x_1, x_2) = g_1(x_1, x_2) + e_2g_2(x_1, x_2)$ is an arbitrary function satisfying $\partial g/\partial x^* = 0$, and

$$\tilde{T}u(x) = -\frac{1}{\pi}\iint_{E_R}\frac{u(y^*)}{y_* - x_*}d\sigma_{y^*}, \tag{3.20}$$

where $E_R = \{(x_1, x_2, 0) \mid x_1^2 + x_2^2 \mid < R^2\}$.

Proof From (3.16), it follows that

$$v(x) = \int_0^{x_3} 2\frac{\partial}{\partial x_*}u dx_3 + w(x_1, x_2).$$

Substituting the above expression into (3.15) and letting $x_3 = 0$, we obtain

$$\frac{\partial}{\partial x_3}u(x_1, x_2, x_3)|_{x_3=0} + 2\frac{\partial}{\partial x^*}w(x_1, x_2) = 0,$$

and then

$$w(x_1, x_2) = -\frac{1}{2}\tilde{T}(\frac{\partial}{\partial x_3}u(x_1, x_2, x_3)|_{x_3=0}) + g(x_1, x_2)$$

can be obtained, in which $g(x_1, x_2)$ satisfies $\partial g/\partial x^* = 0$. Thus formula (3.19) holds.

Moreover, if $v(x)$ is given by (3.19), we can see that $u(x)$ and $v(x)$ satisfy (3.16). Since $u(x)$ is a harmonic function and

$$4\frac{\partial}{\partial x^*}\frac{\partial}{\partial x_*}u = (\frac{\partial^2}{\partial x_1^2} + \frac{\partial^2}{\partial x_2^2})u = -\frac{\partial^2}{\partial x_3^2}u,$$

it is easy to verify that $u(x)$ and $v(x)$ satisfy (3.17).

Lemma 3.4 *Suppose that $u(x) = u_1(x) + e_2u_2(x)$ is a harmonic function in $G = \{|x| < R\}$ in Clifford analysis and $u(x)$ is continuous on $\overline{G}$. Then $u(x)$ can be expressed as*

$$u(x) = -\int_{|y|=R}\frac{\partial}{\partial n}(\frac{1}{4\pi}\frac{1}{|x-y|} - \frac{1}{4\pi}\frac{R}{|x||\tilde{x}-y|})u(y)dS_y, \tag{3.21}$$

where $x = x_1 + x_2e_2 + x_3e_3$, $y = y_1 + y_2e_2 + y_3e_3$, $\tilde{x} = R^2x/|x|^2$, $\partial/\partial n$ *denotes the exterior normal derivative with respect to* y *on the sphere* $|y| = R$, *and* dS_y *is the area element of* $|y| = R$.

Proof Due to $\Delta u = 0$ in G and because $u(x)$ is continuous on $\overline{G}$, from the Poisson formula for harmonic functions in a ball, it follows that the formula (3.21) holds.

Now, we find the conjugate harmonic functions of $1/|x-y|$ and $R/|x||\tilde{x}-y|$ with respect to x. Noting that

$$\int_0^{x_3} 2\frac{\partial}{\partial x_*}\frac{1}{|x-y|}dx_3 = \int_0^{x_3}\left(-\frac{x^*-y^*}{|x-y|^3}\right)dx_3$$

$$= -\frac{(x^*-y^*)(x_3-y_3)}{|x^*-y^*|^2(x-y)} - \frac{(x^*-y^*)y_3}{|x^*-y^*|^2[(x_1-y_1)^2+(x_2-y_2)^2+y_3^2]^{1/2}},$$

and

$$-\frac{1}{2}\frac{\partial}{\partial x_3}\frac{1}{|x-y|}\bigg|_{x_3=0} = -\frac{1}{2}\frac{y_3}{[(x_1-y_1)^2+(x_2-y_2)^2+y_3^2]^{3/2}}$$

$$= \frac{\partial}{\partial x^*}\frac{(x^*-y^*)y_2}{|x^*-y^*|^2[(x_1-y_1)^2+(x_2-y_2)^2+y_3^2]^{3/2}},$$

we obtain

$$-\frac{1}{2}\tilde{T}\left(\frac{\partial}{\partial x_3}\frac{1}{|x-y|}\bigg|_{x_3=0}\right)$$

$$= \frac{(x^*-y^*)y_2}{|x^*-y^*|^2[(x_1-y_1)^2+(x_2-y_2)^2+y_3^2]^{1/2}} + w(x_1,x_2),$$

where $w(x_1,x_2)$ is a function satisfying $\frac{\partial}{\partial x^*}w = 0$. On the basis of Lemma 3.3, we know that the conjugate harmonic function of $1/|x-y|(x \neq y)$ with respect to x possesses the form

$$-\frac{(x^*-y^*)(x_3-y_3)}{|x^*-y^*|^2|x-y|} + c(x_1,x_2) + w(x_1,x_2).$$

In particular, choosing $c(x_1,x_2) = -w(x_1,x_2)$, it is easy to see that $-(x^*-y^*)(x_3-y_3)/|x^*-y^*|^2|x-y|$ is a conjugate harmonic function of $1/|x-y|$. Similarly, we can find that a conjugate function of $R/|x||\tilde{x}-y|$ with respect to y possesses the form

$$-\frac{R(y^*-\tilde{x}^*)(y_3-\tilde{x}_3)}{|x||y^*-\tilde{x}^*|^2|y-\tilde{x}|} = -\frac{R}{|x|}\frac{(y^*-R^2x^*/|x|^2)(y_3-R^2|x_3|/|x|^2)}{|y-R^2x^*/|x|^2|^2|y-R^2x/|x|^2|}$$

$$= -\frac{R|x|(|x|^2y^*-R^2x^*)(|x|^2y_3-R^2x_3)}{|x_1^2y^*-R^2x^*|^2||x|^2y-R^2x|},$$

and a conjugate function of $R/|x||\tilde{x}-y|$ with respect to x possesses the form

$$-\frac{R|y|(|y|^2x^* - R^2y^*)(|y|^2x_3 - R^2y_3)}{||y|^2x^* - R^2y^*|^2||y|^2x - R^2y|}.$$

Denoting

$$g(x,y) = \frac{1}{4\pi}\left(\frac{1}{|x-y|} - \frac{R}{|x|}\frac{1}{|\tilde{x}-y|}\right),$$

$$h(x,y) = -\frac{1}{4\pi}\left(\frac{(x^*-y^*)(x_2-y_2)}{|x^*-y^*|^2|x-y|}\right.$$
$$\left.-\frac{R|y|(|y|^2x^* - R^2y^*)(|y|^2x_3 - R^2y_3)}{||y|^2x^* - R^2y^*|^2||y|^2x - R^2y|}\right),$$

then for $x \neq y$, the function $s(x,y) = g(x,y) + e_3 h(x,y)$ satisfies $\bar{\partial} s = 0$ with respect to x. Since

$$\frac{\partial}{\partial n} = \frac{\partial}{\partial y_1}\frac{y_1}{|y|} + \frac{\partial}{\partial y_2}\frac{y_2}{|y|} + \frac{\partial}{\partial y_3}\frac{y_3}{|y|},$$

by calculation, we obtain

$$\begin{aligned}
-\frac{\partial}{\partial n}s(x,y)|_{|y|=R} = \frac{1}{4\pi R}&\left\{\frac{R^2-|x|^2}{|x-y|^3}\right.\\
+e_3&\left[-\frac{2(x^*(x_3-y_3)+x_3(x^*-y^*)}{|x^*-y^*|^2|x-y|}\right.\\
&+\frac{4(x^*-y^*)(x_3-y_3)(|x^*|^2-(x_1y_1+x_2y_2))}{|x^*-y^*|^4|x-y|}\\
&\left.\left.+\frac{(x^*-y^*)(x_3-y_3)(|x|^2-R^2)}{|x^*-y^*|^2|x-y|^3}\right]\right\}.
\end{aligned} \tag{3.22}$$

Theorem 3.5 *Let $u(x) = u_1(x) + e_2u_2(x)$ be a continuous function on the sphere $|x| = R$. Then there exists a continuous function $f(x)$ on $\overline{G}$, which satisfies the system of first order equations*

$$\bar{\partial} f = 0 \text{ in } G = \{|x| < R\} \tag{3.23}$$

and the boundary condition

$$\operatorname{Re} f = u(x) \text{ on } \partial G = \{|x| = R\}, \tag{3.24}$$

and $f(x)$ can be expressed as

$$f(x)=\frac{1}{4\pi R}\int_{\partial G}\left\{\frac{R^2-|x|^2}{|x-y|^3}+e_3\left[-\frac{2(x^*(x_3-y_3)+x_3(x^*-y^*))}{|x^*-y^*||x-y|}\right.\right.$$
$$+\frac{4(x^*-y^*)(x_3-y_3)(|x^*|^2-(x_1y_1+x_2y_2))}{|x^*-y^*|^4|x-y|}$$
$$\left.\left.+\frac{(x^*-y^*)(x_3-y_3)(|x|^2-R^2)}{|x^*-y^*|^2|x-y|^3}\right]\right\}u(y)dS_y+e_3c(x_1,y_2),\tag{3.25}$$

where $c(x_1,x_2)=c_1(x_1,x_2)+e_2c_2(x_1,x_2)$ is an arbitrary function satisfying

$$\frac{\partial}{\partial x^*}c(x_1,x_2)=0,$$

where $x^*=x_1+x_2e_2$.

Proof From the above discussion, we see that the function expressed by (3.25) is a solution of the Dirichlet boundary value problem (3.23) and (3.24). Conversely, if the Dirichlet problem (3.23) and (3.24) has a solution $f(x)$, we denote the integral on the right-hand side of (3.25) by $F(x)$, i.e.

$$F(x)=\frac{2}{4\pi R}\int_{\partial G}\left\{\frac{R^2-|x|^2}{|x-y|^3}+e_3\left[-\frac{2x^*(x_3-y_3)+x_3(x^*-y^*)}{|x^*-y^*||x-y|}\right.\right.$$
$$+\frac{4(x^*-y^*)(x_3-y_3)(|x^*|^2-(x_1y_1x_2y_2))}{|x^*-y^*|^4|x-y|}$$
$$\left.\left.+\frac{(x^*-y^*)(x_3-y_3)(|x|^2-R^2)}{|x^*-y^*|^2|x-y|^3}\right]\right\}u(y)dS_y.$$

Then $\bar{\partial}F(x)=0$ in G and $\operatorname{Re}F|_{|x|=R}=u(x)$. Hence $\Delta(\operatorname{Re}f-\operatorname{Re}F)=0$ in G, and $(\operatorname{Re}f-\operatorname{Re}F)|_{|x|=R}=0$. This implies that $\operatorname{Re}f=\operatorname{Re}F$ on $\overline{G}$. According to Lemma 3.2, we obtain

$$\frac{\partial}{\partial x^*}(\operatorname{Im}f-\operatorname{Im}F)=0,\ \frac{\partial}{\partial x_3}(\operatorname{Im}f-\operatorname{Im}F)=0.$$

Thus $\operatorname{Im}f-\operatorname{Im}F=c(x_1,x_2)$ and $\partial c(x_1,x_2)/\partial x^*=0$. The theorem is proved (see [4]).

4 Oblique Derivative Problems for Regular Functions and Elliptic Systems in Clifford Analysis

First of all, we consider the case of the Clifford algebra $\mathcal{A}_3$ over the space $\mathbf{R}^3$. Let G be a ball, i.e. $G = \{\sum_{j=1}^3 |x_j|^2 < R^2(<\infty)\}$, and for convenience, let e_2e_3 be denoted by e_4. Any point x in G may be written as $x = \sum_{j=1}^3 x_j e_j$, and any function $f(x)$ in G with values in the Clifford algebra may be denoted by $w(x) = \sum_{j=1}^4 w_j(x)e_j$.

4.1 Oblique Derivative Problems for Generalized Regular Functions in $\mathbf{R}^3$

A generalized regular function $w(x)$ in G is defined as a solution $w(x) = \sum_{j=1}^4 w_j(x)e_j$ $(\in C^2(G))$ for the elliptic system of first order equations in the form

$$\bar{\partial} w = aw + b\bar{w} + c \text{ in } G, \tag{4.1}$$

where

$$a(x) = \sum_{j=1}^4 a_j(x)e_j,\ b(x) = \sum_{j=1}^4 b_j(x)e_j,$$
$$c(x) = \sum_{j=1}^4 c_j(x)e_j \in C^1_\alpha(G), 0 < \alpha < 1.$$

Problem P The oblique derivative problem for system (4.1) is to find a solution $w(x) \in C^1_\alpha(\bar{G}) \cap C^2(G)$ of (4.1) satisfying the boundary condition

$$\frac{\partial w_j}{\partial \nu_j} + \sigma_j(x)w_j(x) = \tau_j(x) + h_j,\ x \in \partial G, w_j(R) = u_j, j = 1,2, \tag{4.2}$$

where $\sigma_j(x), \tau_j(x) \in C^1_\alpha(\partial G)$, $\sigma_j(x) \geq 0$ on $\partial G, j = 1,2,$ $h_j (j = 1,2)$ are unknown real constants to be determined appropriately, $u_j (j = 1,2)$ are real constants, $\nu_j (j = 1,2)$ are vectors at the point $x \in \partial G$, $\cos(\nu_j, \mathbf{n}) \geq 0, j = 1,2,$ $\mathbf{n}$ is the outward normal at $x \in \partial G$, and $\cos(\nu_j, \mathbf{n}) \in C^1_\alpha(\partial G)$.

If $w(x)$ is a solution of Problem P for system (4.1), then we can verify that $[w_1(x), w_2(x)]$ is a solution of the system of second order

$$\begin{cases} \Delta w_1 = \sum_{j=1}^{3}(A_j + B_j)w_{1x_j} + B_5 w_1 + B_6, \\ \Delta w_2 = 2\sum_{j=1}^{3} B_j w_{2x_j} + B_7 w_2 + (A_2 w_1)_{x_1} - (A_1 w_1)_{x_2} \\ -(A_4 w_1)_{x_3} - B_2 w_{1x_1} + B_1 w_{1x_2} + B_4 w_{1x_3} + B_8 w_1 + B_9 \end{cases} \quad (4.3)$$

satisfying the boundary condition (4.2), where $A_j = a_j + b_j$, $B_j = a_j - b_j$, $j = 1, ..., 4$, B_j is only a function of x_j, $j = 1, 2, 3$, B_4 is a real constant, and

$$B_5 = -\sum_{j=1}^{4} A_j B_j + \sum_{j=1}^{3} A_{jx_j},\ B_6 = -\sum_{j=1}^{4} B_j C_j + \sum_{j=1}^{3} C_{jx_j},$$

$$B_7 = \sum_{j=1}^{3} B_{jx_j} - \sum_{j=1}^{4} B_j^2,\ B_8 = A_1 B_2 - A_2 B_1 - A_3 B_4 + A_4 B_3,$$

$$B_9 = -C_{1x_2} + C_{2x_1} - C_{4x_3} - B_1 C_2 + B_2 C_1 + B_3 C_4 - B_4 C_3.$$

In fact, it follows from (4.1) that

$$\begin{cases} w_{1x_1} - w_{2x_2} - w_{3x_3} = A_1 w_1 - B_2 w_2 - B_3 w_3 - B_4 w_4 + C_1, \\ w_{2x_1} + w_{1x_2} + w_{4x_3} = A_2 w_1 + B_1 w_2 - B_4 w_3 + B_3 w_4 + C_2, \\ w_{3x_1} - w_{4x_2} + w_{1x_3} = A_3 w_1 + B_4 w_2 + B_1 w_3 - B_2 w_4 + C_3, \\ w_{4x_1} + w_{3x_2} - w_{2x_3} = A_4 w_1 - B_3 w_2 + B_2 w_3 + B_1 w_4 + C_4. \end{cases} \quad (4.4)$$

When $B_j = B_j(x_j), j = 1, 2, 3$, and B_4 is a real constant, from (4.4) we can derive the first equation in (4.3), and then the second equation in (4.3) can be obtained.

Conversely, if the following conditions hold:

$$B_5 \geq 0,\ B_7 \geq 0 \text{ on } \bar{G}, \quad (4.5)$$

then according to Theorem 1.6, the boundary value problem (4.3),(4.2) has a solution $[w_1(x), w_2(x)]$. Afterwards, if

$$B_3 = B_4 = 0 \text{ on } \bar{G}, \quad (4.6)$$

then we can find $w_3(x), w_4(x)$ by the following integrals:

$$\begin{cases} w_3(x)=\int_0^{x_3}[w_{1x_1}-w_{2x_2}-A_1w_1+B_2w_2-C_1]dx_3+\phi_3(x_1,x_2), \\ w_4(x)=\int_0^{x_3}[-w_{2x_1}-w_{1x_2}+A_2w_1+B_1w_2+C_2]dx_3+\phi_4(x_1,x_2), \end{cases} \quad (4.7)$$

where $\phi_3(x_1,x_2), \phi_4(x_1,x_2)$ satisfy the conditions

$$\begin{cases} \phi_{3x_1}-\phi_{4x_2}-B_1\phi_3+B_2\phi_4-C_3=\psi_3(x_1,x_2), \\ \phi_{3x_2}+\phi_{4x_1}-B_2\phi_3-B_1\phi_4-C_4=\psi_4(x_1,x_2), \end{cases} \quad (4.8)$$

in which

$$\psi_3(x_1,x_2)=[-w_{1x_3}+A_3w_1]|_{x_3=0}, \psi_4(x_1,x_2)=[w_{2x_3}+A_4w_1]|_{x_3=0},$$

and A_j, B_j, C_j satisfy some conditions such that the integrals in (4.7) are single-valued. Suppose that $\phi=\phi_3+\phi_4e_2=\phi_3+\phi_4i$ satisfies the Riemann-Hilbert boundary conditions:

$$\begin{cases} \mathrm{Re}[\overline{\lambda(t)}\phi(t)]=r(t)+h(t), t=t_1+it_2\in\Gamma=\partial G\cap\{x_3=0\}, \\ \mathrm{Im}[\overline{\lambda(d_j)}\phi(d_j)]=g_j,\ j=1,...,2K+1 \text{ for } K\ge 0; \end{cases} \quad (4.9)$$

where $|\lambda(t)|=1$, $\lambda(t), r(t)\in C^2_\alpha(\Gamma)$, $d_j\ (j=1,...,2K+1$ for $K\ge 0)$ are distinct points on Γ, $g_j\ (j=1,...,2K+1$ for $K\ge 0)$ are known real constants, and

$$h(t)=\begin{cases} 0 \text{ on } \Gamma \text{ for } K=\frac{1}{2\pi}\Delta_\Gamma\arg\lambda(t)\ge 0, \\ h_0+\mathrm{Re}\sum_{m=1}^{-K-1}(h^+_m+ih^-_m)t^m \text{ on } \Gamma \text{ for } K<0; \end{cases}$$

herein $h_0, h^\pm_m(m=1,...,-K-1)$ are undetermined real constants. According to Theorems 4.1 and 4.6, Chapter 2 in [81], the functions $\phi_3(x_3,x_4), \phi_4(x_1,x_2)$ may be uniquely determined, and then $w_3(x), w_4(x)$ may be also uniquely determined. Then we obtain

Theorem 4.1 *If the coefficients of the system (4.1) satisfy the conditions (4.5),(4.6), and $W(t)=w_3(t)+iw_4(t)$ satisfies the boundary conditions*

$$\mathrm{Re}[\overline{\lambda(t)}W(t)]=r(t)+h(t),\ t\in\Gamma=\partial G\cap\{x_3=0\}, \quad (4.10)$$

$$\mathrm{Im}[\overline{\lambda(d_j)}W(d_j)]=g_j,\ j=1,...,2K+1 \text{ for } K\ge 0, \quad (4.11)$$

where $\lambda(t), r(t), h(t), d_j, g_j$ and K are as stated in (4.9), then Problem P has a unique solution $w(x)=\sum_{j=1}^4 w_j(x)e_j\in C^1_\alpha(\bar G)\cap C^2(G)$, $0<\alpha<1$ (see [3]).

4.2 Oblique Derivative Problem for a Degenerate Elliptic System of First Order in $\mathbf{R}^3$

Now, we discuss the degenerate elliptic system of first order equations

$$\begin{cases} w_{1x_1}-w_{2x_2}-x_3^{\mu}w_{3x_3}=A_1w_1-B_2w_2-B_3x_3^{1+\mu}w_3-B_4w_4+C_1,\\ w_{2x_1}+w_{1x_2}+w_{2x_3}=A_2w_1+B_1w_2-B_4x_3^{\mu}w_3+B_3w_4+C_2,\\ x_3^{\mu}w_{3x_1}-w_{4x_2}+w_{1x_3}=A_3w_1+B_2w_2+B_1x_3^{\mu}w_3-B_2w_4+C_3,\\ w_{4x_1}+x_3^{\mu}w_{3x_2}-w_{2x_3}=A_4w_1-B_3w_2+B_2x_3^{\mu}w_3+B_1w_4+C_4, \end{cases} \tag{4.12}$$

where $A_j(x)$, $B_j(x), C_j(x)(j=1,...,4)$ are known functions as stated in (4.4), and $C_1(x)=x_3^{1+\mu}c_1(x)$, $C_j(x)=x_3^{\mu}c_j(x)$, $j=2,3,4$, $c_j(x)\in C^1_\alpha(\overline{G})$, $j=1,...,4$, μ is a non-negative constant, and G is a domain in the upper halfspace $x_3>0$ with the boundary $\partial G=D_1\cup D_2\in C^2_\alpha$, where D_1 is in $x_3>0$ and D_2 in $x_3=0$, $\Gamma=\overline{D}_1\cap\{x_3=0\}$. The Riemann-Hilbert boundary value problem (Problem A) for (4.12) is to find a bounded solution $w(x)=\sum_{j=1}^4 w_j(x)e_j\in C^1_\alpha(\overline{G})\cap C^2(G)$ of (4.12) satisfying the boundary condition

$$\begin{cases} \dfrac{\partial w_3}{\partial\nu_3}+\sigma_3(x)w_3(x)=\tau_3(x),\ x\in D_1,\\ \dfrac{\partial w_4}{\partial\nu_4}+\sigma_4(x)w_4(x)=\tau_4(x),\ x\in\partial G, \end{cases} \tag{4.13}$$

$$\begin{cases} \operatorname{Re}[\overline{\lambda(t)}(w_1(t)+iw_2(t)]=r(t)+h[\zeta(t)],\ t=x_1+ix_2\in\Gamma,\\ \operatorname{Im}[\overline{\lambda(d_j)}(w_1(d_j)+iw_2(d_j))]=g_j,\ j=1,...,2K+1 \text{ for } K\geq 0, \end{cases} \tag{4.14}$$

where $\cos(\nu_j,n)>0$, $j=3,4$, n is the outward normal at $x\in\partial G$, and $\sigma_j(x)>0$, $\tau_j(x)\in C^1_\alpha(\partial D_2)$, $j=3,4$, $\lambda(t),r(t),h(t),d_j,g_j$ and K are similar as in (4.9), $\zeta(t)$ is a conformal mapping from the unit disk $\{t|<1\}$ onto D_2. Similarly to before, if B_j is only a function of $x_j, j=1,2,3$, B_4 is a real constant, and $A_j=B_j, j=1,...,4$, then the solution $w(x)=\sum_{j=1}^4 w_j(x)e_j$ of Problem O for (4.12) is also a solution of the following boundary value problem for the degenerate system of second order equations:

$$\begin{cases} \Delta w_3=2B_1w_{3x_1}+2B_2w_{3x_2}+A_5x_3^{-1}w_{3x_3}+A_6w_3+A_7,\\ \Delta w_4=\displaystyle\sum_{j=1}^{3}(A_j+B_j)w_{4x_j}+A_8w_4+A_9w_3+A_{10} \end{cases} \tag{4.15}$$

with the boundary condition (4.13), in which

$$A_5=-\mu+B_3x_3(1+x_3),\ C_5\in C^2(\overline{G}),$$

$$A_6=(1+\mu)B_3-\sum_{j=1}^{4}B_j^2+\sum_{j=1}^{3}B_{jx_j}+(B_3^2-B_{3x_2})(1-x_3),$$

$$A_7=(-B_1C_3-B_2C_4+B_3C_1+B_4C_2-]C_{1x_3}+C_{3x_1}+C_{4x_2})x_3^{-\mu},$$

$$A_8=-\sum_{j=1}^{4}B_j^2+\sum_{j=1}^{3}B_{jx_j},\ A_9=A_4x_3^{\mu-1}(B_4\mu+A_3x_3-B_3x_3^2),$$

$$A_{10}=-A_1C_4+A_2C_3-A_3C_2+A_4C_1+C_{2x_3}-C_{3x_2}+C_{4x_1}.$$

Conversely, if the following conditions hold:

$$A_6\geq 0,\ A_8\geq 0\ \text{ in }\ \overline{G}, \tag{4.16}$$

then by means of Corollary 2.3, the boundary value problem (4.15),(4.13) is solvable. Moreover, if

$$A_1=A_2=0\ \text{ in }\ \overline{G}, \tag{4.17}$$

then $[w_1(x),w_2(x)]$ can be found by the following integrals:

$$\begin{cases} w_1(x)=\int_0^{x_3}[-x_3^{\mu}w_{3x_1}+w_{4x_2}+B_1x_3^{\mu}w_3-B_2w_4+C_3]dx_3+\phi_1(x_1,x_2),\\ w_2(x)=\int_0^{x_3}[x_3^{\mu}w_{3x_2}+w_{4x_1}-B_2x_3^{\mu}w_3-B_1w_4-C_4]dx_3+\phi_2(x_1,x_2), \end{cases} \tag{4.18}$$

in which $\phi_1(x_1,x_2)$, $\phi_2(x_1,x_2)$ satisfy the conditions

$$\begin{cases} \phi_{1x_1}-\phi_{2x_2}-A_1\phi_1+B_2\phi_2=\psi_1(x_1,x_2),\\ \phi_{2x_1}+\phi_{1x_2}-A_2\phi_1-B_1\phi_2=\psi_2(x_1,x_2), \end{cases}$$

where

$$\psi_1(x_1,x_2)=[x_3^{\mu}\mu w_{3x_3}+C_1]|_{x_3=0},\ \psi_2(x_1,x_2)=[-w_{4x_3}+C_2]|_{x_3=0}, \tag{4.19}$$

and A_j, B_j, C_j satisfy some conditions.

Besides, we require that $\phi=\phi_1+\phi_2e_2=\phi_1+\phi_2i$ satisfies the boundary condition (4.14), i.e.

$$\begin{cases} \operatorname{Re}[\overline{\lambda(t)}\phi(t)]=r(t)+h(t),\ t\in\Gamma,\\ \operatorname{Im}[\overline{\lambda(d_j)}\phi(d_j)]=g_j,\ j=1,\ldots,2K+1 \text{ for } K\geq 0. \end{cases} \tag{4.20}$$

Similarly to (4.10),(4.11), the function $[\phi_1(x_1,x_2),\phi_2(x_1,x_2)]$ is uniquely determined, and hence $[w_1(x),w_2(x)]$ is also uniquely determined. Thus we have

Theorem 4.2 *Suppose that the coefficients of system (4.12) satisfy conditions (4.16),(4.17) etc. stated as before. Then Problem A for degenerate elliptic system (4.12) has a solution* $w(x)=\sum_{j=1}^{4} w_j(x)e_j \in C^1_\alpha(G\cup D_1)\cap C^2(G)$, $0<\alpha<1$ (see [80]6)).

We mention that when the boundary condition (4.13) is replaced by

$$w_3(x)=\tau_3(x),\ w_4(x)=\tau_4(x),\ x\in\partial G, \tag{4.21}$$

we can similarly discuss the solvability of the boundary value problem (4.12) and (4.21).

4.3 Oblique Derivative Problem for an Elliptic System of First Order in $\mathbf{R}^4$

Now, we discuss the elliptic system of first order equations in a domain $Q\subset\mathbf{R}^4$. The so-called regular function $w(x)=\sum_{j=1}^{8} w_j(x)e_j$ is indicated as a solution of the system of first order equations

$$\begin{cases} w_{1x_1}-w_{2x_2}-w_{3x_3}-w_{4x_4}=0,\ w_{1x_2}+w_{2x_1}+w_{5x_3}+w_{6x_4}=0,\\ w_{1x_3}+w_{3x_1}-w_{5x_2}+w_{7x_4}=0,\ w_{1x_4}+w_{4x_1}-w_{6x_2}-w_{7x_3}=0,\\ -w_{2x_3}+w_{3x_2}+w_{5x_1}-w_{8x_4}=0,\ -w_{2x_4}+w_{4x_2}+w_{6x_1}+w_{8x_3}=0,\\ -w_{3x_4}+w_{4x_3}+w_{7x_1}-w_{8x_2}=0,\ w_{5x_4}-w_{6x_3}+w_{7x_2}+w_{8x_1}=0, \end{cases} \tag{4.22}$$

in Q, in which for convenience we denote $e_5=e_2e_3$, $e_6=e_2e_4$, $e_7=e_3e_4$, $e_8=e_2e_3e_4$. The generalized regular function $w(x)=\sum_{j=1}^{8} w_j(x)$ is indicated as a solution of the system of first order equations

$$\overline{\partial}w=Aw+B\overline{w}+C \ \text{ in } \overline{Q}, \tag{4.23}$$

where $A(x)=\sum_{j=1}^{8} A_j(x)$, $B(x)=\sum_{j=1}^{8} B_j(x)$, $C(x)=\sum_{j=1}^{8} C_j(x)$, $\overline{w(x)}=w_1(x)-\sum_{j=2}^{8} w_j(x)$. Let $Q=\{|x_j|<R\,(j=1,2,3,4),\ 0<R<\infty\}$. The oblique derivative problem (Problem P) of (4.22) in Q is to find a bounded solution $w(x)=\sum_{j=1}^{8} w_j(x)e_j\in C^1_\alpha(\overline{Q})\cap C^2(Q)$ of (4.22) satisfying the boundary conditions

$$\begin{cases} \dfrac{\partial w_j}{\partial \nu_j}+\sigma_j(x)w_j(x)=\tau_j(x)+h_j,\ x\in\partial Q,\\ w_j(R)=u_j,\ j=1,2,3,5, \end{cases} \tag{4.24}$$

where $\sigma_j(x), \tau_j(x) \in C^1_\alpha(\partial Q)$, $\sigma_j(x) \geq 0\,(j = 1, 2, 3, 5)$, $h_j(j = 1, 2, 3, 5)$ are undetermined constants, and $u_j(j = 1, 2, 3, 5), \alpha$ are real constants, $\nu_j(j = 1, 2, 3, 5)$ are vectors at every point $x \in \partial Q$ and $\cos(\nu_j, n) > 0$, n is the outward normal at the point $x \in \partial Q$, and $\cos(\nu_j, n) \in C^1_\alpha(\partial Q)$, $j = 1, 2, 3, 5$.

We first find the partial derivatives for the first four equations in (4.22) with respect to x_1, x_2, x_3, x_4, and then add the equations, thus

$$\Delta w_1 = w_{1x_1^2} + \cdots + w_{1x_4^2} = 0. \tag{4.25}$$

Similarly we can obtain

$$\Delta w_j = w_{jx_1^2} + \cdots + w_{jx_4^2} = 0,\ j = 2, 3, 5. \tag{4.26}$$

On the basis of Theorem 1.6, there exists a solution $[w_1(x),\ w_2(x),\ w_3(x),\ w_5(x)]$ of the boundary value problem (4.25),(4.26),(4.24), where $w_j(x) \in C^1_\alpha(\overline{Q}) \cap C^2(Q)$, $j = 1, 2, 3, 5$. Substituting the functions into (4.22), we have

$$\begin{cases} w_{4x_4} = w_{1x_1} - w_{2x_2} - w_{3x_3},\ w_{6x_4} = -w_{1x_2} - w_{2x_1} - w_{5x_3}, \\ w_{7x_4} = -w_{1x_3} - w_{3x_1} + w_{5x_2},\ w_{8x_4} = -w_{2x_3} + w_{3x_2} + w_{5x_1}. \end{cases} \tag{4.27}$$

From the above system, it follows that

$$\begin{aligned} w_4 &= \int_0^{x_4} [w_{1x_1} - w_{2x_2} - w_{3x_3}]dx_4 + g_4(x_1, x_2, x_3), \\ w_6 &= \int_0^{x_4} [-w_{1x_2} - w_{2x_1} - w_{5x_3}]dx_4 + g_6(x_1, x_2, x_3), \\ w_7 &= \int_0^{x_4} [-w_{1x_3} - w_{3x_1} + w_{5x_2}]dx_4 + g_7(x_1, x_2, x_3), \\ w_8 &= \int_0^{x_4} [-w_{2x_3} + w_{3x_2} + w_{5x_1}]dx_4 + g_8(x_1, x_2, x_3), \end{aligned} \tag{4.28}$$

and $g_j(x_1, x_2, x_3)\ (j = 4, 6, 7, 8)$ satisfy the system of first order equations

$$\begin{cases} g_{4x_1} - g_{6x_2} - g_{7x_3} = -w_{1x_4}|_{x_4=0} = a_1, \\ g_{4x_2} + g_{6x_1} + g_{8x_3} = w_{2x_4}|_{x_4=0} = a_2, \\ g_{4x_3} + g_{7x_1} - g_{8x_2} = w_{3x_4}|_{x_4=0} = a_3, \\ g_{6x_3} - g_{7x_2} - g_{8x_1} = w_{5x_4}|_{x_4=0} = a_4. \end{cases} \tag{4.29}$$

Let g_4, g_6, g_7, g_8 be replaced by w_1, w_2, w_3, w_4, then the system (4.29) can be rewritten as

$$\begin{cases} w_{1x_1} - w_{2x_2} - w_{3x_3} = a_1,\ w_{1x_2} + w_{2x_1} + w_{4x_3} = a_2, \\ w_{1x_3} + w_{3x_1} - w_{4x_2} = a_3,\ w_{2x_3} - w_{3x_2} - w_{4x_1} = a_4, \end{cases} \tag{4.30}$$

i.e. $w(x) = \sum_{j=1}^{4} w_j(x_j)e_j$ satisfies the equation

$$\overline{\partial} w = e_1 w_{x_1} + e_2 w_{x_2} + e_3 w_{x_3} = a,\ a = \sum_{j=1}^{4} a_j e_j,$$

and (4.30) is just a special case of (4.1) with $a(x) = b(x) = 0$, $c(x) = a(x)$. By Theorem 4.1 and the result in [38], the system (4.30) has a solution $w(x) = \sum_{j=1}^{4} w_j(x)e_j \in C_\alpha^1(\overline{D}) \cap C^2(D)$; herein $D = Q \cap \{x_4 = 0\}$. Thus the functions $g_4(x), g_6(x), g_7(x), g_8(x)$ are found. Let these functions be substituted into (4.28). Then $w_j(x)(j = 4, 6, 7, 8)$ are determined, and the function $w(x) = \sum_{j=1}^{8} w_j(x)e_j$ is just the solution of Problem P for (4.22). Hence we have the following theorem.

Theorem 4.3 *Problem P of the elliptic system of first order equations (4.22), i.e. the regular functions in the domain Q, has a solution $w(x) = \sum_{j=1}^{8} w_j(x)e_j \in C_\alpha^1(\overline{Q}) \cup C^2(Q)$, where $0 < \alpha < 1$* (see [80]6)).

As for the generalized regular functions, under certain conditions, we can also prove the existence of solutions of Problem P for system (4.23) in Q. By a similar method as stated in Subsection 4.2, we can prove the solvability of Problem P for the corresponding degenerate elliptic systems of first order equations with some conditions in the domain G.

Finally, we consider the case of arbitrary dimension n. Let $\mathcal{A}_n(R)$ be a real Clifford algebra and Q be a polycylinder $Q_1 \times \cdots \times Q_n$ in the space $\mathbf{R}^n$, $Q_j = \{|x_j| \le R_j\}, 0 < R_j < \infty, j = 1, ..., n$.

Problem P' The oblique derivative problem for generalized regular functions in $Q \subset \mathbf{R}^n$ is defined to be the problem of finding a generalized regular function

$$w(x) = \sum_{j=1}^{2^{n-1}} w_j(x)e_j = \sum_A w_A(x)e_A \text{ in } G,$$

$$A = \{j_1, ..., j_k\} \subset \{1, ..., n\}, e_A = e_{j_1} ... e_{j_k}, 1 \le j_1 < \cdots < j_k \le n,$$

satisfying the boundary condition

$$\frac{\partial w_j}{\partial \nu_j} + \sigma_j(x) w_j(x) = \tau_j + h_j,\ x \in \partial Q,\ w_j(R) = u_j, j = 1, ..., 2^{n-2}. \tag{4.31}$$

Here $j = A$ if $j \leq n$ and if A includes at least two integers greater than 1, then $j(n < j \leq 2^{n-1})$ denotes one of integers system A such that j and A possess a one to one relation, ν_j is a vector at the point $x \in \partial Q$, $\cos(\nu_j, n) \geq 0$, $\sigma_j(x) \geq 0$, $x \in \partial Q$, $h_j\,(j = 1, ..., 2^{n-1})$ are undetermined constants.

Theorem 4.4 *A function*

$$w(x) = \sum_A w_A(x) e_A \text{ in } Q$$

is a generalized regular function if and only if the $w_A(x)$ *satisfy the real elliptic system of first order equations*

$$\sum_{k=1}^{n} \delta_{\overline{kB}} w_{Bx_k} = \sum_{d=1}^{n} (a_d + b_d) w_M \delta_{\overline{dM}} + \sum_{d=1}^{n} (a_d - b_d) w_N \delta_{\overline{dN}} + c_A, \quad (4.32)$$

where $\overline{kB} = A$, $\overline{dM} = A$, $\overline{dN} = A$ *and* $\delta_{\overline{bB}}$ *etc. are proper signs* (see [80]10)).

Proof Since $w(x)$ is a generalized regular function, it is clear that $w(x)$ satisfies the elliptic system of first order equations

$$\bar{\partial} w = aw + b\bar{w} + c. \quad (4.33)$$

Moreover we have

$$\bar{\partial} w = \sum_A \sum_{k=1}^{n} \delta_{\overline{kB}} w_{Bx_k} e_A = \sum_{k,A} \delta_{\overline{kB}} w_{Bx_k} e_A.$$

Thus

$$aw + b\bar{w} + c = \textstyle\sum_{d,A} (a_d + b_d) w_M \delta_{\overline{dM}} e_A$$

$$+ \textstyle\sum_{d,A} (a_d - b_d) w_N \delta_{\overline{dN}} e_A + \sum_A c_A e_A,$$

where $\overline{kB} = A$, $\overline{dM} = A$ and $\overline{dN} = A$, which shows that $w_A(x)$ satisfies (4.32).

Under certain conditions, by introducing various quasi-permutations as stated in Chapter I, from (4.32) we can obtain

$$\Delta w_k = \sum_{m=1}^{n} d_{mk} w_{kx_m} + f_k w_k + w_k,\ k = 1, ..., 2^{n-2}. \quad (4.34)$$

On the basis of the result in Section 1, a solution $[w_1(x), ..., w_{2^{n-2}}(x)]$ of (4.34) satisfying the boundary condition (4.31) can be found. Moreover, we rewrite (4.32) in the form

$$W_{k\overline{x_k}} = F_{kl}(z_1, ..., z_m, W_{2^{n-3}+1}, ..., W_{2^{n-2}}),$$
$$k = 2^{n-3}+1, ..., 2^{n-2},\ l = 1, ..., m, \tag{4.35}$$

where

$$n = 2m,\ x_{2k-1} + ix_{2k} = z_k,\ w_{2k-1} + iw_{2k} = W_k,\ k = 1, ..., 2^{n-2}.$$

If Q is a polycylinder, under some conditions, we can obtain $[W_{2^{n-3}+1}(x), ..., W_{2^{n-2}}(x)]$ satisfying

$$\mathrm{Re}[\overline{z_1}^{K_1}...\overline{z_m}^{K_m} W_j(z_1, ..., z_m)] = r_j(z_1, ..., z_m) + h_j,$$
$$z(z_1, ..., z_m) \in \partial Q_1 \times \cdots \times \partial Q_m,\ j = 2^{n-3}+1, ..., 2^{n-2}. \tag{4.36}$$

The above result can be written as

Theorem 4.5 *Under some conditions, Problem P' for (4.14) has a solution $[w_1(x), ..., w_{2^{n-1}}(x)]$, where Q is a polycylinder* (see [31]2)).

References

[1] Alkhutov Yu. A., Mamedov I. T., *The first boundary value problem for nondivergence second order parabolic equations with discontinuous coefficients.* Math. USSR Sbornik, 1988, 59:471-495.

[2] Avanissian V., *Cellule d'Harmonicité et Prolongement Analytigve complexe.* Travaux en Cours, Hermann, Paris, 1985.

[3] Begehr H., Wen Guo-chun, *Nonlinear elliptic boundary value problems and their applications.* Addison Wesley Longman, Harlow, 1996.

[4] Begehr H., Xu Zhen-yuan, *Nonlinear half-Dirichlet problems for first order elliptic equations in the unit ball of* $\mathbf{R}^m (m \geq 3)$. Appl. Anal., 1992, 45:3-18.

[5] Bernstein S., *Operator calculus for elliptic boundary value problems in unbounded domains.* Zeitschrift für Analysis und ihre Anwendungen, 1991, 10(4):447-460.

[6] Brackx F., Delanghe R., Sommen F., *Clifford analysis.* Research Notes in Mathematics 76, Pitman, Boston, 1982.

[7] Brackx F., Pincket W., *Two Hartogs theorems for nullsolutions of overdetermined systems in Euclidean space.* Complex Variables: Theory and Application, 1985, 4:205-222.

[8] Cartan É., *Sur les domaines bornés homogénes de léspaace de n variables complexex.* Abh. Math. Sem. Univ. Hamburg, 1936, 11:106-162.

[9] Chai Zhiming, Huang Sha, Qiao Yuying, *A nonlinear boundary value problem for biregular function vectors with values in real analysis.* Acta Mathematica Scientia, 2000, 20(1):121-129 (Chinese).

[10] Chen Yezhe, *Krylov's methods of a priori estimates for full nonlinear equations.* Adv. in Math. 1986, 15:63-101 (Chinese).

[11] Cordes H. O., *Über die Randwertaufgabe bei quasilinearen Differentialgleichungen zweiter Ordnung in mehr als zwei Variablen.* Math. Ann., 1956, 131:278-312.

[12] Delanghe R., 1) *On regular analysis function with values in a Clifford algebra.* Math. Ann., 1970, 185:91-111.

2) *Some remarks on the principal value kernal in* R^m. Complex Varibles, 2002, 47:653-662.

[13] Delanghe R., Sommen F., Souček V., *Clifford algebra and spinor-valued functions, A function theory for the Dirac operator.* Math. and Its Appl. 53, Kluwer Academic Publishers, Dordrecht, Boston, London, 1992.

[14] Dong Guangchang, 1) *Initial and nonlinear oblique boundary value problems for fully nonlinear parabolic equations.* J. Partial Differential Equations, Series A, 1988, 2:12-42.
2) *Linear partial differential equations of second order.* Zhejiang University Press, 1991 (Chinese).
3) *Nonlinear second order partial differential equations.* Amer. Math. Soc., Providence, RI, 1991.

[15] Eriksson S.-L. and Leutwiler H., 1) *Hypermonogenic functions.* Clifford Algebrbs and their Applications in Mathematics Physis, Vol.2, Birkhuser, Boston, 2000, 287-302.
2) *Hypermonogenic function and Möbius transformation.* Advances in Applied Clifford Algebras, Mexico, 2001, 11(s2):67-76.

[16] Fox C., *A Generalization of the Cauchy principal value.* Canadian J. Math., 1957, 9:110-119.

[17] Gürlebeck K., Sprössig W., *Quaternionic and Clifford Calculus for Physicists and Engineers.* John Wiley and Sons, Chichester, 1997.

[18] Gilbarg D., Trudinger N. S., *Elliptic partial differential equations of second order.* Springer-Verlag, Berlin, 1977.

[19] Gilbert R. P., Buchanan J. L., *First order elliptic systems, A function theoretic approach.* Academic Press, New York, 1983.

[20] Gong Sheng, *Singular integrals in several complex variables.* Shanghai Science and Technology Publishing House, Shanghai, 1982 (Chinese).

[21] Hasse H., *Automorphe function von mehreren veränderlichen und Dirichletsche Reihen.* Abh. Math. Sem. Univ. Hamburg, 1949, 16, 3/4:72-100.

[22] Haseman C., *Anwendung der Theoric Integralgleichungen Aufeinige Randwertaufgaben.* Göttingen, 1907.

[23] Hekolaeshuk A. M., *About the stability of Malkushevech boundary value problem.* Ukraine J. Math., 1974, 26(4):558-559.

[24] Hino Y., *Almost periodic solutions of a linear volterra System.* J. Differential Equations and Integral Equations, 1990, 3:495-501.

[25] Hu Chuangan, Yang Chongzun, *Vector-valued functions and their applications.* Kluwer Academic Publishers, Dordrecht, 1992.

[26] Hua Luogeng, 1) *Harmonic analysis for several complex variables in the classical domains.* Science Press, Beijing, 1958 (Chinese).
2) *Starting with the unit circle.* Springer-lerlag, Berlin, 1981.

[27] Hua Luogeng, Lu Qikeng, *Theory of harmonic functions in classical domains.* Scientia Sinica, 1958, 8:1031-1094 (Chinese).

[28] Huang Liede, *The elliptic systems of first order partial differential equations in the space and its boundary value problem.* Tongji University Publishing House, Shanghai, 1981 (Chinese).

[29] Huang Sha, 1) *A nonlinear boundary value problem with Haseman shift in Clifford analysis.* J. Sys. Sci. and Math. Sci., 1991, 11:336-346 (Chinese).
2) *Nonlinear boundary value problem for biregular functions in Clifford analysis.* Science in China (Series A), 1996, 39(11):1152-1164 (Chinese).
3) *Existence and uniqueness of solutions for a boundary value problem with a conjugate value in real Clifford analysis.* Kexue Tongbao (A Monthly Jour. of Sci.), 1991, 36(9):715-716 (Chinese).
4) *Two boundary value problems for regular functions with values in a real Clifford algebra in the hyperball.* J. Sys. Sci. and Math. Sci., 1996, 9(3):236-241 (Chinese).
5) *A boundary value problem in real Clifford analysis.* J. Sys. Sci. and Math. Sci., 1993, 13(1):90-96 (Chinese).
6) *Quasi-permutation and P-R-H boundary value problem in real Clifford analysis.* J. Sys. Sci. and Math. Sci., 2000, 20(3):270-279 (Chinese).
7) *A kind of boundary value problems for the hyperbolically harmonic funcions in the Clifford analysis.* J. Sys. Sci. and Math. Sci., 1996, 16(1):60-63 (Chinese).
8) *A (linear) nonlinear boundary value problem for generalized biregular function in Clifford analysis.* Acta. Math. Sci., 1997, 40(6):914-920 (Chinese).
9) *The regularization theorem of the singular integral equations on*

characteristic manifold in Clifford analysis. Acta Mathematica Scientia, 1998, 18(3):257-263 (Chinese).
10) *Three kinds of high order singular integrals and nonlinear differential integral equations in the real Clifford analysis.* Advances in Mathematics, 2000, 29(3):253-268 (Chinese).
11) *The Cauchy's estimates of three integrals depending on a parameter in Clifford analysis.* Proceedings of Systems and Systems Engineering, Culture Publish Co., Great Wall(H.K.), 1991, 161-168 (Chinese).
12) *Poincaré-Bertrand transformation formulas of the singular integrals in Clifford analysis.* Acta mathematica Sinica, 1998, 41(1):119-126 (Chinese).
13) *The inverse formula of singular integrals in Clifford analysis.* J. Sys. Sci. and Math. Sci., 1999, 19(2):246-250 (Chinese).

[30] Huang Sha, Jiao Hongbing, Qiao Yuying, Chen Zhenguo, *Nonlinear boundary value problems for several unknown functions vectors in Clifford analysis.* Acta mathematica Scientia, 1998, 41(6):1185-1192 (Chinese).

[31] Huang Sha, Li Sheng-xun, 1) *On oblique derivative problem for generalized regular functions with values in a real Clifford algebra.* Proceedings of the Integral Equations and Boundary Value Problems, World Scientific Publishing, Singapore, 1991, 80-85.
2) *On the structure of functions with values in complex Clifford algebra.* Pure and Applied Math., 1990, 6(6):27-35 (Chinese).

[32] Huang Sha, Qiao Yuying, 1) *A new hypercomplex structure and Clifford analysis.* J. Sys. Sci. and Math. Sci., 1996, 16(4):367-371 (Chinese).
2) *Harmonic analysis in classical domains and complex Clifford analysis.* Acta Mathematica Sinica, 2001, 44(1):29-36 (Chinese).

[33] Iftimie V., *Functions hypercomplexes.* Bull. Math. de la Soc. Sci. Math. de la R. S. R., 1965, 9(57):279-332.

[34] Kähler U., *Elliptic boundary value problems in bounded and unbounded domains.* Dirac Operators in Analysis, Pitman Research Notes in Mathematics series, 394, 1998, 122-140.

[35] Kandmanov R. M., *Bochner-Martinelli integrals and their application.* Sciences Siberia Press, 209-213 (Russian).

[36] Kimiro Sano, *The derivatives of Cauchy's kernel and Cauchy's estimate in Clifford analysis.* Complex Variables, 1989, 11:203-213.

[37] Krylov N. V., *Nonlinear elliptic and parabolic equations of the second order.* D. Reidel Publishing Co., Dordrecht, 1987.

[38] Ladyshenskaja O. A. and Ural′tseva N. N., *Linear and quasilinear elliptic equations.* Academic Press, New York, 1968.

[39] Le Huang Son, *Cousin problem for biregular functions with values in a Clifford algebra.* Complex Variables, 1992, 20:255-263.

[40] Leray J. and Schauder J., *Topologie et équations fonczionelles.* Ann. Sci. École Norm. Sup. 1934, 51:45-78; YMH 1946, 1:71-95 (Russian).

[41] Leutwiler H., *Modified Clifford analysis.* Complex Variables, 1992, 17:153-171.

[42] Li Mingde, Qin Yuchun, *On boundary value problems for singular elliptic equations.* Hangzhoudaxue Xuebao (Ziran Kexue), 1980, 2:1-8 (Chinese).

[43] Li Mingzhong, Cheng Jin, *Riemann-Hilbert boundary value problem for overdetermined system with several variables and several unknown functions.* Acta. Math. Sci. 1988, 31(5):671-679 (Chinese).

[44] Li Yuching, Qiao Yuying, *Several properties of a system in complex Clifford analysis.* Journal of Mathematics, 2001, 21(4):391-396.

[45] Lu Jianke, *Selected works about analytic functions and singular integral equations.* Wuhan University Publishing House, 1998 (Chinese).

[46] Lu Jianke, Zhong Shoutao, *Integral equations.* Higher Education Press, 1990 (Chinese).

[47] Lu Qikeng, 1) *An analytic invariant and its characteristic properties.* Sci. Record (N.S.), 1957, 7:370-420 (Chinese).
2) *Classical manifolds and classical domains* Shanghai Science and Technology Publishing House, Shanghai, 1963.
3) *Slit space and extremal principle.* Science Record (N.S.), 1959, 3:289-294 (Chinese).

[48] Malonek H., *A new hypercomplex structure of the Euclidean space* $\mathbf{R}^{m+1}$ *and the concept of hypercomplex differentiability.* Complex Variables, 1990, 14(1):25-33.

[49] McIntosh A., *Clifford algebras, Fourier Theory, Singular Integrals, and harmonic functions on Lipschitz domains.* Clifford Algebras in

Analysis and Related Topics. CRC Press, Boca Raton New York London Tokyo, 1996, 33-88.

[50] Miranda C., *Partial differential equations of elliptic type.* Springer-Verlag, Berlin, 1970.

[51] Mitrea M., *Clifford Wavelets, Singular Integrals and Hardy Spaces.* Lecture Notes in Mathematics 1575, Springer-Verlag, Berlin Heidelberg New York, 1994.

[52] Monakhov V. N., *Boundary value problems with free boundaries for elliptic systems.* Amer. Math. Soc., Providence, RI, 1983.

[53] Mshimba A. S. A. and Tutschke W., *Functional analytic methods in complex analysis and applications to partial differential equations.* World Scientific, Singapore, 1990.

[54] Mushelishvili N. I., 1) *Singular integral equations.* Noordhoff, Groningen, 1953.
2) *Some basic problems of the mathematical theory of elasticity.* Nauka, Moscow, 1946 (Russian); Noordhoff, Groningen, 1953.

[55] Nirenberg L., 1) *On nonlinear elliptic partial differential equations and Hölder continuity.* Comm. Pure Appl. Math., 6(1953), 103-156.
2) *An application of generalized degree to a class of nonlinear problems.* Coll. Analyse Fonct. Liége, 1970, Vander, Louvain, 1971, 57-74.

[56] Norguet F., *Intégrales de formes differentilles exterieures non fermees.* Rendiconti Matematica, 1991, 20:335-371.

[57] Obolashvili E. I., *Higher order Partial differential equations in Clifford Analysis.* Birkhäuser, Boston Basel Berlin, 2003.

[58] Oleinik O. A., *On equations of elliptic type degenerating on the boundary of a region.* Dokl. Akad. Nauk SSSR (N.S.) 1952, 87:885-888.

[59] Panejah B. P., *On the theory of solvability of the oblique derivative problem.* Math. Sb. (N.S.), 1981, 114:226-268 (Russian).

[60] Berjiske-Sabilof E. E., *Geometry of classical domains and theory for automorphic functions.* Nauka, Moscow, 1961 (Russian).

[61] Protter M. H. and Weinberger H. F., *Maximum principles in differential equations.* Prentice-Hall, Englewood Cliffs, N.J., 1967.

[62] Qian T., 1) *Singular integrals with monogenic kernels on the m-torus and their Lipschits perturbations.* Clifford Algebras in Analysis and Related Topics. CRC Press, Boca Raton New York London Tokyo, 1996, 157-171.
2) *Singular integrals associated with holomorphic functional calculus of the spherical Diracoperator on star-shaped Lipschitz surfaces in the quaternionic space.* Math. Ann., 310, 1998, 601-630.
3) *Fourier analysis on starlike Lipschitz surfaces.* Journal of Functional Analysis, 183, 2001, 370-412.

[63] Qian T., Hempfling T., McIntosh A., Sommen F., *Advances in Analysis and Geometry.* Birkhäuser Verlag, Basel Boston Berlin, 2004.

[64] Qiao Yuying, 1) *A nonlinear boundary problem with a shift for generalized biregular functions.* J. Sys. Sci. and Math. Sci., 2002, 22(1):43-49 (Chinese)
2) *A nonlinear boundary value problem with a Haseman shift for biregular functions.* J. Sys. Sci. and Math. Sci., 1999, 19(4):484-489 (Chinese).
3) *The oblique derivative boundary value problem for generalized regular functions with value in quaternions.* Acta Mathematica Scientia, 1997, 17(4):447-451 (Chinese).

[65] Qiao Yuying, Huang Sha, *Some problems for six kinds of high order singular integrals of Quasi-Bochner-Martinelli type in the real Clifford analysis.* J. Sys. Sci. and Math. Sci., 2002, 22(2):180-191 (Chinese).

[66] Qiao Yuying, Huang Sha, Zhao Hongfang, Chen Zhenguo, *A nonlinear boundary value problem in Clifford analysis.* Acta Mathematica Scientia, 1996, 16(3):284-290.

[67] Qiao Yuying, Xie Yonghong, *A nonlinear boundary value problem in Clifford analysis.* Advances in Applied Clifford Algebras, Mexico, 2001, 11(11):260-276.

[68] Ryan J., 1) *Complexified Clifford analysis.* Complex Variables, 1982/1983, 1:119-149.
2) *Special functions and relations within complex Clifford analysis.* Complex Variables, 1983, 2:177-198.
3) *Dirac operators on spheres and hyperbolae*, Bolletin de la Sociedad Matematica a Mexicana, 3, 1996, 255-270.
4) *C^2 extensions of analytic functions defined in the complex plane.*

Advances in Applied Clifford Algebra, 11, 2002, 137-145.
5) *Clifford algebras in Analysis and Related Topics.* CRC Press, Boca Raton, 1996.

[69] Schauder L., *Der fixpunktsatz in Funktionalräumen.* Studia Math. 1930, 2:171-181.

[70] Schwarz L., *Théorie des distributions* I, II. Actualitiés Sci., Paris, 1950.

[71] Sobolev S. L., *Applications of functional analysis in mathematical physics.* Amer. Math. Soc., Providence, RI, 1963.

[72] Sommen F., 1) *Monogenic differential forms and homology theory.* Proc. R. Irish. Acad., 1984(2):87-109.
2) *Some connections between Clifford analysis and complex analysis.* Complex Variables, 1982/1983, 1:97-118.
3) *Spherical monogenic functions and analytic functions on the unit sphere.* Tokyo J. Math., 1981, 4:427-456.

[73] Spivak M., *Calculus on manifolds.* A modern approch to classical theorems of advanced caluclus, The Benjamin/Cummings Publishing Company, 1968.

[74] Trudinger N. S., *Nonlinear oblique boundary value problems for nonlinear elliptic equations.* Trans. Amer. Math. Soc., 1986, 295: 509-546.

[75] Tutschke W., *Boundary value problems for generalized analytic functions of several complex variables.* Ann. Polon. Math., 1981, 39:227-238.

[76] Vahlen K. T., *Über Bewegungen und complexe zahlen.* Math. Ann., 1902, 55:585-593.

[77] Vekua I. N., *Generalized analytic functions*, Pergamon, Oxford, 1962.

[78] Wang Chuanrong, *The Hadamard principal value of singular integral* $\int L\frac{f(\tau)d\tau}{(\tau-t)^{n+1}}$. Math. Ann., 1982, 3(2):195-202 (Chinese).

[79] Wang Xiaoqin, *Singular integrals and analyticity theorems in several complex variables.* Doctoral Dissertation, Uppsala University, Sweden, 1990.

[80] Wen Guochun, 1) *Modified Dirichlet problem and quasiconformal mappings for nonlinear elliptic systems of first order.* Kexue Tongbao (a Monthly Journal of Science), 1980, 25: 449-453.
2) *The mixed boundary value problem for nonlinear elliptic equations of second order in the plane.* Proc. 1980 Beijing Sym. Diff. Geom. Diff. Eq., Beijing, 1982, 1543-1557.
3) *Oblique derivative boundary value problems for nonlinear elliptic systems of second order.* Scientia Sinica, Ser. A, 1983, 26:113-124.
4) *Linear and nonlinear elliptic complex equations.* Shanghai Science Techn. Publ., Shanghai, 1986 (Chinese).
5) *Some boundary value problems for nonlinear degenerate elliptic complex equations.* Lectures on Complex Analysis, World Scientific, Singapore, 1988, 265-281.
6) *Oblique derivative problems in Clifford analysis.* J. Yantai Univ. (Natur. Sci. Eng.), 1990, 1:1-6 (Chinese).
7) *Conformal mappings and boundary value problems.* Amer. Math. Soc., Providence, RI, 1992.
8) *Approximate methods and numerical analysis for elliptic complex equations.* Gordon and Breach Science Publishers, Amsterdam, 1999.
9) *Linear and nonlinear parabolic complex equations.* World Scientific, Singapore, 1999.
10) *Clifford analysis and elliptic system, hyperbolic systems of first order equations.* World Scientific, Singapore, 1991, 230-237.
11) *Linear and quasilinear complex equations of hyperbolic and mixed type.* Taylor & Francis, London, 2002.

[81] Wen Guochun, Begehr H., *Boundary value problems for elliptic equations and systems.* Longman, Harlow, 1990.

[82] Wen Guochun, Tai Chungwei, Tain Mao-ying, *Function theoretic methods of free boundary problems and their applications to mechanics.* Higher Education Press, Beijing, 1996 (Chinese).

[83] Wen Guochun, Yang Guang-wu and Huang Sha, *Generalized analytic functions and their generalizations.* Hebei Education Press, Hebei, 1989 (Chinese).

[84] Wendland W., *Elliptic systems in the plane.* Pitman, London, 1979.

[85] Xie Yonghong, Huang Sha, Qiao Yuying, *A nonlinear boundary value problem with conjugate value and a kind of shift for generalized biregular functions.* Journal of Mathematical Research and Exposition, 2001, 21(4):567-572.

[86] Xu Zhenyuan, 1)*On Riemann problem for regular functions with values in Clifford Algebra.* Science Bulletin, 1987, 32(23):476-477 (Chinese).
2) *A function theory for the operator D-λ.* Complex Variables, 1991, 16(1), 37-42.

[87] Zhao Zhen, *Singular integral equations.* Beijing Normal University Press, Beijing, 1982 (Chinese).

[88] Zhong Tongde, 1) *Boundary nature of Cauchy's type integral for functions of several complex variables.* Acta. Math. Sinica, 1965, 15(2):227-241 (Chinese).
2) *Transformation formula of several dimensional singular integrals with Bochner-Martinelli kernel.* Acta Mathematica Sinica, 1980, 23(4):554-565 (Chinese).
3) *Integral expressions in complex analysis of several variables and several dimensional singular integral equations.* Xiamen University Press, Xiamen, 1986 (Chinese).
4) *Plemelj formula and its application.* Advances in Mathematics, 1994, 23(3):205-211 (Chinese).

[89] Zhong Tongde, Huang Sha, *Complex analysis of several variables.* Hebei Education Publishing House, Shijiazhuang, 1990 (Chinese).

[90] Zhou Yulin, 1) *Some problems for quasilinear elliptic and parabolic equations.* Beijingdaxue Xuebao (Acta Sci. Natur. Univ. Peki.) 1959, 283-326 (Chinese).
2) *Nonlinear partial differential equations.* Acta. Math. Sinica 11(1961), 181-192 (Chinese).
3) *On the nonlinear boundary value problems for quasilinear elliptic and parabolic equations.* Acta Ji-lin Univ. (Natur. Sci.) 1980, 1:19-46 (Chinese).

Index